A series of student texts in

CONTEMPORARY BIOLOGY

General Editors:
Professor E. J. W. Barrington, F.R.S.
Professor Arthur J. Willis

The Comparative Endocrinology

of the Invertebrates

Second Edition

Kenneth C. Highnam
M.Sc., Ph.D.

Reader in Invertebrate Endocrinology,
University of Sheffield

and

Leonard Hill
B.Sc., Ph.D.

Senior Lecturer in Zoology,
University of Sheffield

University Park Press

© Kenneth C. Highnam and Leonard Hill, 1977

First published 1969
by Edward Arnold (Publishers) Ltd.,
25 Hill Street, London W1X 8LL

Second edition 1977

First published in the USA in 1977 by
University Park Press
233 East Redwood Street
Baltimore, Maryland 21202

Library of Congress Cataloging in Publication Data

Highnam, Kenneth Charles.
    The comparative endocrinology of the invertebrates.

    (A Series of student texts in contemporary biology)
    Includes bibliographical references and index.
    1. Endocrinology, Comparative.   2.   Invertebrates—
    Physiology.   I.   Hill, Leonard, joint author.   II.   Title.
    [DNLM: 1. Endocrine glands.   2. Invertebrate hormones.
    3.   Invertebrates.   QL868.H638c]
    QP187.H54   1977   592'.01'4   77-15979
    ISBN 0-8391-1193-2

Printed in Great Britain

Preface

From its beginning endocrinology has been an experimental science. Yet because of legal restrictions (particularly in the United Kingdom) upon the use of vertebrate animals for experimental purposes, the undergraduate student has been forced to consider vertebrate endocrinology a largely theoretical subject, a confusing mass of text-book information to be absorbed and unravelled without the benefits of personal laboratory investigations.

In this book we have attempted to show that the principles of hormonal integration and co-ordination are as well exemplified in the invertebrate animals as in the vertebrate. These principles can be demonstrated practically, using species which are not precluded by the vivisection regulations, and which are often more suitable than vertebrates for class use. We have therefore incorporated a certain amount of practical detail into the text to illustrate the relatively simple experiments and techniques involved. It is hoped that this will encourage teachers and students to initiate their own programmes of endocrinological investigations.

Insects, in particular, are excellent animals to use for such experiments. Many species can be reared easily in the laboratory, and their endocrine systems, although simpler than those of any vertebrate, are sufficiently complex to pose suitably satisfying problems of technique and interpretation. It is not difficult to make extracts of the corpora cardiaca of a locust or a cockroach and to test their effects upon the rate of heart beat or upon diuresis. But then further experiments are necessary to decide whether the activities of the extracts are due to the intrinsic hormones of the corpora cardiaca, or to hormones produced elsewhere and stored temporarily in the glands. Ecdysone, the insect moulting hormone, is now available commercially, together with plant steroids such as ponasterone and inokosterone which mimic its effects. Other compounds with

juvenile hormone activity can also be purchased. The effects of these substances upon the development and reproduction of laboratory colonies of insects can thus be tested, even when facilities are not available for surgical operations upon the endocrine systems of the insects. The large proportion of this book devoted to insect endocrinology reflects not only the amount of information on the subject now available, but also our feeling that laboratory investigations on insect hormones should be of considerable importance in view of the current emphasis upon experimental method in biology in both schools and universities.

Neurosecretory hormones dominate this account of invertebrate endocrinology. This is inevitable when many of the groups appear to possess only this particular class of hormones. But more importantly, it is our opinion that in almost all animals neurosecretory mechanisms are of the greatest significance in relating the development and reproduction of an individual to its changing environment, and in this book we have tried to give reasons for this view. The suggestion that vertebrate animals are no different from the invertebrates in this regard explains the presence of some vertebrate endocrinology within our pages. This concept of the role of neurosecretion owes a very great deal to the writings of Professors Ernst and Berta Scharrer.

Since the first edition of this book was published, major advances have been made in invertebrate endocrinology, particularly in the hormonal control of reproduction in molluscs and annelids, and in the chemistry and biochemistry of hormones in arthropods and echinoderms. The chapters dealing with these topics have consequently been completely or extensively revised. Further contributions to the endocrinology of the less highly organized invertebrates have also been incorporated, together with new information about the 'metabolic' hormones in insects and crustaceans, and the endocrine control of insect reproduction. Other parts of the book have been rewritten to emphasize more recent interpretations of the available information.

We have assumed throughout some knowledge of the structural organization and biology of the animals whose endocrinology is discussed, although more detail has been given where it was considered necessary. Other volumes in this series should do much to make up any deficiencies.

Sheffield,   K. C. H.
1976   L. H.

Table of Contents

I

Nervous and Chemical Co-ordination

In all Metazoa, the nervous and endocrine systems so co-ordinate the activities of the various organs and tissues in the body that the animals function as individuals. The nervous system serves for rapid communication, essential, for example, in the systematic contractions of muscles during locomotion. For this, complicated chains of interconnected neurones are necessary for the transmission of transient impulses, together with the highly localized production of chemicals such as adrenaline and acetylcholine which are rapidly destroyed. The endocrine system uses circulating body fluids to carry its chemical messengers to more or less specific target organs. These chemicals take time to build up to an effective concentration, and consequently must have a longer biological life than the chemicals of the nervous system before they are eventually destroyed or excreted. Hormones are consequently well suited to exert their effects over extended periods of time, and the endocrine system controls long-term processes within the body, such as the co-ordinated growth of organs or the maintenance of appropriate metabolite concentrations in the blood and tissues.

An animal derives information about the chemical and physical features of its environment through sensory receptors. This information passes to the nervous system where it is collated and interpreted, and motor impulses are initiated which cause the animal to react appropriately. Short-term environmental changes, such as the appearance of predators or prey, the scents produced by possible mates, or the hazards associated with a sudden downpour of rain, each evoke rapid and particular responses in individuals through nervous activities. But such

reactions are not necessarily rigidly fixed. If the stimuli are repeated many times, the sensory receptors may adapt and no longer respond. Or the animal may habituate to the stimuli, learning to disregard them. The same stimulation may evoke different reactions when the animal is feeding, moving or engaging in courtship. Moreover, a particular activity can have a marked after effect when it is completed, considerably modifying the immediately subsequent behaviour.[168, 169] Behavioural variability is the result of complicated neuronal interactions within the central nervous system.

It must not be supposed that the nervous system functions quite independently of the endocrine system. A mature female grasshopper will move towards and mate with a courting male. An immature female avoids him, and will even fight him if he persists in his courtship.[193] The hormone complements of the two females cause their central nervous systems to interpret the same information differently, and to initiate quite opposite reactions. It is likely that central nervous activity in most animals is strongly affected by hormones most of the time.

The opposite is also true: hormone production and release is dependent upon nervous activity. Nearly all animals have to respond developmentally to environmental changes, particularly seasonal fluctuations, throughout the year. Unfavourable periods are avoided by dormancy or migration, or overcome by other changes in habit or physiology. Advantage is taken of favourable seasons for rapid development and multiplication. Even transient fluctuations, such as temporary shortage of food, or the absence of suitable mates, can have dramatic effects upon development. The act of mating in a female insect can accelerate the development of her eggs; the changing length of day can control the onset of metamorphosis in annelids; colour change in many crustaceans is brought about by alterations in the shade of their immediate environment. All these developmental or physiological responses result from changes in the concentrations of circulating hormones caused by nervous impulses originating in the stimulation of particular sense organs. The nervous and endocrine systems are strictly interdependent: the normal activity of either requires the functional presence of the other.

Paradoxically, there are very few endocrine organs which are richly innervated. The adrenal medulla of the Vertebrates is an obvious exception, innervated by pre-ganglionic fibres from the coeliac plexus, part of the sympathetic nervous system. But the cells of the adrenal medulla are considered to be modified post-ganglionic sympathetic neurones, specialized for the production of large amounts of adrenaline and nor-adrenaline—the chemicals produced by normal post-ganglionic neurones —to be poured into the circulation at the appropriate times of fight or flight. The adrenal medulla is consequently a rather atypical endocrine

gland. Other endocrine organs in vertebrate animals may contain nerve fibres, but these are usually associated with the blood supply to the glands. Endocrine activity may be partly controlled by altering the blood flow through glands, but the mechanism is not sufficiently all-embracing to explain the often close correlation between hormonal processes and environmental events. In particular, the anterior pituitary has been long known to produce a number of hormones which control the activities of other endocrine glands, but it contains very few nerve fibres indeed.

How then can nervous activity control endocrine function? More than fifty years ago, Stefan Kopec suggested that the larval brain of the gipsy moth, *Lymantria dispar*, produced a hormone which induced pupation.[177] At this time, both nervous and endocrine mechanisms, particularly in vertebrate animals, were being intensively investigated. For many reasons, these researches followed quite separate paths and it was never seriously considered that nervous systems could produce hormones. Kopec's experiments attracted very little attention, especially when it was subsequently shown that in other insects epithelial endocrine glands in the head, the corpora allata, apparently played an important role in the control of moulting and metamorphosis (Chapter 6). In the butterflies and moths, the corpora allata lie close to the brain, and it was assumed that Kopec had removed these glands unintentionally, together with the brain.

But during the last twenty years, opinions about the possible endocrine function of nervous tissue have been completely reversed. It is now generally accepted that most, if not all, nervous systems *do* have the ability to produce hormones. This new point of view results from the experimental demonstration that the central nervous systems of arthropods and vertebrates contain particular endocrine cells. These cells are morphologically similar to neurones, with axons, dendrites, Nissl granules and neurofibrillae. They are also able to transmit nervous impulses, but they differ from other neurones in two important respects: their axons do not innervate effector organs such as muscles, nor make synaptic connections with other neurones; and they manufacture materials, often visible in stained sections of nervous tissue, which are released from the ends of the axons and exert a biological effect some distance away. In effect, the cells are neurones which also produce hormones and are consequently called neurosecretory cells (Fig. 1.1).

In many animals, the neurosecretory cells are often clumped into groups which are very conspicuous features of the central nervous system after appropriate staining. The ends of the neurosecretory cell axons are usually swollen, and the material manufactured by the cell can be stored here before being released. The swollen axon terminals lie outside the nervous system, usually closely associated with the circulatory system

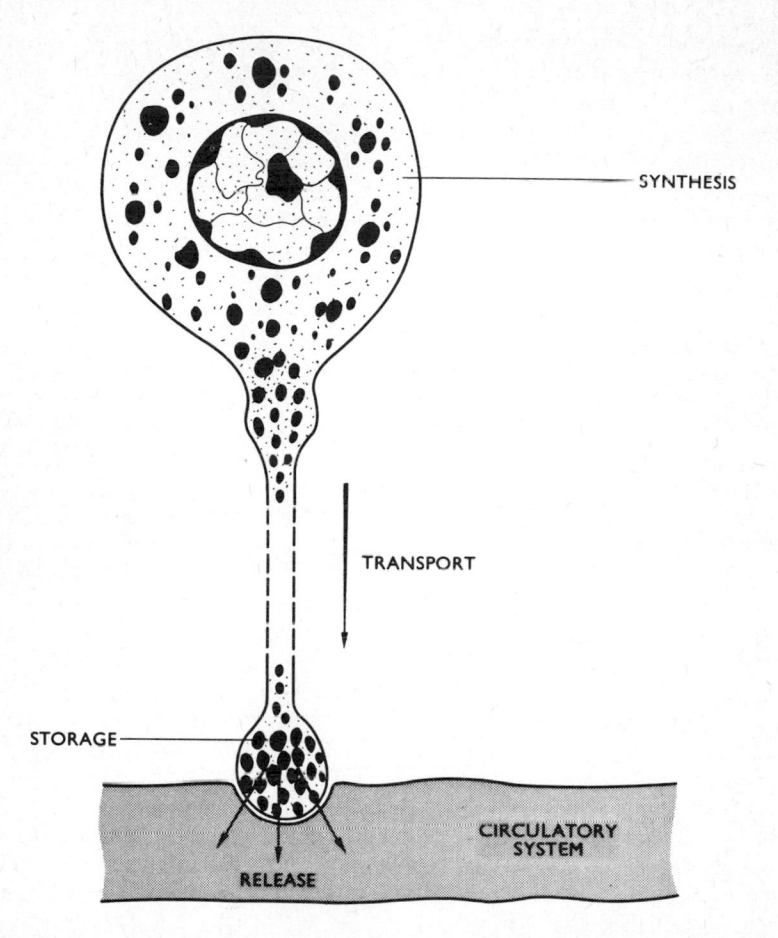

SYNTHESIS

TRANSPORT

STORAGE

CIRCULATORY SYSTEM

RELEASE

Fig. 1.1 Diagram of a neurosecretory cell. Globules of neurosecretory material, which can be stained and made visible with the light microscope, are synthesized in the cell body, or perikaryon, and transported along the axon to the swollen terminal where they are stored or released into the circulatory system. In vertebrate animals, it is likely that synthesis can also occur within the axon.

which carries the neurosecretory hormones around the body. Where numbers of neurosecretory cells are grouped together, their axon terminals can form well developed organs outside the nervous system, and since these structures are combined with blood vessels, they are called **neurohaemal organs.**

Neurosecretory cells from invertebrates and vertebrates are essentially

similar, although it is possible that the invertebrate cells may lack dendrites and it is not yet *conclusively* proved that they can transmit nervous impulses.[266, 293] The material elaborated by neurosecretory cells is largely protein and reacts histologically and histochemically in much the same way whether it is produced by invertebrate or vertebrate neurosecretory cells. Ultrastructurally, the resemblance is even closer: the neurosecretory particles are usually electron-dense spheres between 1000 to 3000 Å in diameter, surrounded by a thin membrane (Fig. 1.2). Some

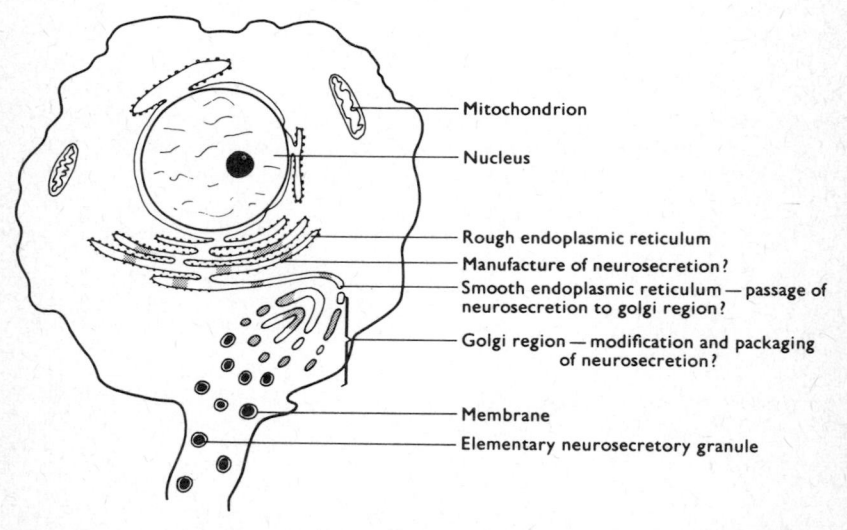

Fig. 1.2   Diagram of part of the ultrastructural organization of a neurosecretory cell. Protein, which makes up at least part of each elementary neurosecretory particle, is synthesized upon ribosomes associated with the endoplasmic reticulum. The protein material is passed to the Golgi apparatus where it is organized and provided with a membrane covering.

neurosecretory cells in both invertebrates and vertebrates produce electron-transparent vesicles. The dense spheres originate in the Golgi apparatus of the cell bodies,[14] in a manner similar to protein droplet formation in other endocrine and ordinary gland cells. But in addition, synthesis, or perhaps reorganization, of neurosecretory material can take place in the axons of many neurosecretory cells. The protein material synthesized by neurosecretory cells, called **neurophysin** in the vertebrates, is considered to be a carrier for the actual hormones. There is good evidence for this from vertebrate neurosecretory systems; whether the same applies to all invertebrate neurosecretions is problematical.

The discovery of neurosecretory cells has added a new dimension to the study of endocrine systems. Previously, endocrine glands were

divided into two categories: those which were derived from embryonic ectoderm, mesoderm and endoderm layers, called epithelial endocrine glands; and those which were derived from nervous tissue. The first category included all the endocrine glands and tissues of a typical vertebrate, except the adrenal medulla and the posterior pituitary which developed from a transformed sympathetic ganglion and part of the brain respectively, and were therefore placed in the second category. But it is now known that the posterior pituitary is a neurohaemal organ for groups of neurosecretory cells in the hypothalamus of the brain. The posterior pituitary hormones—all octapeptides like the oxytocin and vasopressin of mammals—are manufactured by the hypothalamic neurosecretory cells and released into the blood from the pars nervosa, perhaps after preliminary storage there (Fig. 1.3). The posterior pituitary is thus quite a different endocrine organ from the adrenal medulla. Other neurosecretory hormones are released from the median eminence into the

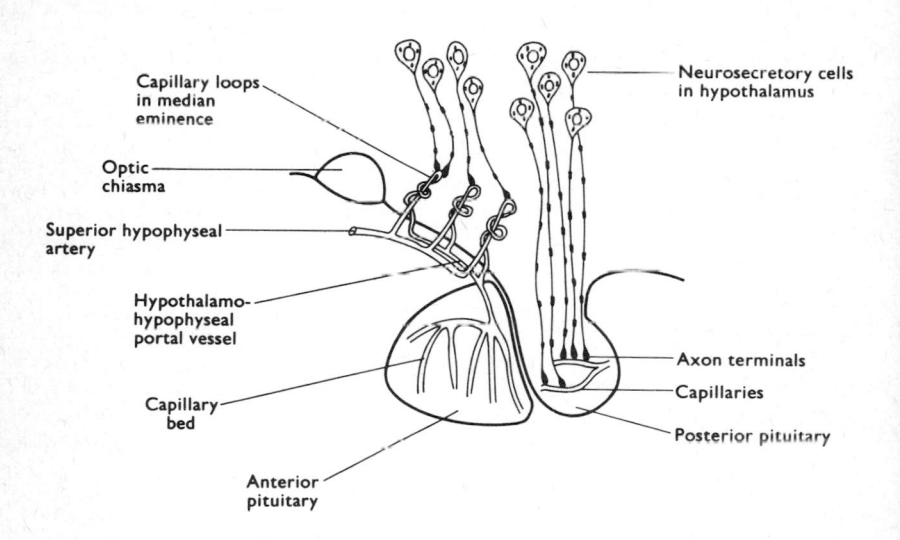

Fig. 1.3 The relationships between neurosecretory cells in the hypothalamus of the brain and the pituitary in vertebrates. Some neurosecretory cells send their axons directly to blood capillaries within the posterior pituitary, and release the octapeptides oxytocin and vasopressin. Other neurosecretory cell axons terminate upon capillaries in the anterior part of the infundibular stalk, the median eminence. These capillaries pass to the anterior pituitary, carrying blood containing neurosecretory factors which control the release and production of anterior pituitary hormones.

capillaries of the hypothalamo-hypophyseal portal system[66] (Fig. 1.3). These *releasing factors* are carried to the anterior pituitary to control the release and synthesis of the anterior pituitary hormones.[123] In some vertebrates a second major neurosecretory system is situated at the posterior end of the spinal cord.[103] The neurohaemal organ for this caudal neurosecretory system is the urophysis:[15] the functions of the hormones produced here have not so far been precisely determined.[16, 292]

In the invertebrate animals, epithelial endocrine glands, endocrine organs developed from transformed nerve ganglia, and neurosecretory cells often with well formed neurohaemal organs are all present, though not necessarily together in the same taxonomic group (Chapter 2). Even in complicated animals like the arthropods, the number of epithelial endocrine glands is much less than in the vertebrates, and in most other invertebrate groups they are absent altogether. Endocrine mechanisms in the invertebrates are consequently usually simpler than the intricate interactions between numerous endocrine glands producing even larger numbers of hormones which characterize vertebrate endocrinology. Because of the relative paucity of epithelial endocrine glands in the invertebrates, neurosecretory mechanisms assume great importance.

Their dual properties make neurosecretory cells the most likely candidates for transforming nervous into endocrine information.[236] Where this function is not complicated by the presence of numerous other endocrine mechanisms, the significance of neurosecretory processes in co-ordinating developmental events within an animal with environmental fluctuations may be more clearly demonstrated. The vertebrate hypothalamo-hypophyseal neurosecretory system plays the same part as the simpler neurosecretory systems in the invertebrate animals: a knowledge of these systems must surely precede any study of the much more exact but more complicated mechanisms found in the vertebrates.

However, many problems are associated with the determination of the functions of neurosecretory systems. The classical endocrinological experiment involves the removal of a suspected endocrine gland followed by its subsequent reimplantation to a different site in the body. If the consequences of removal are reversed and brought back to normal on reimplantation, then it is established that a hormonal mechanism is involved. But when a neurohaemal organ is removed, the cut ends of the neurosecretory axons remaining within the animal may still release hormones, sometimes in an uncontrolled manner. Or a new neurohaemal organ may be rapidly regenerated, so that the effects of a deficit of neurosecretory hormones may not be obvious. Moreover, the reimplantation of the neurohaemal organ may introduce several hormones into the animal in concentrations, and in relative proportions, which are never normally attained. When the neurosecretory cells which supply the

neurohaemal organ are destroyed or removed, the stored hormones within the organ may be released, or may leach out, for some time after the operation. Again, hormonal deficiency may not be immediately apparent. In some animals, the neurosecretory cells may be scattered singly or in small numbers throughout the central nervous system (Chapter 2), and a well defined neurohaemal organ may be lacking. Some neurosecretory hormones are not released to the circulatory system, but are transported axonally directly to their target organs.[156] It is difficult to determine the exact function of such neurosecretory mechanisms because of the impossibility of extracting and testing biological materials from individual cells. In many instances, the neurosecretory nature of such cells is inferred simply because the cells are histologically and histochemically similar to those in other animals whose function has been determined experimentally.

The invertebrates include groups which possess epithelial endocrine glands whose functions can be elucidated by the classical techniques of extirpation and reimplantation (Chapter 2). A very small number of hormones from such glands have been purified, their chemical composition determined and they can now be synthesized (Chapter 12). This allows much more sophisticated experiments in which the effects of different concentrations can be established, the ways in which combinations of hormones can interact, how the introduction of particular hormones at unnatural times can affect development, and so on. The well developed neurosecretory systems of some invertebrates allow detailed investigations into the problems of the production and release of combinations of neurosecretory hormones. Other invertebrates possess neurosecretory mechanisms whose functions can only be inferred from circumstantial evidence. In subsequent chapters, an attempt will be made to analyse the evidence for endocrine function in a broad range of invertebrate animals. Although the evidence may be slight in many instances, it would be unreasonable to exclude it: it could provide the foundation for much fruitful research.

2

Invertebrate Endocrine Systems

THE PORIFERA, COELENTERATA, TURBELLARIA, NEMERTEA AND NEMATODA

In the less highly organized invertebrates, epithelial endocrine glands are apparently absent, and neurosecretions are therefore the only hormonal coordinators. Cells with the histological characteristics of neurosecretion are found in the simplest invertebrate animals. The presence of a nervous system in members of the Porifera is uncertain, although cells with the morphological appearance of neurones have been described. These bipolar and multipolar cells are particularly numerous in the inner aspect of the rim around the osculum in *Sycon ciliatum* and some contain droplets which stain with chrome haematoxylin and alcian blue. Bulbous swellings along the cell processes often stain or contain reactive vesicles.[189b]

In *Hydra* (Coelenterata), neurosecretory cells are present in the epidermal nerve net predominantly in the hypostome and tentacle bases, but also in the peduncle and adhesive base[29, 75d] (Fig. 2.1). Vesicles can often be seen in the axons, and peristaltic waves carry the vesicles along the lengths of the axons to the tip, where they are released.[31] Neurosecretory cells are also present in the gastrodermis of *Hydra*.[75c] *Hydra* is unusual in that its nervous system is continually regenerating. The animal renews its tissues every three to five weeks by cell divisions in a meristematic region just behind the hypostome, and so new nerves are continually formed to replace those which are eventually sloughed off at the extremities of the body column. The new nerves develop from interstitial cells and only two

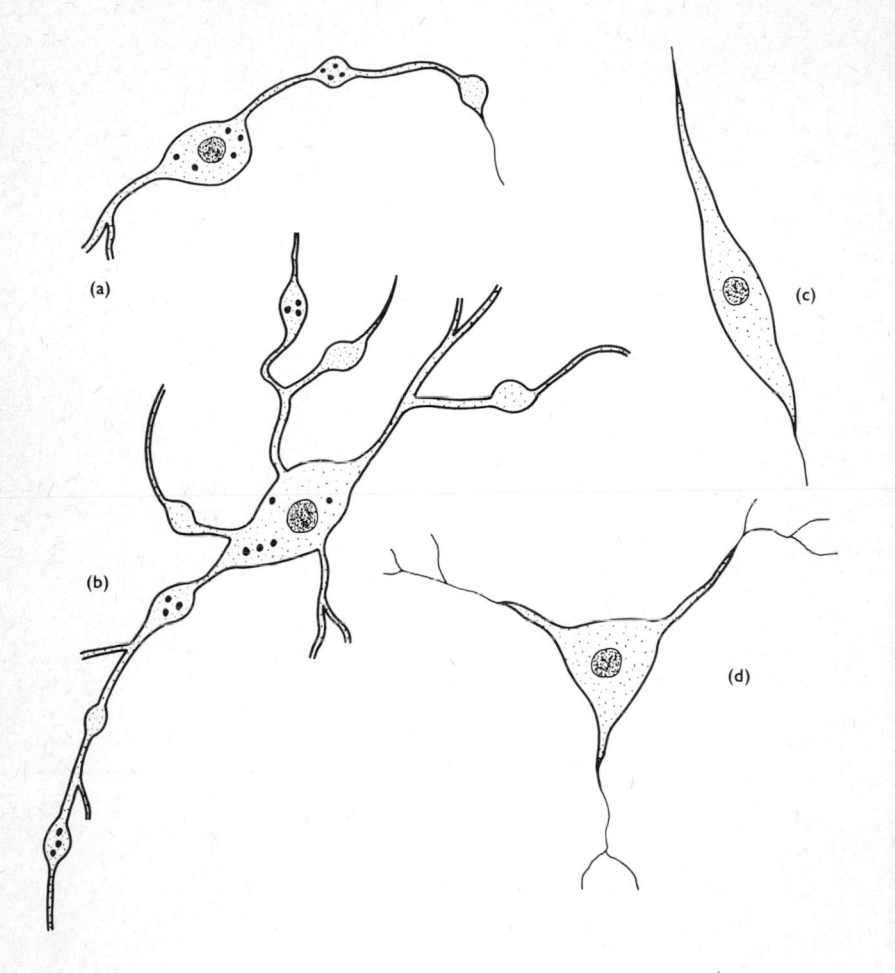

Fig. 2.1 Neurones from the sub-hypostomal region of *Hydra*. (a) and (b) contain stainable granules in the perikaryon and within vesicles arranged along the dendritic processes. Some vesicles are devoid of granules. (c) and (d) are bi- and tri-polar neurones respectively, containing no vesicles or granules. (After Burnett and Diehl[29])

types of neurone are formed: peripheral sensory cells and central multi-functional neurones.[268e] The latter type is a primitive neurone which combines several functions—motor, neurosensory and neurosecretory. Individual neurones thus have the potential to function in neurosecretion, neurotransmission and photoreception. The existence of such primitive

neurones with both secretory granules and synaptic vesicles is of evolutionary interest (p. 31).

In the Turbellaria, neurosecretory cells are present in the cerebral ganglia of several species of Platyhelminthes. *Polycelis nigra* and *Dugesia gonocephala* have neurosecretory cells in the ventral part of the brain which increase in number posteriorly, and are present also in the posterior parts of the ventral nerve cords.[266a] Ultrastructural studies have revealed a plexus of neurosecretory cells near the ovaries and testes in both *Polycelis* and *Dugesia*: neurosecretory granules are discharged from the endings of the nerve fibres by exocytosis into the spaces of the parenchyma next to the gonads.[118h, 118i, 118j]

The cerebral ganglion of species of Nemertea also contains neurosecretory cells.[186] In *Cerebratulus marginatus*, large numbers of neurosecretory cells are present in the medio-dorsal region of the cerebral ganglion around the cephalic clefts. Axons from these cells run singly and in bundles towards the neuropile and terminate there in close contact with the lateral blood vessel. This area may therefore be a very primitive neurohaemal organ.[17e, 17f] The cerebral organ of nemerteans, a structure composed of glandular and nervous elements, has for long been compared with neurosecretory structures in other groups.[232] In different nemertean groups, the cerebral organ is progressively incorporated into the central nervous system, which was thought at one time to suggest the way in which neurosecretory structures originated (p. 31). It is now clear that there is no relation between the cerebral organ, which is probably sensory in function, and neurosecretion.[17g, 191a] Within the Nematoda, *Ascaris lumbricoides* has neurosecretory cells in the lateral ganglia on the anterior nerve ring (Fig. 2.2), and *Phocanema decipiens* has similar cells in the dorsal and ventral ganglia[74] (Fig. 2.3). Many of the primary sense organs on the lips of these two species also stain in the same way as the ganglionic neurosecretory cells.

Experiments to determine the functions of these neurosecretory cells have been carried out in all these less well organized invertebrates (Chapter 3). But because of the relatively small numbers of cells present, their dispersion within the nerve ganglia, and the complete absence of any neurohaemal organs, or their analogues, in practice it is impossible to relate any experimental results precisely to factors produced by the neurosecretory cells.

THE ANNELIDA

Groups of scattered neurosecretory cells are found in the brain, or supra-oesophageal ganglion, and in the ganglia of the ventral chain in both polychaetes and oligochaetes[100] (Fig. 2.4). In the polychaete families

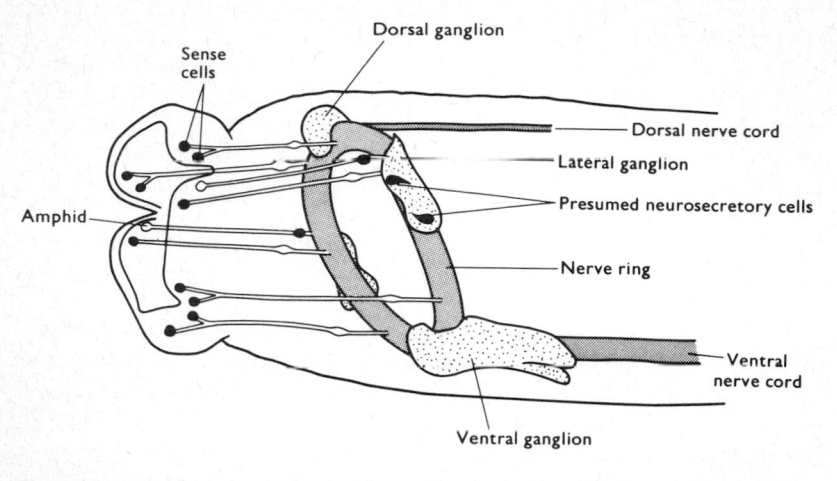

Fig. 2.2 The anterior nervous system of the nematode *Ascaris lumbricoides*. Neurosecretory cells (shown in black) are present in the lateral ganglia on each side of the nerve ring. Many of the sense cells on the lips also stain in the same way as neurosecretory cells. (After Davey[74])

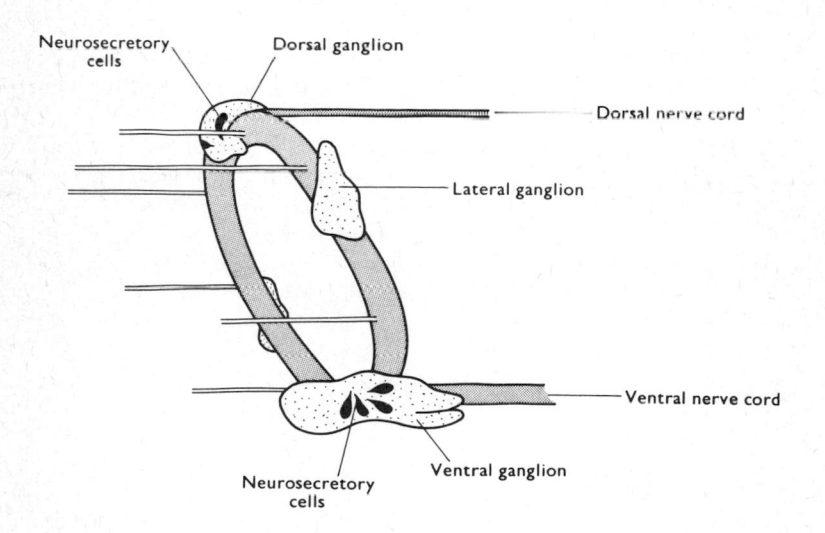

Fig. 2.3 The anterior nervous system of the nematode *Phocanema decipiens*. In this species neurosecretory cells (shown in black) occur in the dorsal and ventral ganglia of the nerve ring. (After Davey[74])

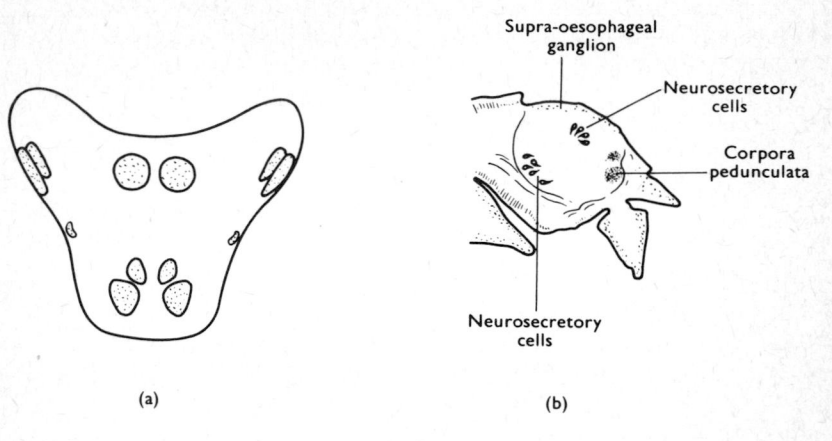

Fig. 2.4 (a) Plan of the brain of *Nereis diversicolor* showing the ganglionic nuclei (stippled) containing neurosecretory cells which are reported to show histological changes after the amputation of posterior segments of the body. (b) Parasagittal section through the prostomium and brain of *Platynereis dumerilii*. Two major groups of neurosecretory cells are present in the brain which are likely to be the source of the oocyte inhibiting hormone. ((a) After Clark and Bonney[57]; (b) after Hauenschild[125])

Nereidae, Nephtyidae and Polynoidae, the neurosecretory axonal tracts of the brain lead ventrally to the surface of the brain where they form part of a cerebrovascular complex.[8b, 8c, 21, 54, 115a, 115b, 115d] The complex is probably a simple neurohaemal organ, but also contains a glandular component, the infracerebral gland (Fig. 2.5). In nereids, four types of axonal fibres terminate in swollen endings against the brain capsule part of the complex, and these fibres spread out to cover an area of the capsule far greater than the cross-sectional area of the tracts themselves. Some neurosecretory axons do not terminate against the capsule, but pass through and ramify among the cells of the infracerebral gland. It is not clear, however, whether these axons innervate the infracerebral gland, or whether they originate in this region and run centripetally to the brain.[115b] Ultrastructural and cytological indications of the secretory activities of the different cell types composing the infracerebral gland suggest that the gland fulfills one or more functions. In *Nereis limnicola*, the dorsal and ventral parts of the brain (the ventral part containing the infracerebral gland) are each not as effective as a complete brain in controlling maturation (p. 76)[8d] suggesting that the cerebral neurosecretory cells together with the infracerebral gland form a system whose integrity is essential for the secretion at normal levels of its hormone(s).

In the syllid polychaetes, glandular cells have not yet been detected in the proventriculus, but the components of the stomatogastric nervous system present in this organ may have an endocrine function (p. 71).

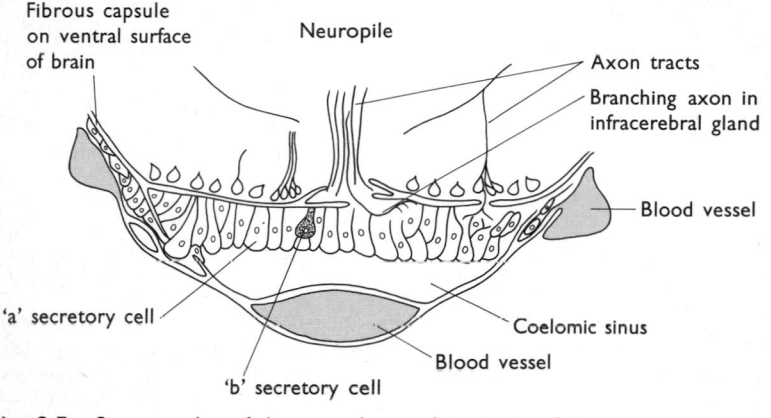

Fig. 2.5   Cross section of the ventral part of the brain of *Nephtys californiensis* showing the pericapsular organ, an area where neurosecretory cells in the posterior part of the brain may release their secretions into the blood. It is by no means certain that *all* the axons from the posterior groups of neurosecretory cells run to this cerebro-vascular complex. There are also anterior groups of neurosecretory cells in the brain which have no connection with the organ; their secretions must be released elsewhere. (After Clark and Olive[58a])

It is likely that many so-called neurosecretory cells in the annelid nervous system are not neurosecretory at all. In some leeches, 50 to 95% of the neurones in the brain have been called neurosecretory cells.[119] It must be realized that the criteria for identifying a neurosecretory cell are often solely histological: the presence of inclusions within the perikarya, the amount varying under different conditions; inclusions along the axons from the cells; and if present, a neurohaemal organ in which the axons terminate. But many ordinary nerve cells may contain inclusions which vary in amount at different times, and may duplicate the staining properties of neurosecretory proteins. Even ultrastructurally, some neurones contain membrane-bound elementary particles, 1000–3000 Å in diameter, which resemble closely elementary neurosecretory granules; recently electron-transparent neurosecretory vesicles have been described which are difficult to differentiate from synaptic vesicles found at axonal synaptic junctions. The only true test of neurosecretory function is the experimental demonstration of endocrine activity: the annelid brain passes this test (p. 57). But even so, endocrine function cannot be attributed to particular neurosecretory cell groups in the brain.

THE MOLLUSCA

Neurones with the histological characteristics of neurosecretory cells occur in almost all ganglia of the nervous system in gastropods and lamellibranchs (Figs. 2.6, 2.7). Histological changes in these cells have been correlated with a variety of developmental events (p. 86). But whether all such cells are actually neurosecretory is still open to question:

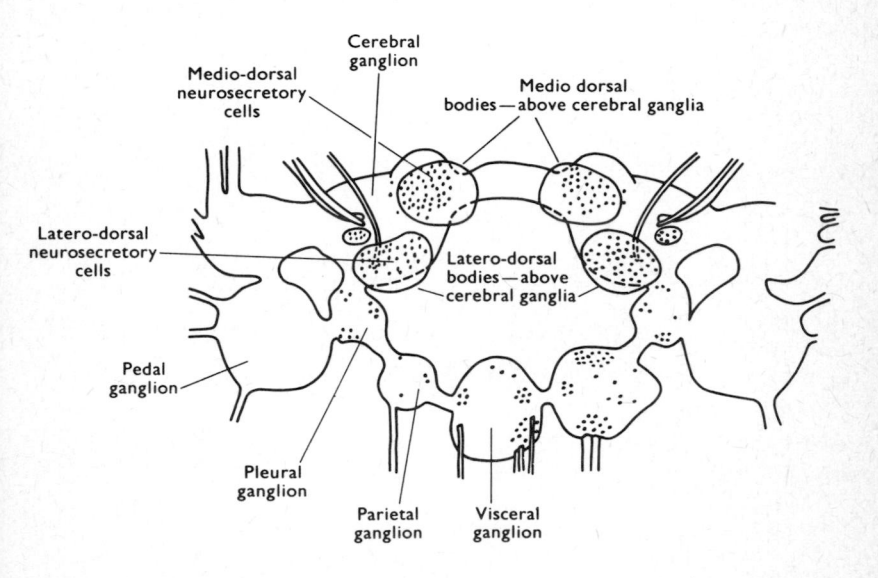

Fig. 2.6 The central ganglia of the pond snail, *Lymnaea stagnalis*, showing the location of scattered neurosecretory cells (black dots). The positions of the medio-dorsal and laterodorsal bodies above the cerebral ganglia are also shown. (After Lever, de Vries and Jager[190])

many of the stainable inclusions could be lysosomes, or glycogen granules, pigmented droplets, etc. (Chapter 5).

Particular regions of the perineurium of nerves adjacent to blood vessels, containing much neurosecretory material (Fig. 2.8), occur in some gastropod molluscs. These could represent the simplest of all neurohaemal organs. There is now evidence that, at least in some species, neurosecretory hormones are stored in these regions. Neurosecretory cells are also said to occur in the tentacles of some pulmonate gastropods, and there is evidence that these cells exert some developmental effect (p. 90).[220, 180] But unfortunately, the neurosecretory nature even of these cells is not definitely proved.

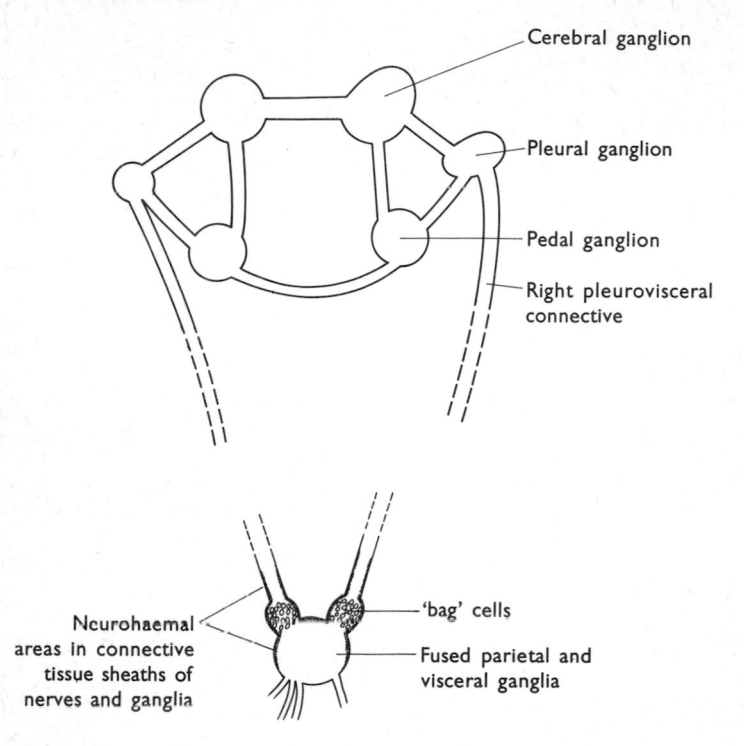

Fig. 2.7 Diagram (not to scale) of the central nervous system of *Aplysia californica* to show the position of the neurosecretory 'bag' cells and their associated neurohaemal organs. The fused parietovisceral ganglion is sometimes called the 'abdominal' ganglion.

In freshwater pulmonate snails, **dorsal bodies** are attached to the perineurium of the cerebral ganglia (Fig. 2.6). There are paired mediodorsal bodies in all species so far examined, and in some of these paired laterodorsal bodies also occur. The suggestion that the dorsal bodies are innervated by axons from neurosecretory cells in the cerebral ganglia[189d] is not borne out by ultrastructural examination.[21a] In *Lymnaea stagnalis*, neurosecretory axons from cells in the cerebral ganglia, although found within the dorsal bodies, are situated close to blood spaces and capillaries within the narrow strands of connective tissues which separate the cell groups of the dorsal bodies, and are therefore likely to be merely part of the perineural neurohaemal system.[268a] Certainly, the dorsal bodies and the neurosecretory cells of the cerebral ganglia have quite different functions (p. 88).

In terrestrial pulmonates, structures comparable with the dorsal bodies

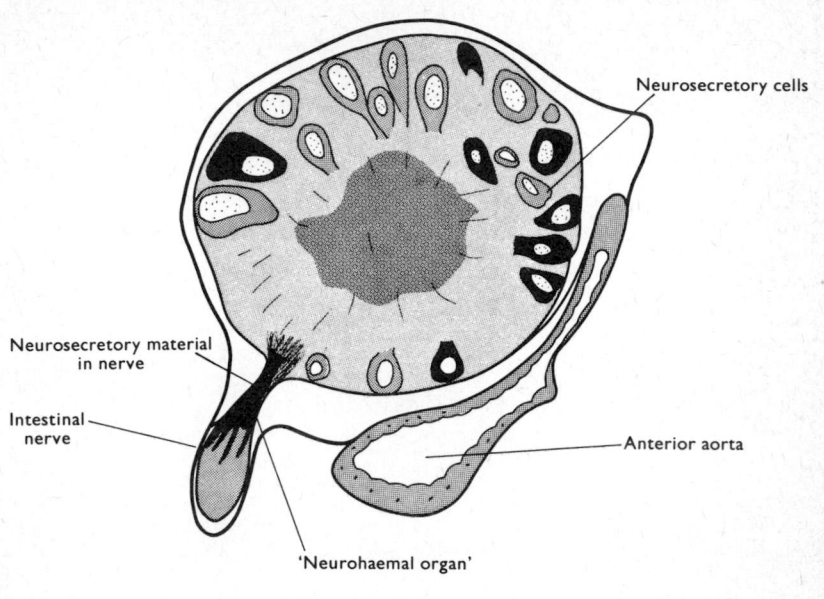

Neurosecretory cells

Neurosecretory material
in nerve

Intestinal
nerve

Anterior aorta

'Neurohaemal organ'

Fig. 2.8 Cross section through the visceral ganglion and intestinal nerve of a gastropod mollusc. The axons of neŭrosecretory cells in the ganglion pass into the intestinal nerve, where their close association with the anterior aorta has suggested that this area of the nerve may be a neurohaemal organ. (After Simpson, Bern and Nishioka[246])

are located in the thick connective tissue sheath around the cerebral ganglia and the cerebral commissure.[229d] The equivalent *'organes juxtaganglion-naires'* in the prosobranch and opisthobranch molluscs are attached to the anterodorsal parts of the cerebral ganglia.[201c, 201d, 266b, 266c, 266d] The polyplacophoran amphineurans have the *'organe juxtacommissural'*, probably homologous with the *'organes juxtaganglionnaires'*, situated between the intercerebral commissures and an anterior blood space.[266e, 266f]

In the cephalopod molluscs (octopus and squids), the optic glands are well developed endocrine organs lying on the optic stalks just internal to the eyes (Fig. 2.9). These glands are intimately involved in the control of reproductive development (p. 95). The glands are possibly nervous in origin, arising from the regions at the sides of the vertical lobe of the brain.[21e] The optic glands are well vascularized and innervated, but without trace of neurosecretory axons. Their possible equivalence with the dorsal bodies of some pulmonate gastropods is suggested both by their position and their function (p. 99). The presence of a neurosecretory system associated with the anterior parts of the vena cava is now well

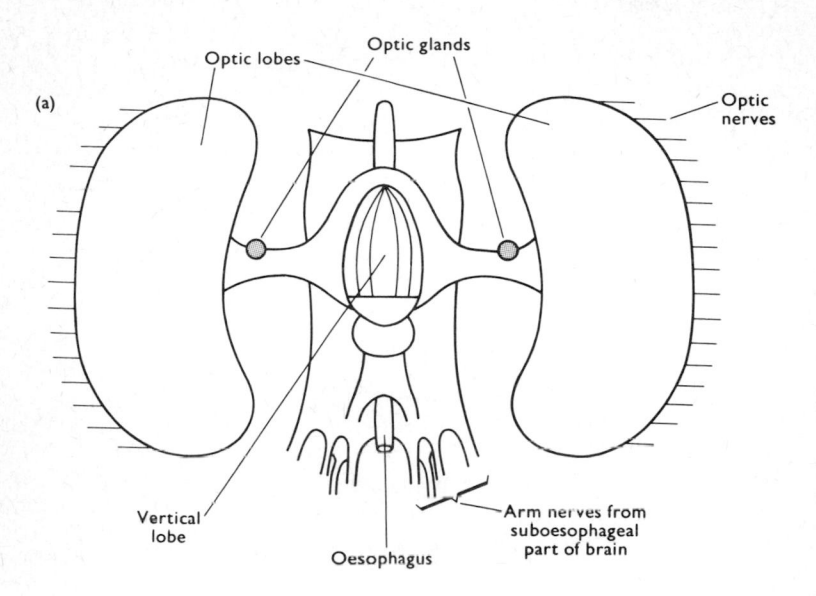

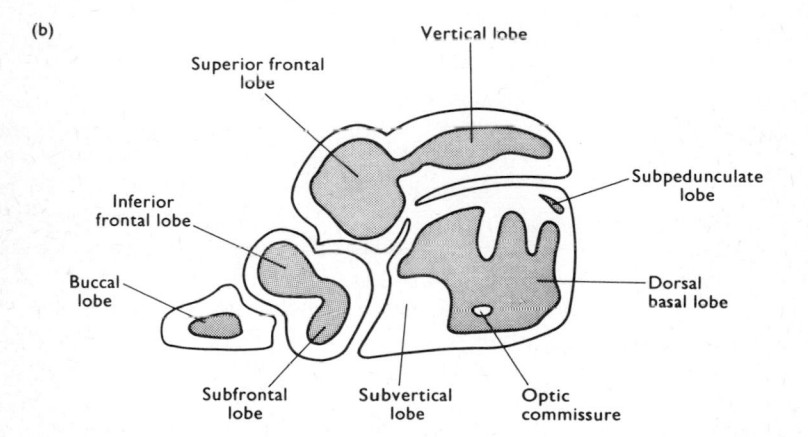

Fig. 2.9 Structure of the brain of *Octopus*. (a) Dorsal view of the brain as it would be seen after its removal from the cartilaginous brain case. The optic glands are situated dorsally on the optic stalks. (b) A longitudinal vertical section of the supra-oesophageal part of the brain showing its different areas. *For further details see text.* (After Wells and Wells[268])

established in a variety of cephalopod molluscs. The neurosecretory cells are located in a layer on the palliovisceral lobe of the brain, and in paired ganglionic trunks which emerge from the visceral lobe. The axons from these neurosecretory cells eventually penetrate the muscular wall of the vena cava, and form a fine neuropile adjacent to the inner surface of the

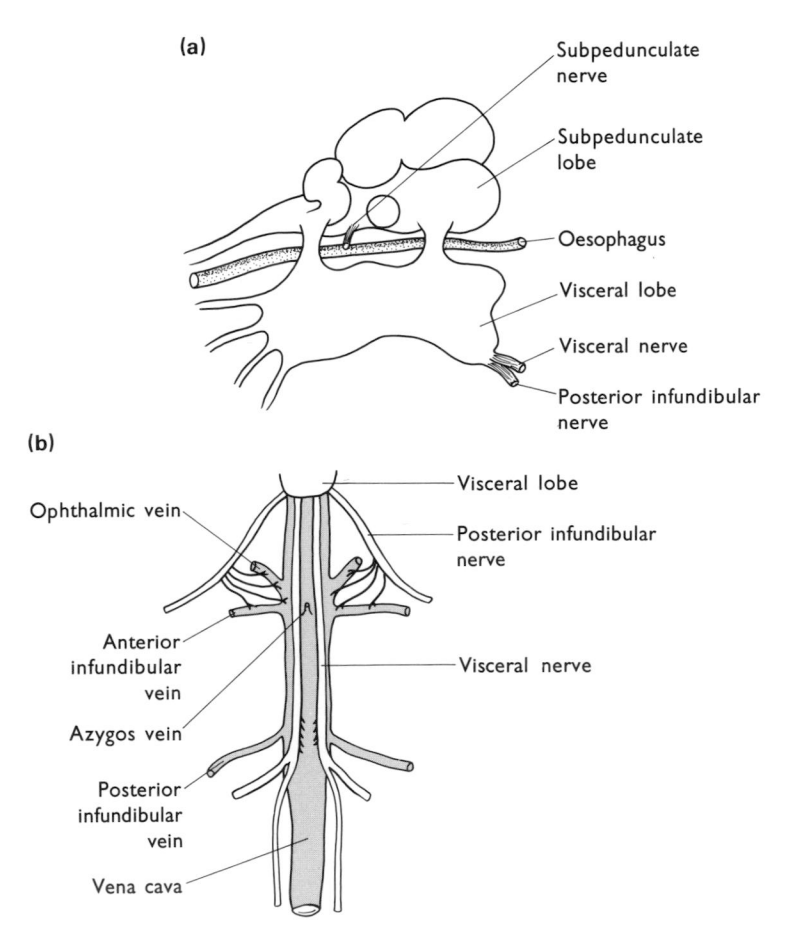

Fig. 2.10 (**a**) Diagrammatic side view of the central nervous system of *Octopus* to show the origins of the nerves (subpedunculate, visceral and posterior infundibular) with which neurosecretory fibres are associated. (**b**) The vena cava neurosecretory system of *Sepia officinalis*. Neurosecretory fibres (black) associated with the visceral and posterior infundibular nerves send branches to the anterior parts of the vena cava and the roots of its anterior tributaries. (**b** after Alexandrowicz[1j])

vessel[11, 1j, 17a, 201b] (Fig. 2.10). A further neurosecretory system has recently been described in *Octopus*: this has cell bodies located in the subpedunculate lobe of the brain, with neurosecretory axons running in the subpedunculate nerve and ending with neurosecretory terminals at the pharyngo-ophthalmic vein.[95d] (Fig. 2.11). The function of the vena cava neurosecretory system is now known (p. 102); that of the subpedunculate lobe–pharyngo-ophthalmic system is less certain (p. 99). The mesodermal branchial glands on the gills were once thought to be endocrine glands, with functions similar to those of the adrenal glands of vertebrates,[258] but are now known to be concerned with the synthesis of the blood pigment haemocyanin.[77d, 206c]

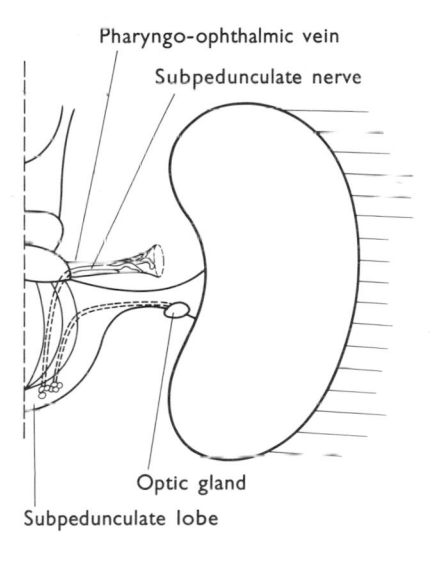

Pharyngo-ophthalmic vein

Subpedunculate nerve

Optic gland

Subpedunculate lobe

Fig. 2.11 Diagram of the right side of the brain of *Octopus* to show the path of neurosecretory fibres in the subpedunculate nerve which terminate upon the pharyngo-ophthalmic vein. The path of the (non-neurosecretory) nerve to the optic gland is also shown. (After Froesch[95d])

THE ARTHROPODA

The arthropods are complicated, highly organized invertebrate animals. In number of species, and in number of individuals, they are the most successful of all animal groups in their colonization of land and sea. The insects and the crustaceans are pre-eminently the dominant groups

among the arthropods. With the exception of the cephalopod molluscs (also highly organized animals), the arthropods are the only invertebrates known to possess epithelial endocrine glands. In combination with well-developed neurosecretory mechanisms, the epithelial endocrine glands form endocrine systems which in arrangement and scope resemble those of the vertebrate animals.

Insecta

The insect endocrine system has four major components: groups of *neurosecretory cells* in the brain, the *corpora cardiaca*, the *corpora allata*, and the *thoracic glands* or their equivalents (Fig. 2.12). The corpora allata and thoracic glands are epithelial endocrine glands, arising embryonically as ectodermal invaginations in the region of the first and second maxillary pouches.[154, 265, 213, 130] The corpora allata are connected by nerves to the corpora cardiaca and to the sub-oesophageal ganglion; the thoracic glands are innervated variously from the sub-oesophageal ganglion and thoracic ganglia (Figs. 2.12, 2.13). Structures homologous with the thoracic glands, the *ventral glands*, are found in some insects: they remain in the head after invagination instead of migrating further backwards into the thorax.[255] But in the cockroaches, both ventral and thoracic glands are present, so the structures must be serially homologous.[226, 227] The corpora cardiaca develop embryonically as invaginations of the foregut at the same time and in the same series as the dorsal sympathetic nervous system, which includes the frontal, hypocerebral and stomachic ganglia, connected by the recurrent nerve[229] (Fig. 2.14). The corpora cardiaca are consequently transformed nerve ganglia, and are still often referred to as oesophageal ganglia. As their name implies, they are usually in close association with the dorsal heart. The corpora cardiaca produce their own intrinsic hormones (Chapter 8), but their major function in most insects is to store and release neurosecretory hormones from the brain (Fig. 2.14). They are thus the most important insect neurohaemal organ; with the neurosecretory cells in the brain, they form the *cerebral neurosecretory system*. The cerebral neurosecretory cells are in two groups in each half of the pars intercerebralis region of the forebrain.[136] The medial groups are often closely apposed: most of their axons decussate in the brain before passing to the corpora cardiaca (Fig. 2.12). The lateral cell groups are usually smaller than the medial; their axons run directly to the ipsilateral corpus cardiacum. The corpora cardiaca thus receive paired inner and outer nerves from the cerebral neurosecretory cells. The size and staining reactions of the medial cells allow the differentiation of four types, called A-, B-, C- and D-neurosecretory cells.[131] It is difficult to ascribe parti-

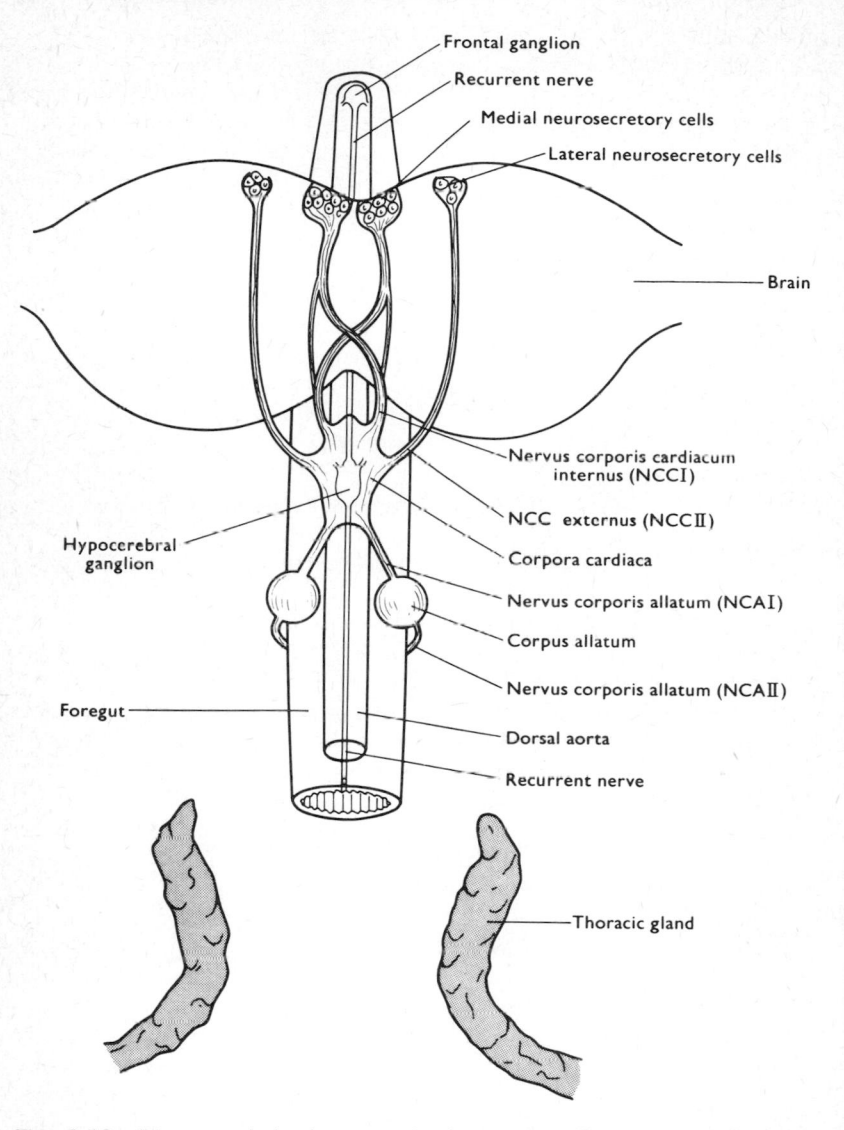

Fig. 2.12 Diagram of the insect endocrine system. Two groups of neuro-secretory cells in each half of the brain send axons to the corpora cardiaca, in close association with the aorta. The majority of the axons from the medial groups of neurosecretory cells decussate before leaving the brain. The corpora cardiaca are connected by nerves with the corpora allata, which also send nerves to the suboesophageal ganglion. The thoracic glands, or their equivalents, are not normally in anatomical connection with the corpora cardiaca and corpora allata.

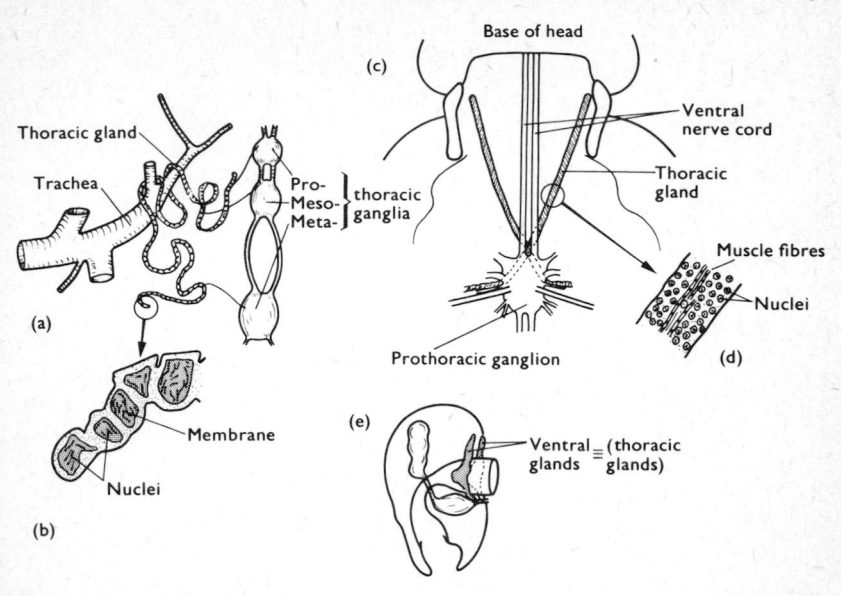

Fig. 2.13 Thoracic glands of the lime hawk moth (a, b), cockroach (c, d) and locust (e). The glands may be innervated from the thoracic ganglia (a, c) although in some insects the glands are found in the head and not the thorax, and are called ventral glands (e). The histology of the glands is variable : in the lime hawk moth, the nuclei are large and deeply folded with little surrounding cytoplasm (b); in the cockroach, the nuclei are small and spherical separated by considerable cytoplasm, and the glands contain small muscle fibres (d). ((c) After Rae and O'Farrell[227])

cular functions to these cell types. Non-neurosecretory axons run with those from the neurosecretory cells to the corpora cardiaca: the neuronal origin of these fibres is unknown.

Other neurosecretory cells occur in the brain and in all the ventral ganglia of the central nervous system[76] (Fig. 2.15). Those in the ventral ganglia are associated with a series of small neurohaemal organs, called the **perisympathetic organs**,[78] found in the vicinity of the small median ventral nerve (Fig. 2.16). With few exceptions (pp. 182, 194) the functions of these cells are unknown.

In many groups of insects the originally paired corpora cardiaca and allata fuse to form single organs. A variety of arrangements of the glands is thus possible[43] (Fig. 2.17). Since the glands originate on either side of the aorta, this vessel is usually closely associated with the fused glands (Fig. 2.17). The greatest degree of condensation of the endocrine system is found in larvae of the Diptera Cyclorrhapha (blowflies, houseflies,

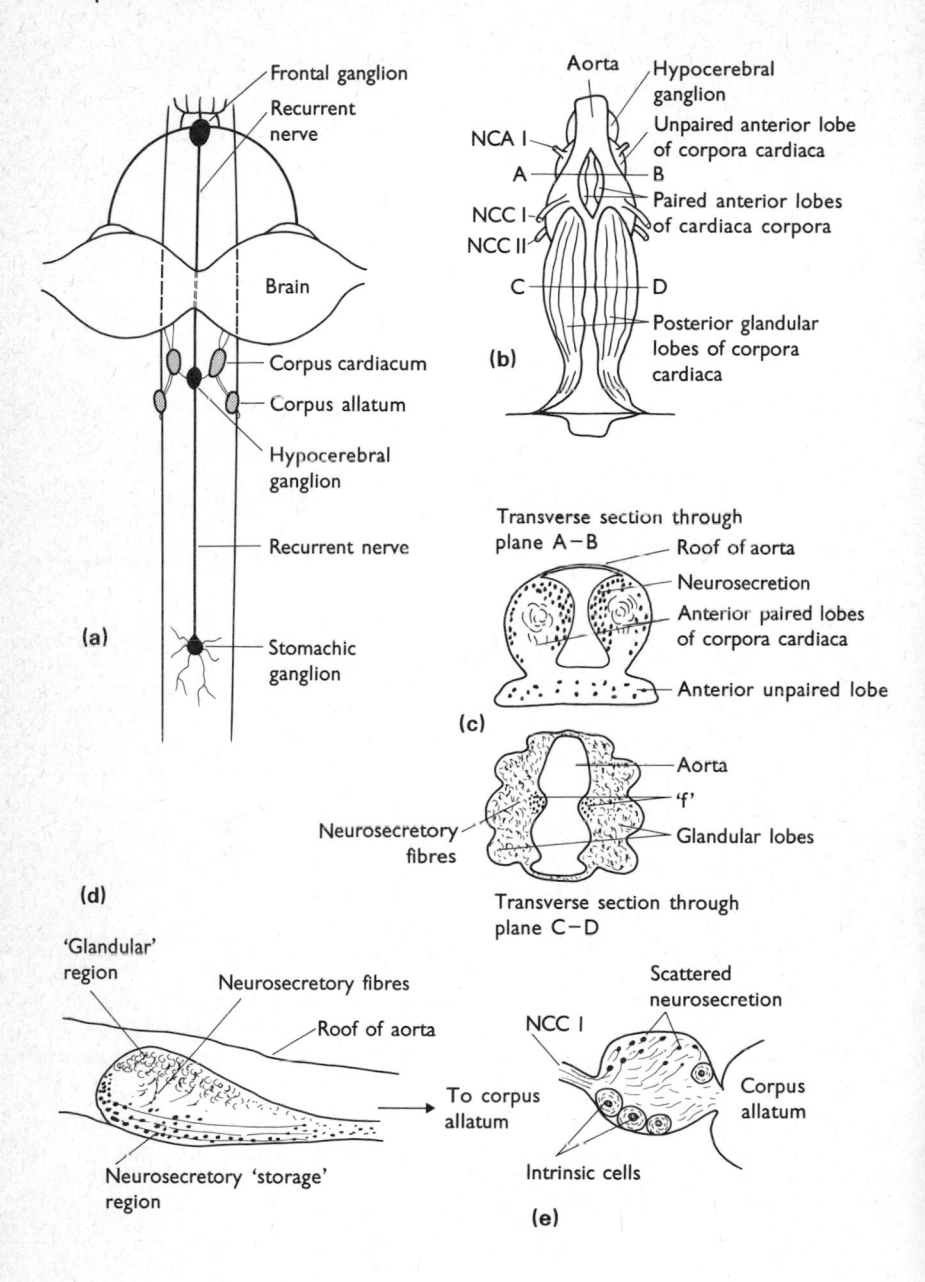

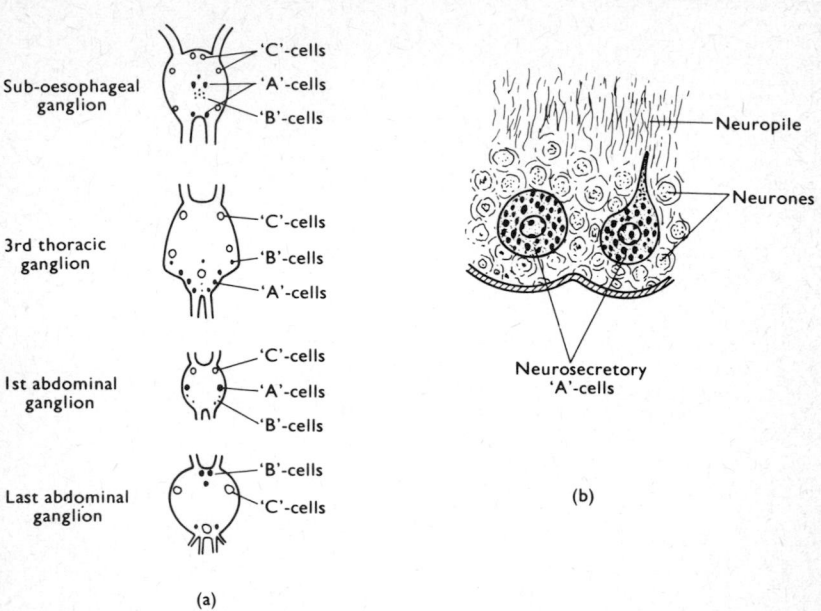

Sub-oesophageal ganglion — 'C'-cells, 'A'-cells, 'B'-cells

3rd thoracic ganglion — 'C'-cells, 'B'-cells, 'A'-cells

1st abdominal ganglion — 'C'-cells, 'A'-cells, 'B'-cells

Last abdominal ganglion — 'B'-cells, 'C'-cells

Neuropile

Neurones

Neurosecretory 'A'-cells

(a)   (b)

Fig. 2.15 (a) Arrangement of individual neurosecretory cells with different staining properties within some of the ventral ganglia of the locust (after Delphin[76]). (b) Drawing of two neurosecretory cells from the third thoracic ganglion of the locust.

fruitflies, etc.) where there is a single corpus allatum dorsal to the aorta, a single corpus cardiacum ventral to the vessel, and the nerves connecting the gland around the sides of the aorta are surrounded by the homologues of the ventral glands (Fig. 2.18). The complex ring-shaped structure is called Weismann's ring after its discoverer.[267]

Fig. 2.14 (a) Relationships between the dorsal sympathetic nervous system (in black), corpora cardiaca and allata (cross hatched) and brain in a generalized insect. (b) corpora cardiaca of the locust, with the aorta cut open to show the anterior lobes. Note the separation of neurosecretory storage from glandular lobes of the corpora cardiaca. (c) Transverse sections through neurosecretory storage lobes and glandular lobes of the locust corpora cardiaca. The glands compose the greater part of the aorta wall in this region. The neurosecretory storage lobes send axons (marked 'f') posteriorly to the glandular lobes, and neurosecretory fibres ramify between the cells of the latter. (d) Longitudinal section through a corpus cardiacum of the cockroach. In contrast to the arrangement in the locust, the glandular region is in intimate contact with the neurosecretory storage region although neurosecretory axons still ramify between its cells. (e) Corpus cardiacum of the lime hawk moth, with large intrinsic cells and scattered neurosecretion.

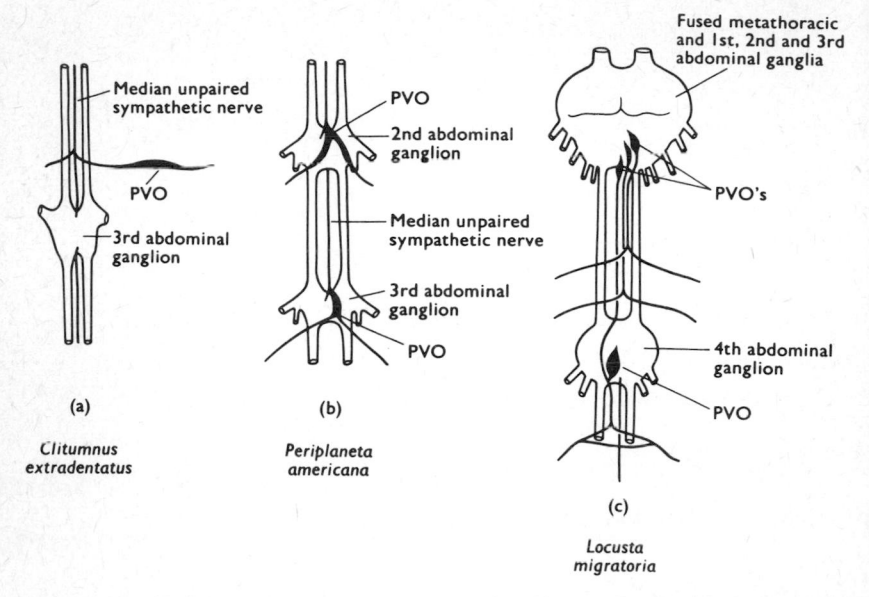

Fig. 2.16   Perivisceral neurohaemal organs (PVO) associated with the ventral ganglia and ventral sympathetic nervous system of a stick insect, a cockroach, and a locust. ((a) After Raabe[78] (b) after de Besse[17b], (c) after Chalaye[44c])

Neurosecretory cells in the central nervous system, corpora cardiaca and allata, and the thoracic glands all originate from ectodermal tissues in the embryo. Recently, the apical tissue of the testes in the glow-worm, *Lampyris noctiluca*, has been shown to produce a hormone[211] (p. 131). Such tissue is mesodermal in origin, as are the pericardial cells around the heart which may produce hormones in some insects (p. 201).

Crustacea

The crustacean endocrine system cannot be categorized so easily as that in the insects. There is a pair of **Y-organs** in the head, which are epithelial endocrine glands analogous to the insect thoracic glands, and in male crustaceans, a pair of **androgenic glands** in the vicinity of the sperm ducts[46] (Fig. 2.19). There are three major neurohaemal organs: the **sinus glands**, in the eyestalks of the stalked-eyed crustacea and within the head of sessile-eyed forms; the **postcommissural** organs, behind the tritocerebral commissure; and the **pericardial organs**, in the vicinity of the heart (Fig. 2.19). The greatest confusion has arisen

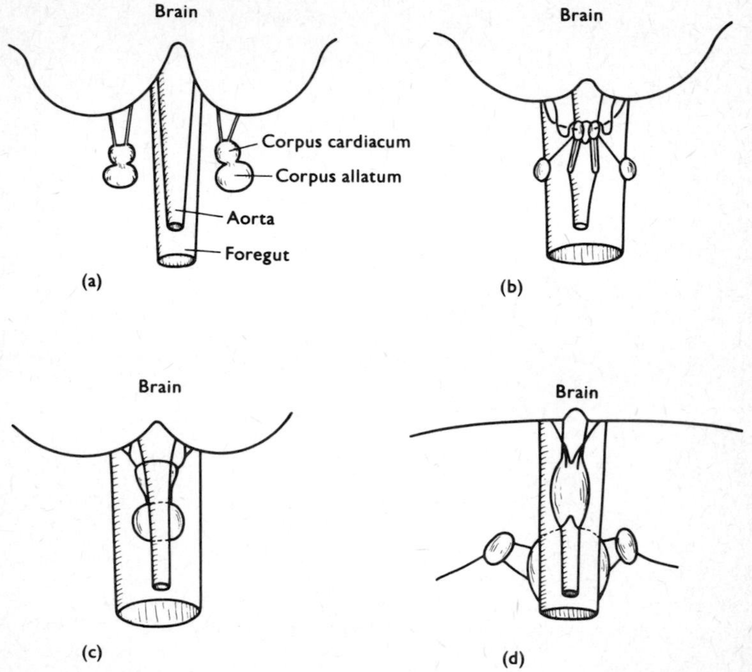

Fig. 2.17 Arrangement of the corpora cardiaca and corpora allata in (a) lime hawk moth, (b) locust, (c) *Pyrrhocoris apterus* (Hemiptera) and (d) *Ephemera danica. Ephemera* is unusual in that the major allatal nerves are those to the sub-oesophageal ganglion. Compare Fig. 2.12.

over the nomenclature of the neurosecretory cells which supply the sinus glands.[118, 41] Many of the cells are within the optic ganglia and are called the **X-organs**, but at least four X-organs, located in different segments of the optic ganglia have been described (Fig. 2.20). Other neurosecretory cell groups in the brain send axons to the sinus gland (Fig. 2.19). Five different axon terminals, differing in shape, dimensions and contained granules, are found in the same gland, and since different granule types are not mixed within individual terminals, they are considered to represent different hormones.[2b] This accords well with what is known of the diversity of hormones which are known to be present in the crustacean eyestalk (Chapters 9 and 11). The postcommissural organs are innervated by neurosecretory cells in the tritocerebrum of the brain; the pericardial organs by at least seven cell groups in the thoracic ganglionic mass.

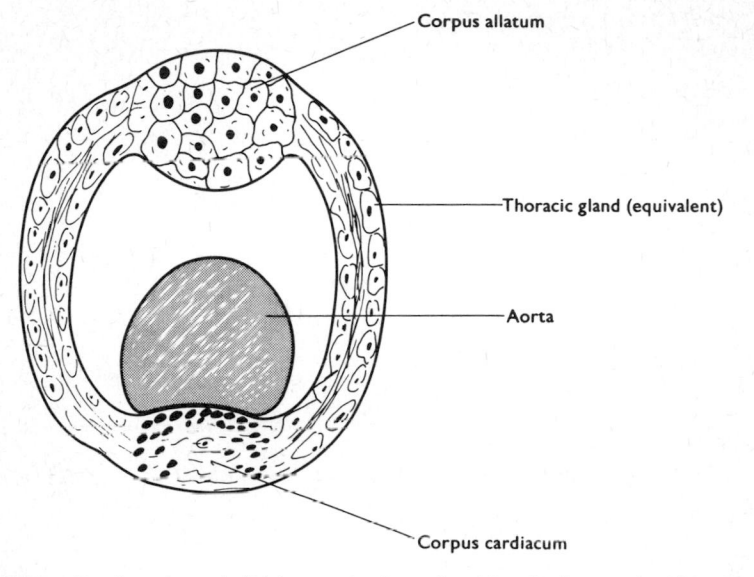

Fig. 2.18 Section through Weismann's ring of a blowfly larva, showing the single corpus allatum dorsal to the aorta, corpus cardiacum ventral to the vessel and the thoracic gland homologues connecting the two.

Other Arthropoda

Protocerebral neurosecretory cells are present in both chilopods and diplopods in the Myriapoda. In these groups, neurohaemal organs analogous to the insect corpora cardiaca and the crustacean sinus glands are also present: these are the ***cerebral glands***[225] (Fig. 2.21). A moulting centre behind the head in *Lithobius forficatus*, controlled by the cerebral glands, suggests the presence of analogues of the insect thoracic glands.[239] Many arachnids also possess cerebral neurosecretory cells, with a variety of neurohaemal organs where neurosecretions are stored and released.[98] In some arachnids, glandular structures are present behind the brain, which show histological signs of secretory activity before the moult and regress in the early adult stages: they consequently could be analogous to the insect thoracic glands[188] (see pp. 115–116). In the King-crabs (Xiphosura) many neurosecretory cells have been described throughout the central nervous system: their density in different regions is correlated with the chromatophorotrophic activity of extracts of these regions.[233] It is likely that the overall endocrine control of moulting in the arthropods is similar to that in insects and crustaceans (see Chapters 6 and 9).

Since little detailed experimental work has been performed on arthropods other than these two major groups, the remainder will not be discussed further.

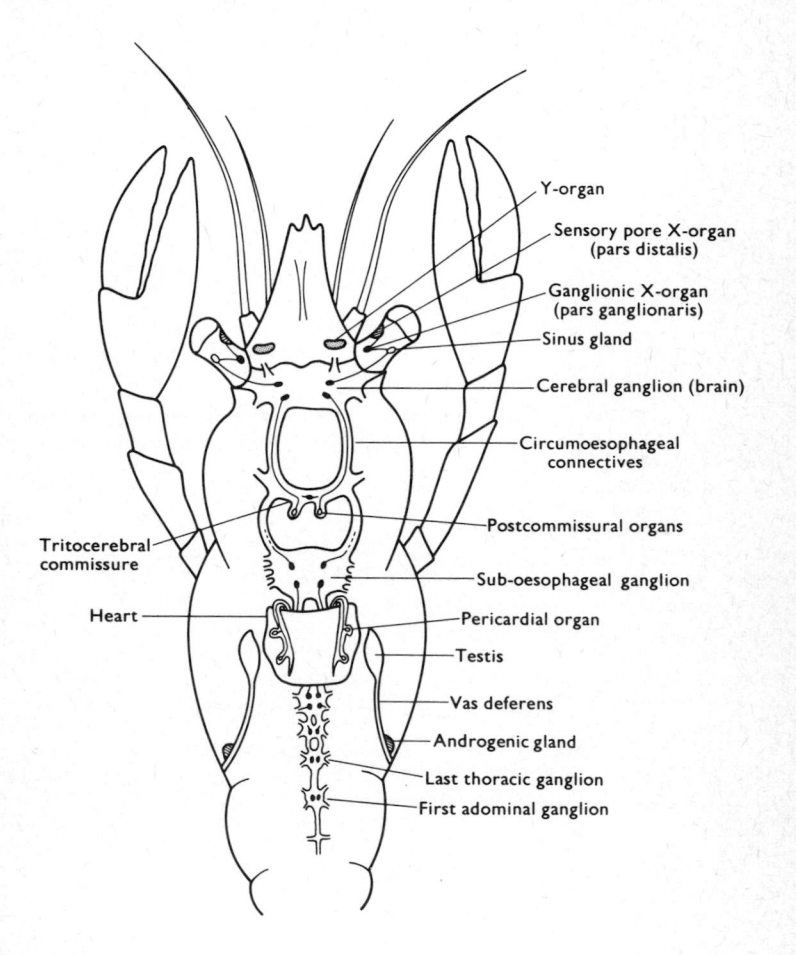

Fig. 2.19 The endocrine system of a generalized male crustacean. Neuro-secretory cells (shown in black) are found in the eyestalks, brain, sub-oesophageal ganglion and throughout the remainder of the central nervous system. The neuro-haemal organs (shown in white) may be supplied by several groups of neuro-secretory cells. The sinus glands receive axons from the ganglionic X-organ and from the brain ; the postcommissural organs receive axons from the brain ; and the pericardial organs, axons from the thoracic ganglia. Epithelial endocrine glands (cross hatched) are the Y-organs and the androgenic glands. (After Gorbman and Bern[118])

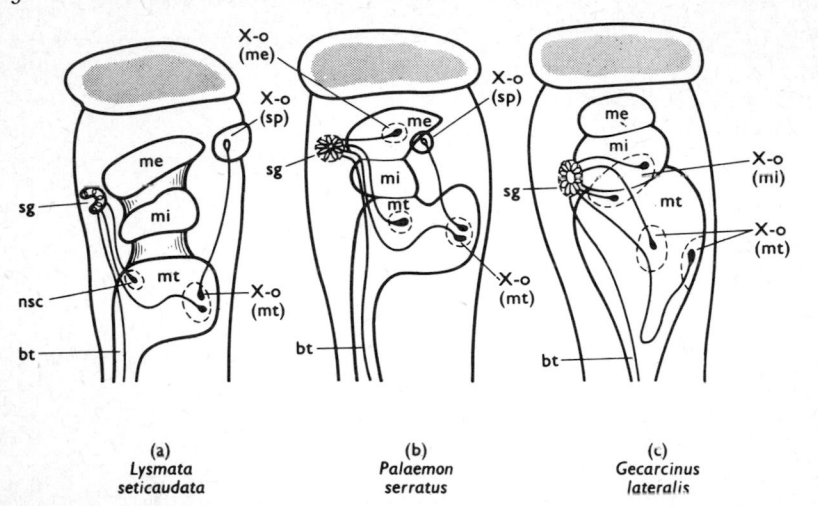

<table>
<tr><td>(a)
*Lysmata
seticaudata*</td><td>(b)
*Palaemon
serratus*</td><td>(c)
*Gecarcinus
lateralis*</td></tr>
</table>

Fig. 2.20   Arrangements of neurosecretory cell groups within the eyestalks of different crustaceans. Dark areas represent neurosecretory cells, light ovals their axon terminals.

(a) *Lysmata seticaudata*. The X-organ consists of neurosecretory cells in the medulla terminalis, with the sinus gland as its major neurohaemal organ. The sinus gland also receives axons from neurosecretory cells in the brain. But some neurosecretory cells in the medulla terminalis send axons to the sensory pore 'X-organ' which, despite its name, is another neurohaemal organ.

(b) *Palaemon serratus*. There are two areas of neurosecretory cells, called ganglionic X-organs, one in the medulla terminalis and one in the medulla externa, both of which send axons to the sinus gland where they are joined by neurosecretory axons from the brain. The sensory pore is absent in this species, but the sensory pore 'X-organ' remains, supplied by neurosecretory axons from the medulla terminalis.

(c) *Gecarcinus lateralis*. In this species there are several ganglionic X-organs in the medulla terminalis and the medulla interna with axons terminating in the sinus gland together with those from the brain.

Abbreviations : bt: neurosecretory axons from the brain terminating in the sinus gland; me, mi, mt: medulla externa, interna and terminalis; nsc: neurosecretory cell group; sg: sinus gland; sp: sensory pore; X-o: X-organ.
(After Gorbman and Bern[118])

THE ORIGIN OF NEUROSECRETORY CELLS

Neurosecretory cells have been defined as neurones which also show the cytological characteristics of gland cells. They receive nervous impulses, but these are not transmitted to other neurones or effector cells: instead the neurosecretory axons terminate in close proximity to parts of the circulatory system, and release substances which act on distant

effector organs or upon endocrine glands. These neurosecretory substances, therefore, are themselves hormones.

This definition implies that neurosecretory cells are modified neurones,
with a glandular function superimposed upon their fundamental neuronal
character.[122] But the opposite could equally well be true: neurosecretory
cells may have been originally gland cells which later acquired some of
the properties of neurones.[52] The morphological relationships of the
nemertean cerebral organ (p. 11) have been used to support this view.[232]

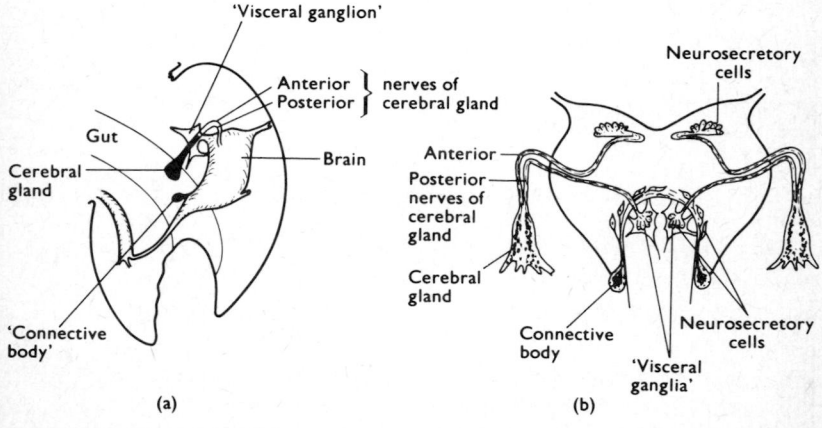

Fig. 2.21 Neurosecretory cells and neurohaemal organs in *Jonespeltis
splendidus* (Myriapoda). (a) From the side, (b) in horizontal section. Groups of
neurosecretory cells in the brain and visceral ganglia send axons to the major
neurohaemal organs, the cerebral glands. Other cells are associated with the
connective bodies, which are likely also to be neurohaemal organs. (After
Prabhu[225])

A similar incorporation of epidermal mucous cells into the posterior lobes
of the brain in the polychaete nephtyids (Fig. 2.22) could be another
example.[53] In the phyllopod crustaceans and copepods there is a sensory
frontal organ often associated with a large secretory cell. The frontal
organ is incorporated into the central nervous system in the malacostracan Crustacea, where it forms an X-organ, again revealing a connection
between an originally epidermal glandular (and sensory) structure and a
later neurosecretory centre.

But evidence for the origin of neurosecretory cells derived from *living*
representatives of modern animal groups must be considered extremely
tentative. Indeed, the widespread occurrence of neurosecretory
cells in animals as diverse as coelenterates and man, together with the
fact that in arthropods and vertebrates the major neurosecretory cell

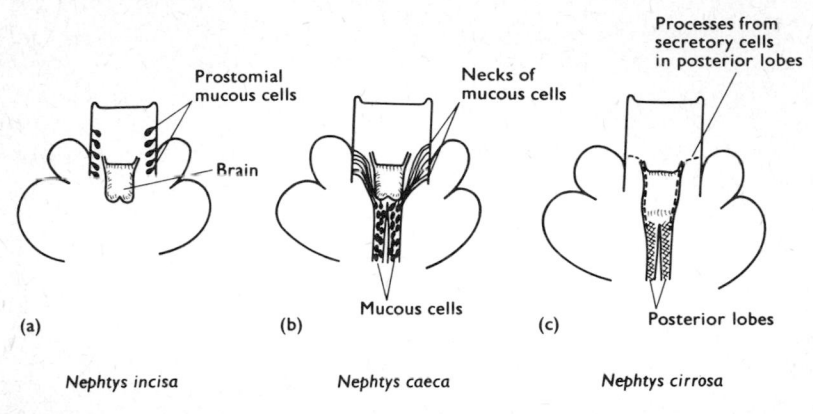

Fig. 2.22 Incorporation of mucous cells into the brain of nephtyids. In (a), the cells are superficial and separate from the brain. In (b), the cells form two lobes posterior to the brain, but their necks open externally. In (c), the posterior lobes of the brain contain neurones and gland cells, although processes still extend from the cells to the epithelium. (After Clark[53])

groups are found in the phylogenetically oldest parts of the brain, suggest a very ancient origin for the cells. Arguments for either a neuronal or a glandular origin of neurosecretory cells become largely irrelevant if it is assumed that the primaeval metazoan nervous system had two major functions, enabling the animal to react quickly to short-term fluctuations in its surroundings and also to co-ordinate developmental and other processes with much more extended environmental change. The neurosecretory cell would thus be as old as the neurone, and speculation about its nervous or glandular origin as unrewarding as that concerning the chicken and the egg.

Much that is known about the function of neurosecretory cells in many animals (Chapter 14) supports this point of view. The morphological similarities between neurosecretory cells and neurones would merely reflect a common origin. Indeed, functionally there is less difference between neurosecretory cells and neurones than is commonly accepted. Neurones are as glandular as neurosecretory cells, manufacturing and secreting chemicals: the essential difference between them is in the kind of chemical produced. Neurones produce materials such as adrenaline and acetylcholine which have a local action at the synapse and are rapidly destroyed. Neurosecretory cells produce a range of peptide or small molecular weight proteins which are comparatively long-lived and act at a distance. But all these substances are secretory products of the cells. In many insects, neurosecretory axons terminate in the corpora allata (Fig. 2.12), and their secretions have a local action upon the cells of the

glands.[235, 214] In other insects, it is claimed that *all* the neurosecretory axons terminate within the organs they control[156] (Fig. 2.23). And in many animals, including the vertebrates, some neurosecretory axons terminate in different parts of the neuropile of the central nervous system perhaps affecting transmission across neuronal synapses.[13] So the distinc-

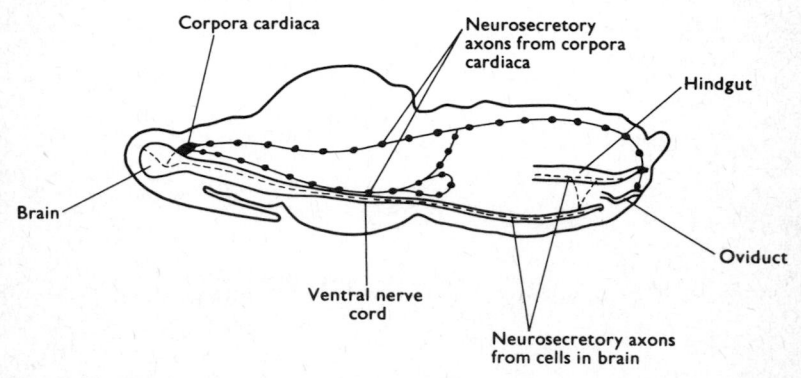

Fig. 2.23   Part of the closed neurosecretory system of the aphid. Neurosecretory axons from the brain and corpora cardiaca run to many parts of the body. Those innervating the hindgut and oviduct are shown here (After Johnson[156]).

tion between neurosecretory cells and ordinary neurones becomes even more blurred, not unexpectedly if the two kinds of cell have a similar, and ancient, origin. The implications of considering neurosecretions to be the 'oldest' animal hormones will be discussed in Chapter 14.

3

Endocrine Mechanisms in the Coelenterata, Turbellaria, Nemertea, Nematoda and Echinodermata

In the previous chapter, neurosecretory cells were said to occur within the nervous systems of a number of the less highly organized invertebrates. Neurohaemal organs seem to be absent from most species, and epithelial endocrine glands do not occur. The difficulties of ascribing functions to dispersed neurosecretory cells, described in Chapter 1, are encountered immediately in these groups.

COELENTERATA

Growth, regeneration, and the development of sexuality in species of *Hydra* provide the most convincing evidence for the existence of hormones in the Coelenterates.[29, 30, 31] In normal *Hydra pseudoligactis*, the growth region lies just proximal to the hypostome and active cell proliferation occurs here. In this region, many nerve cells are found with long, broad axons which are enlarged at intervals into vesicles which often contain droplets which can be stained with vital dyes[29] (Fig. 2.1). The cells may also be stained in fixed material with paraldehyde-fuchsin —a stain much used to differentiate neurosecretory cells in other animals. But in *Hydra*, paraldehyde-fuchsin also stains nematoblasts, mucous droplets in musculo-epithelial cells, and other materials, and vital staining of the nerve cells is to be preferred.

When the hypostome of *Hydra pseudoligactis* is cut and removed from the body column, the numbers of droplets in the nerve cells increase considerably for up to four hours after the operation, and then decrease during the subsequent two hours. But during this latter period, the droplets appear within the tissues surrounding the nerve cells, apparently being released from the nerve cells. There seems every justification for calling these droplet-containing neurones neurosecretory cells. Rapid cell proliferation and regeneration of the body column follows the induced activity of neurosecretory cells in the excised hypostome.[30]

The presence of neurosecretory cells almost exclusively in the growth region of *Hydra*, and their hyperactivity when the accelerated growth characteristic of regeneration is induced, provides strong circumstantial evidence for the production of a growth hormone by the neurosecretory cells. Moreover, normal growth in *Hydra* is almost invariably accompanied by budding. The development of the bud is initially controlled by the parent's growth centre: it dies if prematurely separated. But just as the tentacle outpushings appear on the bud, the bud's own growth centre is developed—and this coincides exactly with the first appearance of neurosecretory cells in this region of the bud.

But there is more direct evidence for the existence of a growth hormone produced in this region. Extracts of the hypostome will stimulate growth, in the form of supernumerary heads, when applied to any part of the body column (Fig. 3.1). Extracts of other parts of the body do not have this

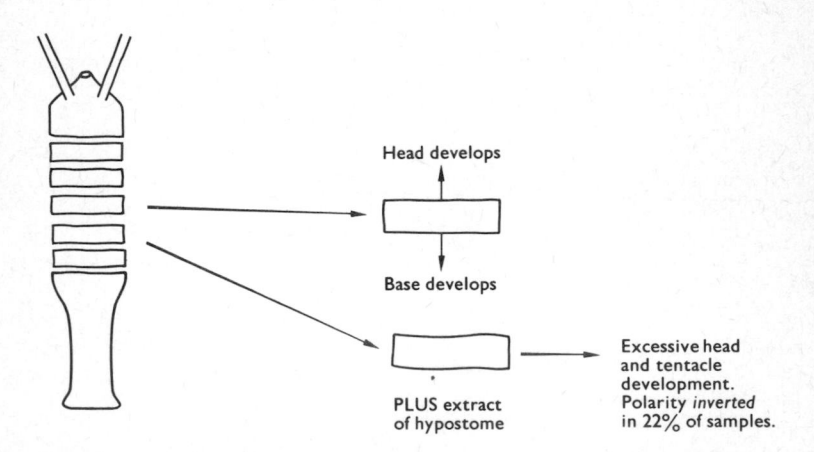

Fig. 3.1 Any part of the column of *Hydra* will reconstitute an individual when isolated. But when hypostomal extract is added, many supernumerary heads and tentacles are formed and in a proportion of samples these develop on the base side of the sections. (After Burnett, Diehl and Diehl[30])

effect. It must be emphasized that it is impossible to be sure that the hypostomal neurosecretory cells are the exact source of the active factor in the extracts, but such an assumption is supported strongly by the observations described above.

Neurosecretory cells in the sub-hypostomal growth region of *Hydra pseudoligactis* are scarce or absent in fully sexual forms. Hypostomal extracts from sexual *Hydra pirardi* will not induce growth when applied to the body columns of other individuals. It seems likely, therefore, that the disappearance of the neurosecretory cells, with the consequent absence of growth hormone, is associated with sexual development.[31]

When the hypostome of a growing *Hydra pirardi* is grafted onto the body column of a sexual *Hydra fusca*, the testes of the latter do not mature. Instead, they begin to develop into small buds (Fig. 3.2). The growth hormone therefore inhibits the onset of sexuality.

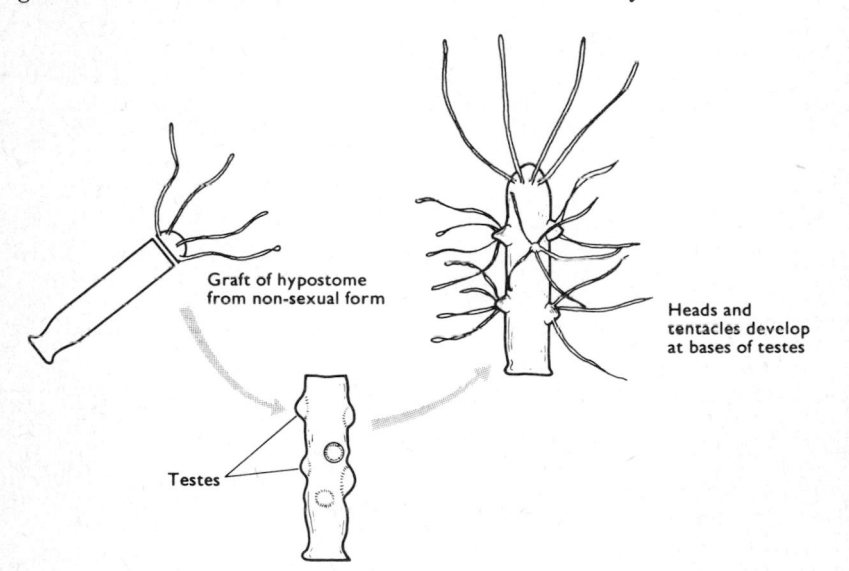

Graft of hypostome
from non-sexual form

Heads and
tentacles develop
at bases of testes

Testes

Fig. 3.2 A sexual *Hydra fusca* transforms its testes into heads and tentacles when its hypostome is replaced with that from a non-sexual individual. The non-sexual hypostome contains a factor which favours somatic rather than reproductive development.

In *Hydra*, the interstitial cells can develop into nematoblasts, nerve cells, or other somatic structures. But they can also differentiate into gametes, and in species in which there is no large store of interstitial cells, somatic and reproductive differentiation are antagonistic. Thus the

tentacles of the sexual forms of *Hydra pirardi* can be quite devoid of nematoblasts.

When individuals of *Hydra pirardi* with well formed testes are severed through the middle of the body column, they will remain in the sexual state for a considerable time. But when the sub-hypostomal regions of growing individuals are grafted onto the excised surfaces of the sexual forms, then the cells at the bases of the gonads show extensive differentiation into nematoblasts within 24 hours. As long as the interstitial cells have not undergone their first meiotic division, the presence of growth hormone switches their development from gamete formation to nematoblast development (Fig. 3.3), in other words away from reproductive and

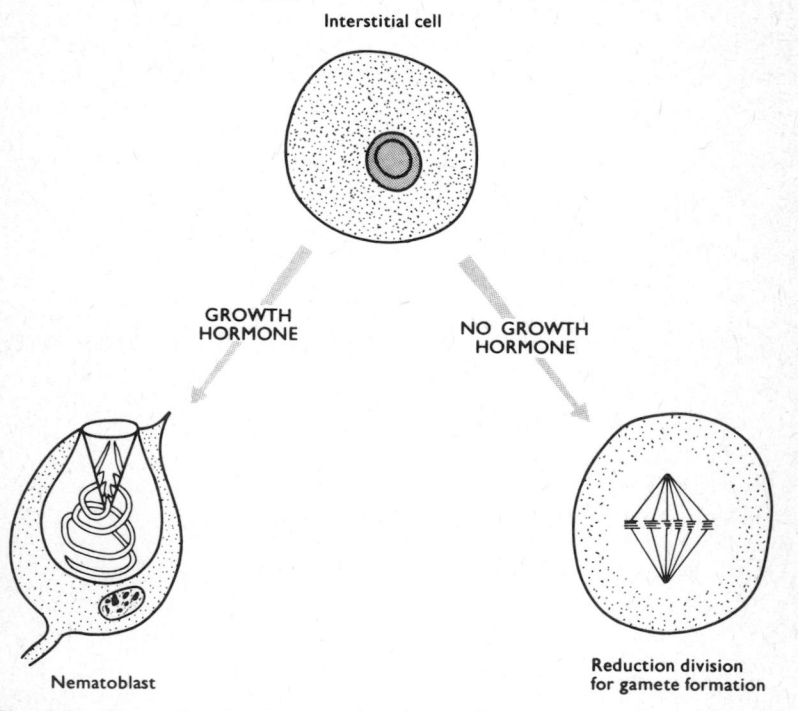

Fig. 3.3 Alternative developmental pathways for an interstitial cell in *Hydra*. In the presence of growth hormone from the hypostomal region, the cell transforms to a nematoblast; in the absence of the hormone, meiosis occurs, leading to gamete formation.

towards somatic development. Very often, the grafted individuals will develop supernumerary heads or tentacles.[30]

The neurosecretory growth hormone produced in the sub-hypostomal

region therefore has two functions: it induces cell proliferation and growth, and it directs interstitial cell development into a somatic course. In the absence of the growth hormone, growth naturally ceases, but in addition, the interstitial cells then develop along a reproductive course.

It follows from this that growth and reproduction in *Hydra* are antagonistic processes, as they are in many other animals. In different species of *Hydra*, various environmental conditions will inhibit growth, and consequently stimulate sexuality. In *Hydra pseudoligactis*, bringing animals from a temperature of 20–21°C to 8°C will induce egg formation within 30 days. In *Hydra pirardi* the same treatment causes testes formation. But reducing the temperature inhibits sexuality in *Hydra littoralis*. In *Hydra pseudoligactis* and *Hydra pirardi*, a progressive shortening of the day length will induce sexuality, although the effect must in some way be combined with temperature because increasing daylength during the Spring will also induce sexuality in *Hydra pseudoligactis*.[31]

Besides temperature and photoperiod changes, sexuality in *Hydra* may be induced by starvation, an increased carbon dioxide concentration, stagnation in the culture medium, and so on. All these factors are likely to interfere with growth, either directly, or indirectly through an effect upon the sub-hypostomal neurosecretory cells. The resulting development of sexual forms, with the eventual production of a resting zygote, is clearly of advantage to the species in overcoming the unfavourable conditions. In some species the reaction to changes in daylength also favours the production of an overwintering zygote. It is of considerable interest that neurosecretory mechanisms are involved in these protective adaptations in Coelenterates in much the same way as they are in much more complicated animals like the insects (see p. 177*ff*).

Neurosecretion has also been implicated in the control of developmental events in another coelenterate group, the Scyphozoa.[64e] In some scyphozoans, the polyp is the asexual form and the medusoid sexual generation is formed by strobilization (transverse fission) of the polyp to produce ephyrae. In *Chrysaora quinquecirrha*, strobilization can be divided into three stages: neck formation, where the polyp has one distinct constriction beneath the base of the tentacles; segmentation, where the polyp has a series of constrictions which divide the body into a number of segments; and metamorphosis, where the polyp possesses mature ephyrae, and the tentacles, septal muscles and nematocysts of the polyp have been resorbed and the new structures of the ephyrae, such as lappets, rhopalia and manubria have appeared. Before and during neck formation, neurosecretory cells between the epithelio-muscular cells can be recognized easily by their content of dense, membrane-bound granules. As segmentation starts, the granules disappear from the perikarya of the neurosecretory cells and there is a dramatic increase in the number of granules in the

axons. During metamorphosis, most of the perikarya and axons are devoid of neurosecretory granules. These observations suggest a movement and release of neurosecretion during segmentation, and thus strobilization could be controlled by the release of a neurosecretory hormone. There is some evidence that strobilization is coordinated with environmental events, and the neurosecretory cell could therefore be the link between the environment and this particular aspect of development, but until these histological interpretations are subjected to the rigours of experimental method, the conclusions must be considered tentative.

TURBELLARIA

There is increasing evidence that regeneration and reproduction in Turbellaria are controlled by neurosecretory hormones. In many triclad worms, reproduction and regeneration are intimately linked, since asexual reproduction, which occurs by transverse fission of the animal, is inhibited during sexual reproduction. After transverse fission, the anterior fragment of the worm which retains the pharynx regenerates a new tail while the posterior fragment regenerates a new head.

In *Dugesia gonocephala*, the number of neurosecretory cells in the brain can be correlated with the asexual fission cycle. At $21\,^{\circ}$C, fission occurs at about 21 day intervals, and the number of neurosecretory cells oscillates with a period coincident with the fission cycle[189a] (Fig. 3.4). During sexual reproduction, when transverse fission is inhibited, the number of neuro-secretory cells in the brain remains high.[188a] These observations suggest that neurosecretion plays some role in the control of fission, but they are not conclusive. Firmer evidence comes from the experimental demonstration that decapitated worms show a much higher rate of fission than either control animals with an equivalent amount of tail tissue removed, or intact controls. Neurosecretion may therefore *inhibit* fission.[17d]

In *Dugesia tigrina*, there is also some interaction between individuals to control population density. Individual worms placed in 50 cm^2 of medium increase in number exponentially if the fission products are removed and isolated in further 50 cm^2 of medium. When 10 worms are placed in 500 cm^2 of medium and the volume of the medium is increased as fission occurs to maintain the same population density, the population growth is much less than exponential—numerical increase of the grouped animals is inhibited. There is no evidence for the release of an inhibitory substance into the medium, but the fact that grouped animals come into direct contact with each other chiefly along their lateral margins suggests that tactile stimuli, or even some kind of pheromonal substance, may affect the control of fission. There is no evidence that neurosecretion is involved at

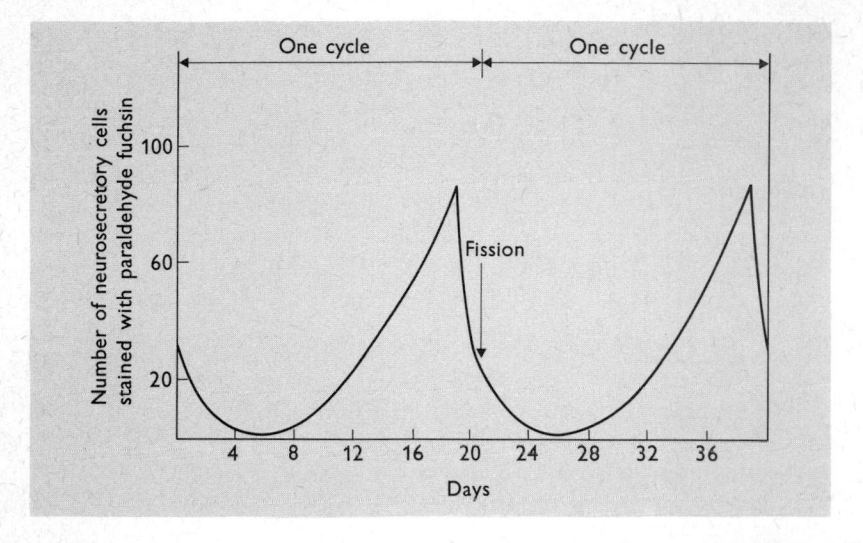

Fig. 3.4 Variations in the number of paraldehyde-fuchsin positive neurosecretory cells in the brain of *Dugesia gonocephala* during the asexual fission cycle. At 21°C the fission cycle takes 21 days : the number of stainable cells rises towards the time of fission and then drops sharply just before fission, suggesting that neurosecretion is released at this time. (After Lender and Zghal[189a])

any stage in this process,[75b] but the evidence for fission control in *D. gonocephala* (see above) suggests this possibility.

Freshwater planarians have long been known to possess quite extraordinary powers of regeneration, in addition to asexual fission, and this ability has been used to study the nature of regeneration and the role of the brain in its control. Undifferentiated cells, the neoblasts, migrate to the vicinity of a wound or an amputation, and are then able to divide and to differentiate into any tissue which has been damaged or removed. The neoblasts are thus both motile and totipotent. If the wound which they repair is extensive, they form a blastema of regenerating tisue rather similar to that found in polychaete worms after segment amputation (p. 56).

What initiates the migration of neoblasts towards the damaged tissues? In *Euplanaria lugubris*, x-irradiation of the anterior part of the body at a level sufficient to destroy most of the tissues is followed by necrosis of the irradiated region and the subsequent death of the individual.[189] But if part of the irradiated region is cut off just before or after x-irradiation, then neoblasts migrate from the posterior, non-irradiated part of the body, and replace not only the amputated section, but the rest of the irradiated region also (Fig. 3.5). The result is the same when the cut is

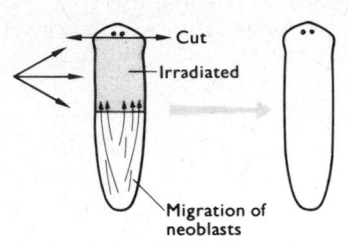

Fig. 3.5 Regeneration in *Euplanaria lugubris*. When an anterior cut is made and part of the body subsequently irradiated, the whole of the irradiated and amputated regions are regenerated from neoblasts which migrate from the posterior parts of the body. Compare Fig. 3.6.

made up to four days before subsequent irradiation; but a longer interval between section and irradiation prevents regeneration. If the cut is made some distance from the anterior border of the individual, and the area is then irradiated, regeneration will not extend anterior to the cut (Fig. 3.6). This region becomes necrotic and dies.

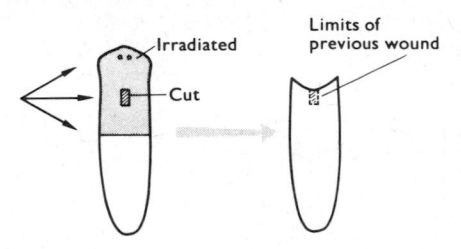

Fig. 3.6 Regeneration in *Euplanaria lugubris*. When a wound is made behind the anterior tip of the body, and the region subsequently irradiated, regeneration does not occur anterior to the wound. The wound is the source of factors (wound hormones) which stimulate the neoblasts to migrate and regenerate the damaged tissues.

The results of these experiments on *Euplanaria lugubris* suggest very strongly that the neoblasts are activated by, and migrate towards, some factor produced by the damaged region of the body. The first steps in regeneration are thus set in motion by, for want of a more precise term, 'wound hormones' produced by the damaged cells and tissues. Little is known about these substances: but their presence in wounded tissues has been inferred not only in planarians, but in worms, insects and vertebrates too.

Having arrived at the site of damage, what controls the differentiation of neoblasts into specific cells and tissues? This problem has been examined in the control of eye regeneration in *Polycelis nigra*. This

planarian, as its name implies, has many (about 80–90) eyes on the anterior borders of the head. If an eye-bearing region is amputated, ocular regeneration occurs in about 7 days at 18°C. But if the brain is removed at the same time, the eyes do not regenerate at all. This absence of regeneration is not due to removal of the ocular nerves, because the retinal and capsule cells begin regeneration, in normal animals, *before* the nerves are formed. Moreover, a section of the anterior border with eyes removed from one individual will regenerate eyes when grafted onto a normal host, in the vicinity of the brain, without any nervous connections between host and graft being established (Fig. 3.7). Thus the brain is

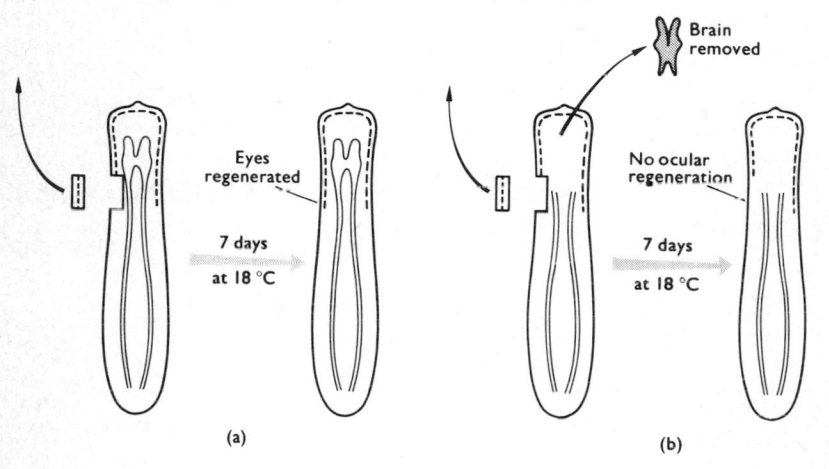

Fig. 3.7   Ocular regeneration in *Polycelis nigra*. In the presence of the brain, eyes are regenerated when part of the ocular region is amputated (a). Where the brain is removed simultaneously with amputation of part of the ocular region, eyes are not regenerated (b). The brain is necessary for ocular regeneration in *Polycelis*.

necessary for ocular regeneration in *Polycelis*, and it exerts its effect at a distance: it would seem that the brain produces a regeneration hormone.[189]

Neurosecretory cells are present in the posterior ventral parts of the brain of *Polycelis*.[189] Homogenates of the anterior parts of planarians added to the water in which they are reared, will induce ocular regeneration in decerebrate individuals. The active principle will withstand temperatures of 60°C for 10 minutes, but is destroyed by boiling for 30 minutes. A homogenate of 5 heads in 10 cm² of water produces only a little ocular regeneration; 10–15 heads in the same volume of water produce normal regeneration; 20 heads accelerate regeneration by 24–48 hours.

When an anterior segment, with eyes removed, is grafted to the posterior part of a normal animal, the eyes do not regenerate[189] (Fig. 3.8). Clearly, the brain cannot influence caudal tissues. But when tails

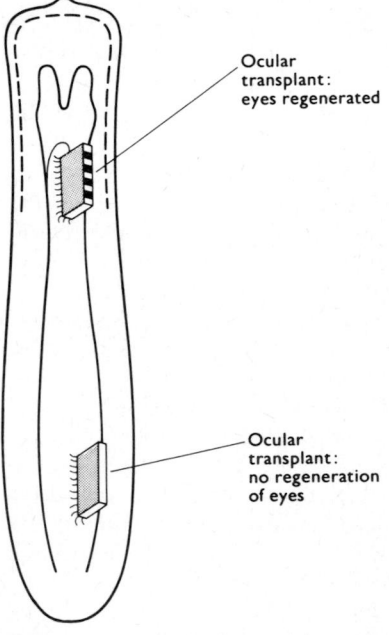

Fig. 3.8 Ocular regeneration in *Polycelis nigra*. Part of the ocular region transplanted from another individual to a site near to the host brain will regenerate eyes. Transplanted to a posterior site, no eyes are regenerated. The influence of the brain in inducing regeneration does not extend very far posteriorly (Figs. 3.5–3.8 after Dubois and Lender[77h]).

are homogenized, and the extract added to water in which decerebrate *Polycelis* with eye segments removed are kept, some ocular regeneration will occur. If the tails are first treated with 70% alcohol, or heated to 60°C for two minutes, or are finely ground with sand, then the extracts are more potent in inducing ocular regeneration. It is concluded that not only is there a gradient of regeneration hormone, high anteriorly and decreasing caudally, but that in addition the cells in the tail region actually absorb or sequester the hormone to reduce its effectiveness in the normal animal.[189]

The regeneration hormone cannot be definitely identified as a product of the neurosecretory cells in the brain. Moreover, although the hormone

is necessary for ocular regeneration, eyes are only formed in those regions where they normally occur. The interaction of some other factor with the regeneration hormone must determine that neoblasts in this region will differentiate into eyes. The situation is thus rather similar to the control of segment regeneration in polychaete worms (p. 62). It must also be emphasized that this property of evoking regeneration is not confined to the *Polycelis* brain. Other substances, particularly extracts of chick embryos, will also induce ocular regeneration. The evidence for the existence of neurosecretory hormones controlling regeneration in flatworms is not yet conclusive, but taken in conjunction with that for the control of fission (p. 39) it makes such a control very probable.

Indirect evidence suggests that neurosecretions are involved in the control of sexual reproduction in *Dugesia lugubris* and *Polycelis nigra*. Ultrastructural studies have shown the presence of neurosecretory fibres near the ovaries and testes (p. 11), and the numbers of granules within these fibres can be correlated with gonadal development. While gamete maturation is in progress, the cellular content of neurosecretory granules is high, and is much reduced when the maturation process ceases. The claim that this indicates a neurosecretory stimulation of gametogenesis[118h, 118i, 118j] must be treated with caution in the absence of experimental evidence, and in view of the difficulties in interpreting histological pictures in terms of the turnover in neurosecretory cells (p. 173). How such possible control of gametogenesis might relate to the control of asexual fission, which is in turn related to sexual activity (p. 39), is problematical.

NEMERTEA

In a few heteronemertean species, experiments have shown that sexual maturation is controlled by a neurosecretory hormone from the cerebral ganglion. In *Lineus ruber*, sexual differentiation begins towards the end of the first year of life; by the end of the second year, the animal is fully mature sexually and subsequently undergoes annual reproductive cycles. Sexual maturation in *Lineus* involves both the maturation of the gonads and somatic changes in cuticular glands and gonoducts.

The removal of the cerebral ganglion from animals less than one year old induces premature sexual differentiation; the transplantation of cerebral ganglia from young animals into such decerebrate individuals prevents the onset of premature maturation. These results indicate that the cerebral ganglia of young *Lineus* produce a hormone which inhibits sexual maturation. Similar experiments with mature animals which are sexually quiescent show that the annual reproductive cycle is also controlled by the cerebral hormone.

Careful experiments involving the removal and reimplantation of different parts of the head complex show that the source of the hormone is the cerebral ganglion and not the cerebral organs (p. 11), and it is concluded that neurosecretory cells in the ganglion produce the hormone.[17i, 17j, 17k] The hormonal control of maturation in *Lineus* thus strikingly resembles the control found in nereid polychaetes (p. 68).

Osmoregulation may also be controlled by neurosecretory hormones in the heteronemertean *Lineus viridis*. When placed in dilute sea water, individuals gain weight rapidly as water enters the body, but within 72 hours the body weight returns to normal. Removal of the cerebral ganglion prevents the efflux of water; reimplantation of the ganglion allows the osmotic response to occur normally. The neurosecretory cells of the cerebral ganglion undergo histological changes suggestive of hyperactivity during the first few hours of osmotic stress[186a] and it is to be supposed that they are the source of a diuretic hormone. The cerebral organs are also involved in the osmotic response, perhaps acting as sensors to monitor the changed osmotic conditions.

Although the problems of precisely defining scattered neurosecretory cells in the central nervous system as particular sources of hormones are acute (p. 8), the classical endocrine experiments—removal and reimplantation of the suspected source of hormones (p. 7)—have been carried out in these heteronemertean species, and the evidence for the involvement of neurosecretory hormones in these regulatory processes thus has a firmer basis than in the Turbellaria.

NEMATODA

Growth, cuticle formation, ecdysis and gonad development are among the postembryonic developmental events in nematodes that, by analogy with other animals, could be hormonally controlled. However, nematodes are difficult animals to work with experimentally: they are entelic, with little or no capacity for regeneration, and their high internal hydrostatic pressure imposes insuperable problems upon surgical experiments which involve disruption of the body wall. Despite these limitations, evidence has been obtained for endocrine regulation of some processes in nematodes.

Nematodes moult their cuticles in a manner very similar to that of the arthropods. A new cuticle is first deposited under the old, and the latter is then partly digested by moulting fluid before splitting along a circumferential line of weakness to allow the emergence of the next developmental stage. One of the enzymes responsible for digestion of the old cuticle is leucine aminopeptidase.[74]

Phocanema decipiens is parasitic in the muscles of the cod, where it occurs as a last stage larva, and in the digestive tract of the seal, where the final moult to the adult stage takes place very soon after ingestion. This final moult can be duplicated in a suitable incubation medium *in vitro*, enabling accurate observations to be made on the control of the moult.

When fourth stage larvae of *Phocanema* are removed from the cod and placed in the incubation medium, a new cuticle begins to be formed within 12 hours, and ecdysis (the moulting of the old cuticle) occurs 3 to 5 days later. Two groups of neurosecretory cells are present in *Phocanema*, one in the dorsal and the other in the ventral ganglion (p. 11). Within 12 hours of being placed in the incubation medium, neurosecretory granules appear in the cells, reaching a peak of accumulation between the second and fifth days. After the sixth day, the granules have disappeared.[74]

Although these histological changes suggest that the neurosecretory cells are active between the second and fifth days of incubation, there is no evidence that neurosecretion is released from the cells during this time. In fact, when the larvae are ligatured to prevent any hormone, if present, from reaching the posterior parts of the body, new cuticle is deposited normally both in front of and behind the ligature. The activity of the neurosecretory cells is not concerned with cuticle deposition.[74]

Leucine aminopeptidase is secreted by the so-called excretory gland. In *Phocanema*, a cycle of synthesis and release of the enzyme occurs at the time of ecdysis. If slight changes are made in the constitution of the incubation medium, the neurosecretory cells do not show their cycle of activity, there is no leucine aminopeptidase production in the excretory gland, and ecdysis is inhibited, although new cuticle formation proceeds normally. Are the neurosecretory cells therefore concerned, not with cuticle deposition, but with ecdysis?

When excretory glands are incubated with saline extracts of the anterior ends of *Phocanema* in which the neurosecretory cells are empty, there is no increase in leucine aminopeptidase in the glands. But incubated with similar extracts of *Phocanema* with full neurosecretory cells, the leucine aminopeptidase activity of the excretory glands increases considerably.[74] It is likely, therefore, that in normal *Phocanema* a hormone from the neurosecretory cells controls the production of moulting fluid enzymes necessary for the casting of the fourth stage larval cuticle. Moreover, conditions must be suitable before the cells are activated to produce their hormone. Presumably the correct conditions are found in the digestive tract of the seal, so that the adult worm eventually emerges from the fourth stage cuticle when its cod host is eaten.

Although parts of the nematode nervous system, suspected to contain neurosecretory cells, cannot be removed by surgery, an attempt has been

made to destroy internal structures by laser microbeam irradiation.[229e] In the free living nematode, *Panagrellus silusiae*, destruction of different body areas has revealed centres controlling growth, ecdysis and gonad development. Irradiation of the area anterior to the nerve ring, which contains the bulk of the cell bodies of the nerve cells, has no effect upon cuticle formation but does block ecdysis, and growth of the nematode, as measured by increase in length, is inhibited.

Irradiation of the hind gut area also inhibits growth and the formation of new cuticle. Irradiation of the nerve ring is followed by incomplete development of the gonads, similar to that resulting from the inhibition of DNA synthesis. These preliminary results are difficult to interpret, but do suggest the possible involvement of neurosecretory centres in the control of growth (including that of the gonads) and ecdysis. But the role of the hind gut area in the control of growth is obscure; an effect of the irradiation upon the transport or uptake of nutrients cannot be ruled out.

Nematodes pass through a number of postembryonic developmental stages before attaining the adult form, and at the end of each stage the body cuticle, and that lining the pharynx, excretory pore and rectum is moulted. This similarity with arthropod moulting has led to the testing of insect hormones upon nematodes. When ecdysterone (p. 267) is added to *in vitro* cultures of the infective larvae of *Trichinella spiralis*, increase in length of the individuals is retarded but the numbers moulting increase significantly. A compound with mass spectra, ultraviolet and infrared absorption data similar to ecdysterone can be extracted from the larvae and has a similar effect upon growth and moulting. Moreover, the addition of a juvenile hormone analogue (p. 284) to the culture results in a significant increase in the length of the larvae compared with the controls.[148c, 227j] These results are analogous to the effects of these developmental hormones upon insects themselves (p. 106) and perhaps point to even greater similarities between the two groups. But until more is known of the complex interactions involved in the control of postembryonic development in nematodes, too much importance should not be attached to such interphyletic effects.

ECHINODERMATA: CONTROL OF SPAWNING AND OOCYTE MATURATION

In the echinoderms, most is known of the control of spawning and maturation in starfish (Asteroidea), although there is some evidence that spawning in sea-urchins (Echinoidea) may also be hormonally controlled.

The gonads of the starfish are paired within each of the five arms,

lying freely in the perivisceral cavity and opening to the exterior by paired gonoducts, the gonopores situated between the junctions of the arms. When ripe, the gametes are shed into the sea: gametes are shed from all the gonads simultaneously, which suggests some regulating mechanism.

When an aqueous extract of the radial nerves is injected into the perivisceral cavity of one arm, *all* the gonads shed gametes about 30 minutes later. The extracts are not sex specific: radial nerve extracts from males will induce the shedding of eggs, and extracts from females induce the shedding of sperm.[44] Nor is the extract species-specific, although exceptions are known: *Henricia leviuscula* and *Othelia tenuispina*, which produce fewer and larger eggs than most starfish, do not respond to shedding extracts prepared from other species.[44] Moreover, radial nerve extract from *Asterina pectinifera* will induce spawning in *Asterias amurensis*, but the converse does not apply.[213d]

The results of these experiments suggest that the control of spawning is hormonal, and since radial nerve extracts alone are effective, that a neurosecretory hormone is involved. When successive layers of the radial nerves are stripped away, either mechanically or by enzymes, only the most ventral part of each nerve is found to contain shedding activity.[44] In starfish with ripe gonads, the ventral layer of the radial nerves contain granules staining with paraldehyde-fuchsin and chrome-haematoxylin-phloxine—which stain neurosecretory material in many other animals. The neurosecretory cells which must be the source of this material in the radial nerves have not yet been identified. But the combination of physiological and histological evidence leaves little doubt about the neurosecretory nature of the hormone.

Isolated ovaries respond to radial nerve extracts by shedding their eggs in the same way as ovaries *in situ*. The hormone must therefore normally act directly upon the gonads, and not through any other endocrine mechanism. The radial nerves are in close proximity to the oral surface and asteroids can absorb amino acids and other substances from sea water.[861] The shedding hormone is a polypeptide (p. 259) and consequently it was once thought that the shedding hormone passed outside the body to be taken up by the peripheral tissues and carried to the coelom where the ovarian effect was induced.[44, 44a] However, although extracts of the tube feet and body wall can have considerable gamete shedding properties, so also does coelomic fluid when the starfish undergo spawning[158q] and a sea water route of entry for the shedding hormone is denied.[158s] Neurosecretion is said to diffuse into the coelom from the radial nerve in the ophiuroids[93b] and the same could occur in asteroids. It is even possible that neurosecretion may pass along nerves directly to the genital ducts and gonadal epithelium.[28a]

In female starfish, spawning involves the dissolution of cementing substances between the follicle cells themselves and between the follicle cells and the oocyte surface: freeing the oocytes from their follicles facilitates spawning under pressure from the expanded ovaries.[158o] At the same time, meiosis, which is arrested during the growth phase of the oocytes, is completed. The shedding hormone thus has two ultimate effects: stripping follicle cells from the oocytes, and inducing the completion of oocyte maturation.

It is now known that the shedding hormone does not exert its dual effects directly (Fig. 3.9). When isolated ovaries are incubated in sea water with shedding hormone added, a small molecular weight, heat stable, non-proteinaceous substance can be extracted from the ovaries which will strip off follicle cells and induce oocyte maturation in other ovaries which have not been incubated with shedding hormone.[158r 244c.] The substance is formed in increasing amounts in proportion to increasing concentrations of shedding hormone in the incubate.[158o 244b.] Moreover, with a constant concentration of shedding hormone, the amount of the substance formed increases with increasing numbers of oocytes with follicles in the incubation medium, and follicle cells alone produce equivalent amounts of the substance; the follicle cells must therefore be the source of the substance.[148a]

The ovarian substance has been identified as 1-methyladenine (p. 259). Synthetic 1-methyladenine is as effective as the natural compound. The follicle cell stripping action and the oocyte maturation effect are brought about by variations in concentration of 1-methyladenine: 1.3×10^{-7} M will induce oocyte maturation, whereas about 10 times this concentration is necessary to strip off follicle cells.[251a]

The way in which the neurosecretory shedding hormone causes the follicle cells to produce 1-methyladenine is unclear. 1-Methyl adenosine (Fig. 12.1) (composed of the purine condensed with a pentose sugar) will produce effects similar to those of 1-methyladenine upon oocytes in the ovary, but has no effects at all upon the maturation of isolated oocytes. 1-Methyladenosine is split into 1-methyladenine and ribose by an enzyme present in ovarian tissue, 1-methyladenosine ribohydrolase.[158u] This suggests that the shedding hormone does not induce the formation of the enzyme, but instead may control the production of 1-methyladenosine by the follicle cells, which is then split by the enzyme already present. This would also explain why exogenous 1-methyladenine when applied to ovaries does not accumulate on oocyte surfaces or within follicle cells.[264b]

When 1-methyladenine is micro-injected into the oocyte, maturation does not ensue.[158p] However, enucleated oocytes will produce fertilization membranes after insemination only when they have been incubated previously with 1-methyladenine.[148b] These results suggest that cytoplasmic

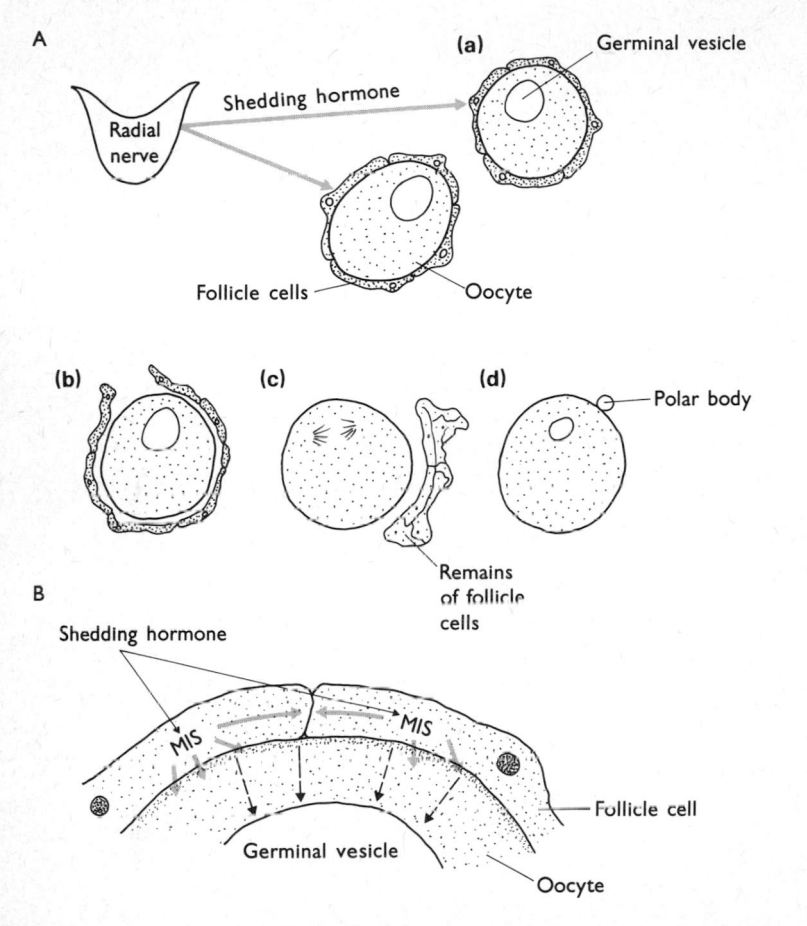

Fig. 3.9 Control of spawning in starfish. *A,* shedding hormone from the radial nerve having penetrated the ovaries (a), causes the follicle cells to rupture at a weak point (b) and to strip away from the oocytes (c); oocyte maturation is completed and a polar body extruded (c and d). *B,* indirect action of the neurosecretory shedding hormone: the hormone induces the follicle cells to produce maturation inducing substance (MIS : 1-methyladenine) which dissolves cementing materials connecting the follicle cells to each other and to the oocyte surface which then induces maturation (dotted arrows).

changes on the oocyte surface are induced by 1-methyladenine, without the participation of the nucleus, and nuclear maturation is consequent upon the cytoplasmic effect.[148b] The importance of 1-methyladenine production by the follicle cells is thus underlined.

High concentrations of radial nerve extracts can *inhibit* the shedding of gametes. This suggests that the radial nerves also contain an inhibitory hormone, with a much higher concentration threshold than the shedding hormone. The two hormones in radial nerve extracts can be separated on Sephadex columns. When radial nerves are assayed throughout the year, the inhibiting hormone is found only in animals with ripe gonads, while the shedding hormone is present continuously.[44] This apparent contradiction is explained by assuming that gamete shedding is primarily controlled by fluctuation in the concentration of the inhibiting hormone. Premature spawning is most likely as the gonads approach maturity: the high level of inhibitory hormone would prevent spawning during this dangerous period. When the concentration of inhibiting hormone begins to decline, the shedding hormone becomes increasingly effective. However, the situation is further complicated by the reported presence of an anti-mitotic substance, apparently the aminoacid L-glutamic acid, which prevents oogonial proliferation and oocyte growth and thus counteracts the action of the shedding hormone early in the breeding season.[153a]

1-Methyladenine is a purine base and is derived from 1-methyladenosine in the follicle cells which surround the oocyte. The compound acts at the surface of the oocyte, and this raises the problem of whether it can properly be called a hormone, since it is not produced by a discrete organ and is not carried any distance in the body fluids. In effect, 1-methyladenine is the means of interaction between two types of cell—the follicle cells and the oocytes—and thus has more in common with the organizer substances which play such an important part in embryogenesis (p. 305) than with hormones *sensu stricto*. The maturation stimulating substance of starfish ovaries may perhaps best be considered as a link between intercellular and purely hormonal communication systems (p. 305) which would emphasize the sequential continuity of such systems in bringing about ordered development at different levels of organization.

In the sea urchin, *Strongylocentrotus purpuratus*, together with a few other species tested, extracts of the radial nerves will induce spawning. The response in the echinoid differs from that in asteroids in that spawning is induced within one minute, whereas it takes about 45 minutes after treatment in the asteroids. Starfish shedding hormone will induce spawning in echinoids within one minute after application, whereas the sea-urchin hormone induces spawning in starfish after 45 minutes[61b] (Fig. 3.10). This suggests that the production of an intermediary substance, e.g. 1-methyladenine, by shedding hormone does not occur in echinoids, and since the gametes lie freely within the gonads in sea-urchins, a follicle stripping substance would seem to be unnecessary. The different responses of sea-urchins and starfish to the shedding hormone would thus lie in the time of onset of gonadal contraction. The concentration of radial nerve

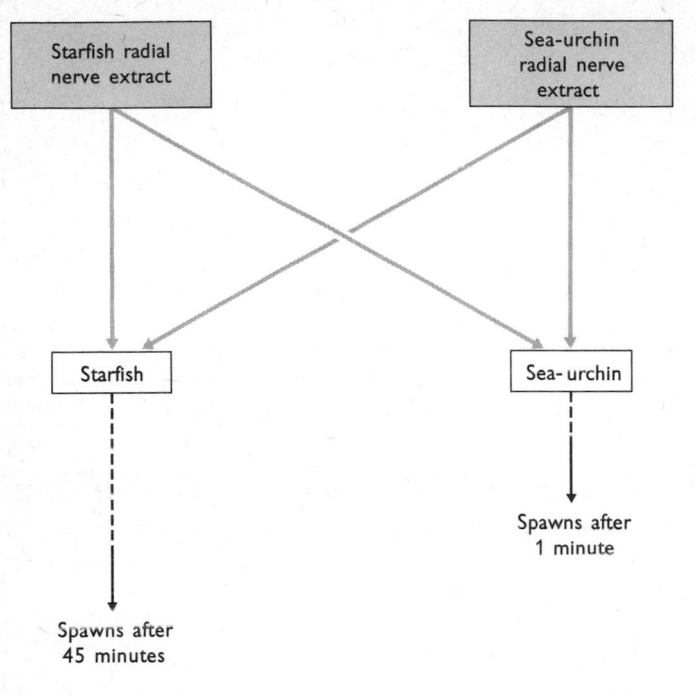

Fig. 3.10 Effect of shedding hormones from starfish and sea-urchins upon spawning. Radial nerve extracts from both starfish and sea-urchins will induce spawning in either group, but spawning always takes longer in starfish, suggesting that the neurosecretory hormone is similar in both groups, but that the indirect action of the hormone in starfish (*via* MIS) is absent in sea-urchins.

hormone in *Strongylocentrotus* increases markedly during the summer spawning season in California, but what controls this seasonal increase is not known.[61b]

THE EVISCERATION FACTOR OF HOLOTHUROIDS

Some species of sea cucumber (Holothuroidea), when subjected to crowded conditions, high temperatures, rough handling, the injection of chemicals into the coelom, and other abnormal stimuli, will expel one or both respiratory trees, the digestive tract and the gonads through a rupture in the body wall. Such evisceration can be followed by the regeneration of the lost parts.

Evisceration in *Thyone briareus* occurs through the anterior end, quantities of coelomic fluid being expelled as the introvert ruptures. A sufficiently large amount of this fluid injected into another animal will induce evisceration in the recipient. In contrast, the coelomic fluid of an intact *Thyone* will never induce the evisceration responses. This suggests that an endogenous evisceration factor is present in the coelomic fluid of eviscerating animals. High concentrations of the factor are present in the haemal system of *Thyone* and ultrastructural studies on *Cucumaria frondosa* have shown the presence of numerous bundles of nerve axons containing membrane bound granules in the walls of the vessels. The evisceration factor may thus be a neurosecretory hormone.[249a]

Whether evisceration is of adaptive value, or indeed whether it occurs under natural conditions, is debatable. The evisceration factor is present in the coelomic fluid of *Cucumaria*—an animal which does not exhibit evisceration. It is likely that this factor has a more general role, and evisceration may only be an incidental, if not accidental, effect. The identity of the evisceration factor is unknown; it is not acetylcholine, 5-hydroxytryptamine, dopamine, 1-methyladenine or histamine.

4

Endocrine Mechanisms in the Annelida

Like the species already discussed no annelid possesses epithelial endocrine glands. Any hormonal control over development is therefore likely to be due to neurosecretions. In 1936, neurosecretory cells were first demonstrated in the supra-oesophageal ganglion, or brain, of a polychaete.[231] But twenty years were to elapse before it was shown that growth, regeneration, asexual reproduction and maturation of the gonads together with epitokal metamorphosis (when present) are all controlled by hormones from the brain. In the oligochaetes and leeches also, neurosecretory cells are present in the brain,[100, 127, 120] although their functions are less well defined than those of the polychaetes.

GROWTH AND REGENERATION IN POLYCHAETES

The typical annelid body consists of a presegmental prostomium and a postsegmental pygidium separated by a variable number of metameric segments (Fig. 4.1). Oligochaetes and leeches hatch from the egg as segmented individuals; the polychaetes become segmented only after metamorphosis from the unsegmented trochophore larva. But in all annelids, growth consists of a proliferation of segments from an actively growing zone in front of the pygidium, followed by enlargement of the new segments. Where segment proliferation continues throughout the life of the annelid, as in many polychaetes, its rate decreases considerably as the individual ages.

The soft-bodied annelid is very vulnerable to attack by predators or

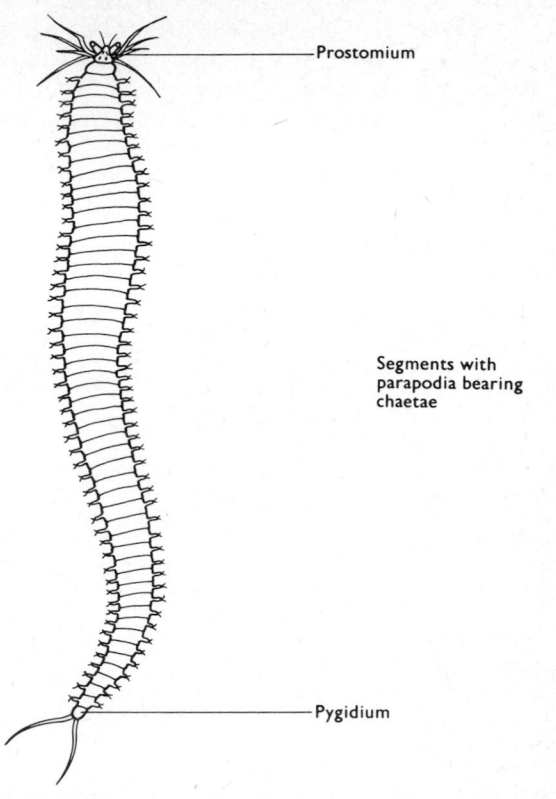

Prostomium

Segments with
parapodia bearing
chaetae

Pygidium

Fig. 4.1 Dorsal view of a polychaete annelid. The body consists of the pre-segmental prostomium behind which are a variable number of metameric segments and, at the posterior end, the post-segmental pygidium. During growth, new segments are formed in the proliferative zone just in front of the pygidium.

damage by physical agents. An ability to regenerate lost parts of the body, including specialized sensory or reproductive organs, is clearly of great advantage to the individual. In annelids, caudal regeneration is wide-spread: the posterior parts of the body are more likely to be lost under natural conditions. Nereid polychaetes readily regenerate a new tail, but only rarely a new head. Sabellid and syllid polychaetes, some earthworms and leeches, will regenerate new heads; and the highly specialized tube-dwelling *Chaetopterus* will regenerate a complete individual from any one of its first fourteen segments.

When segments are lost, the open wound is sealed by the contraction of the body wall muscles and then a blastema is formed, composed of mesoderm with overlying ectoderm, together with neoblast cells of

mesenchymal origin (Fig. 4.2). The blastema produces the ectodermal and mesodermal parts of the regenerating segments, and proliferating endoderm forms the gut. Once a new pygidium has been formed, the blastema takes over the role of the original proliferative region.[129]

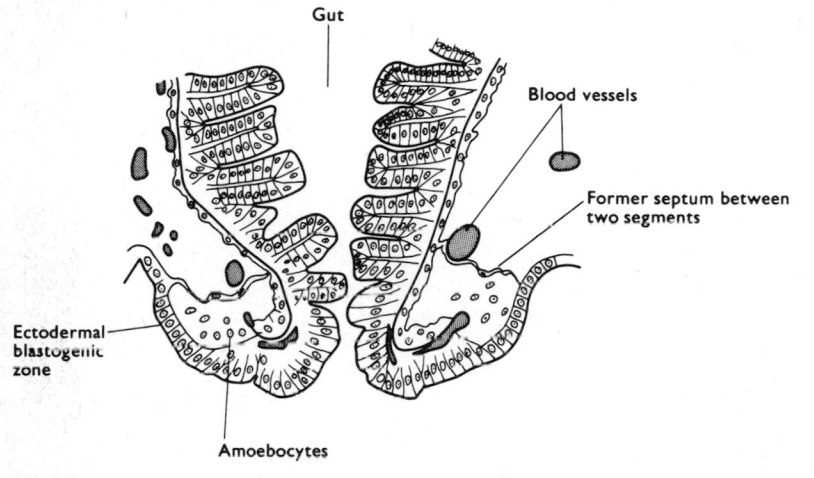

Fig. 4.2 Diagram of a median frontal section through the healing zone of *Nereis diversicolor* eleven days after amputation of posterior segments. The wound has healed by the abutment of the (ectodermal) body wall and the (endodermal) gut wall. The gut remains open. The blastema is beginning to be formed by ectoderm cells (ectodermal blastogenic zone). These cells, and later the mesodermal cells in the area, de-differentiate and form the new region of segment proliferation. Some epithelial cells are histolysed during this process: amoebocytes may aid this breakdown. (After Herlant-Meewls[129])

Growth is stimulated by both local and systemic influences. It is likely that in the annelids **wound hormones** (unspecified chemicals released from damaged cells, see p. 41) play an important part in blastema formation through stimulating the activation and migration of neoblasts.[129] Moreover, the nerve cord exerts a local trophic effect upon regeneration: not only is its presence necessary, but abnormal regeneration will occur if it is deflected from its usual position (Fig. 4.3). In this way, regeneration can be induced along the whole length of the body in *Myxicola aesthetica*. The influence of the nerve cord is non-specific: it induces regeneration, but the *kind* of regeneration which occurs varies according to the region of the body.[129]

Equally important for regeneration is a systemic or hormonal influence. Little is known at present of any possible hormonal regulation of cephalic regeneration. The polychaete *Lycastrus* does not normally

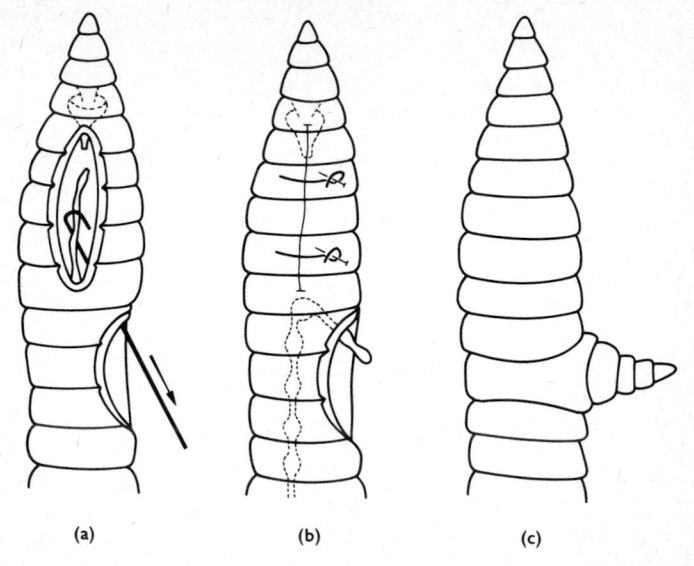

(a)   (b)   (c)

Fig. 4.3 The induction of a lateral head by deflection of the nerve cord in an oligochaete. (a) and (b) show the technique of cutting and deflecting the nerve cord. Such deflection in this cephalic region causes the regeneration of a small supernumerary head (c). If the anterior nerve cord is deflected even further posteriorly, an undifferentiated protruding bud results and not a supernumerary head. Thus although deflection of the nerve cord results in abnormal regeneration, the *kind* of regeneration depends upon the region of the body into which the nerve cord is deflected. (After Herlant-Meewis[129])

regenerate a head after decapitation, but will do so if the headless body is given a brain—which is unlikely to happen in nature.

The hormonal regulation of growth and caudal regeneration is well established. The ragworm *Nereis diversicolor* provides most of the information. Normally, the young worm can regenerate posterior segments. If, however, its brain is removed, growth stops, and amputation of the posterior segments from a decerebrate individual is not followed by regeneration[42, 79, 57] (Fig. 4.4). Subsequent implantation of a brain from a growing worm restores both growth and the power of regeneration.[58] Caudal amputation from a fully grown *Nereis* is not followed by regeneration, even in the presence of the brain.

Proliferation of new segments is the essential feature of both growth and regeneration. The only distinction between the two processes is that segment proliferation is initially very rapid following amputation. Even so, as regeneration proceeds the rapid rate of segment proliferation declines to that of normal growth and it is impossible to determine when

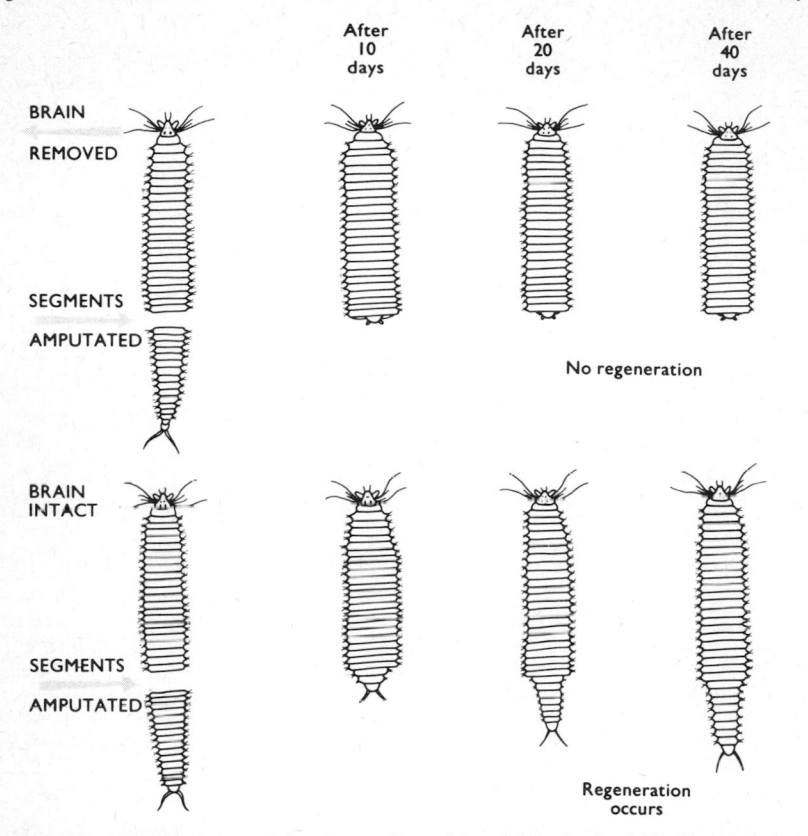

Fig. 4.4 Effect upon caudal regeneration of removing the brain from a growing *Nereis diversicolor*. Amputated posterior segments are regenerated as long as the brain remains *in situ*; when amputation is accompanied by brain removal, segment regeneration is prevented.

regeneration finishes and normal growth recommences[114] (Fig. 4.5). So apart from the initial regeneration of the pygidium, growth and regeneration must be essentially similar, but occurring at different rates.

In *Nereis*, the presence of the brain is necessary for caudal regeneration. But what determines the number of segments to be regenerated? In practice, it is found that the number of segments regenerated more or less equals the number amputated (Fig. 4.6). This is so when the worm's own brain is left in place, or when it is replaced by a brain from another individual.[114] How does the *Nereis* 'know' the number of its segments which have been amputated, particularly when it has been given a new brain, unconnected with any part of its own central nervous system?

One way would be for the brain to produce *more* hormone as more segments are amputated. In other words, there could be a feedback from the region of amputation to the brain, and the rate of segment proliferation would then be directly proportional to the amount of hormone produced.

Until recently, it was thought that segment regeneration in *Nereis* was controlled in just such a manner. The brain was activated to manufacture

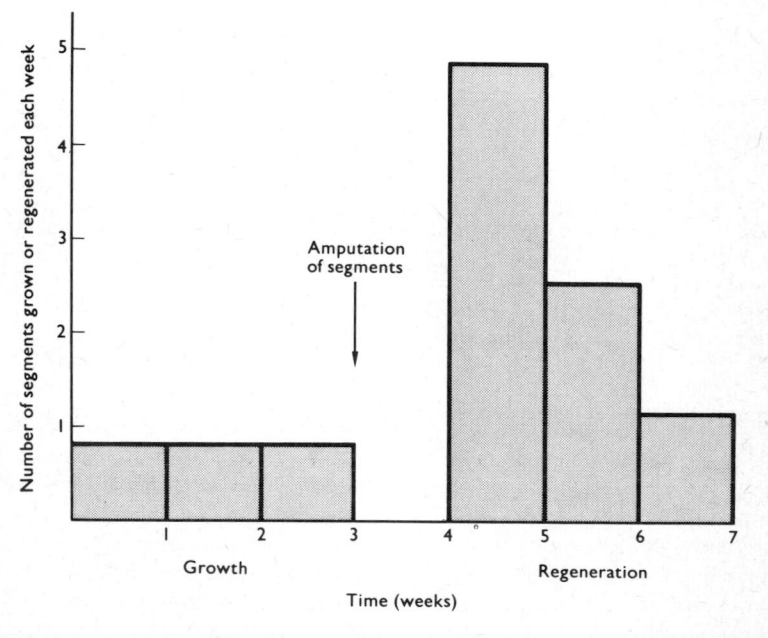

Fig. 4.5 Rates of proliferation of new segments in *Nereis diversicolor* during normal growth and after caudal amputation. After caudal amputation, the rate of segment proliferation increases and then gradually decreases as regeneration proceeds. Thus segment proliferation during normal growth and during regeneration is essentially the same process but occurring at different rates. (After Golding[114])

hormone by the amputation of tail segments, and the hormone accumulated in the brain for the subsequent 3 or 4 days. Then at a critical period, the large amount of hormone was released into the blood, inducing a degree of regeneration proportional to its concentration.[245]

But this hypothesis cannot be valid as the following experiment shows. A brain is taken from a young, intact worm and implanted into a decerebrate, tail-less individual. One day later it is removed, and replaced by

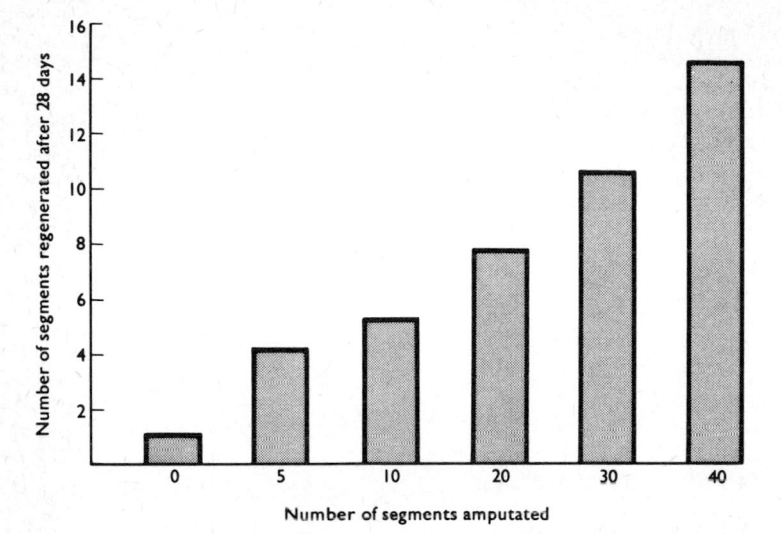

Fig. 4.6   The relationship between number of segments amputated and number regenerated in *Nereis diversicolor*. The worm eventually regenerates a number of segments approximately equal to the number amputated. (After Golding[114])

another similar brain. This is left for one day and replaced, and so on for 10 days. Segment regeneration occurs in all such host individuals, the amount of regeneration being related to the number of amputated segments. But clearly the amount is not dependent upon variable amounts of hormone accumulating in the brain, because *all* animals are given similar numbers of brains for the same short periods of time.[114]

It is possible, of course, that if the implanted brains all secrete hormone at the same rate, the concentration of the hormone in a *Nereis* with many segments amputated would be greater than in one with only a few segments amputated, simply because the volume of body in the first instance would be considerably less than in the second. So even without feedback from the amputated region, the blastema might react to different hormone concentrations. But this possibility is dismissed by the results of the next experiments.

The posterior half of one *Nereis* is grafted on to the body of another, and different numbers of segments are amputated from each. Both the host and the graft regenerate segments appropriate to the number lost (Fig. 4.7), although both are dependent upon the host's brain, and must be reacting to the presence of the same concentration of hormone.[115] Moreover, if an already regenerating *Nereis* is grafted on to a host, which then has segments amputated, the graft *continues* to regenerate at the

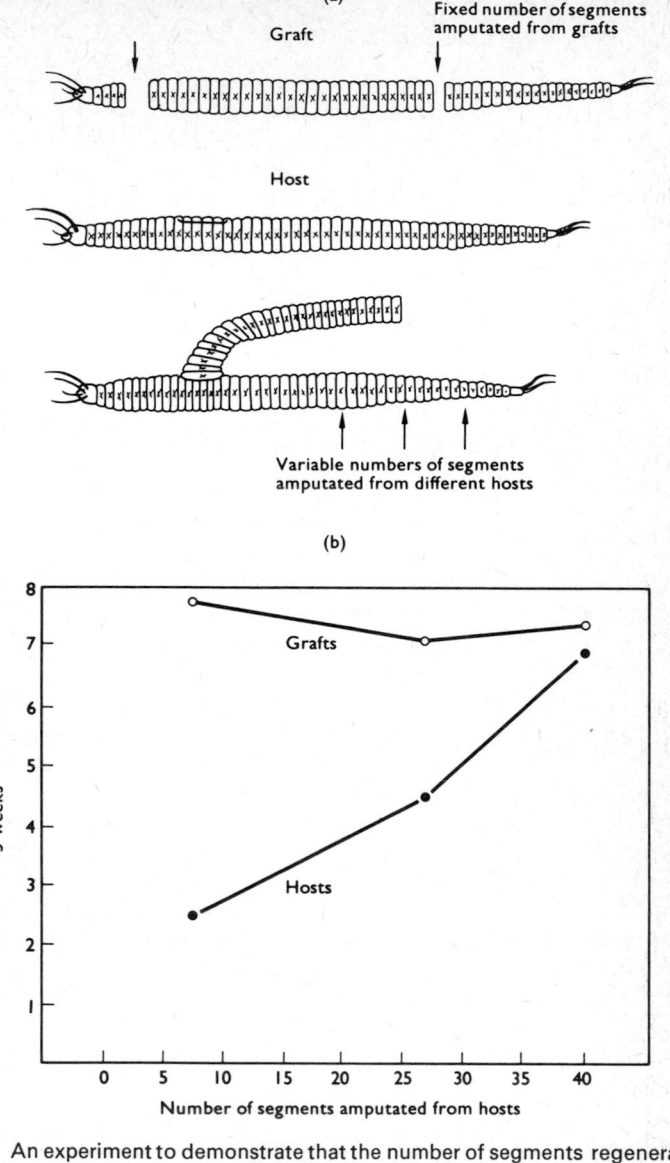

Fig. 4.7 An experiment to demonstrate that the number of segments regenerated is independent of the concentration of brain hormone in the blood of *Nereis diversicolor*. (a) The posterior part of one worm is grafted onto an intact (host) worm. When segments are amputated from both host and graft, regeneration in each will depend upon hormone derived from the host brain: the concentration of brain hormone must be the same in host and graft. (b) The number of segments amputated from the hosts is varied, but the number of segments amputated from the grafts is always the same: although regeneration in the hosts therefore varies, regeneration in the grafts is seen always to remain the same. Consequently, host regeneration cannot be the result of differing hormone concentrations, since the graft is unaffected. (After Golding[115])

rate it had reached when it was joined to the host, and the host regenerates at its usual rate.[115] Thus both the rate of regeneration and the number of segments produced can be very different in an identical hormone environment.[115]

In *Nereis*, it is certain that the brain is necessary for regeneration, and probably for normal growth, but its hormone does not act in any regulating capacity. Instead, the control of segment proliferation must be a local mechanism, the property of the tissues at the wound. The brain secretes hormone at a steady rate, but segment proliferation depends upon a changing competence of the blastema to react to its presence. When more segments have been amputated, the blastema is more reactive to the same level of brain hormone.

In arthropod endocrinology, as in vertebrate, the principle is well established that fluctuating levels of hormones bring about differential changes in their target organs. It is often forgotten that the state of the target itself can modify the apparent effects of hormones. In *Rhodnius prolixus*, for example, epidermal cell division is initiated by developmental hormones (p. 127), but the amount of division is affected by the degree of abdominal stretching brought about by the blood meal, and by other less obvious factors such as the position of epidermal cells in the body. Imaginal buds in holometabolous insects may not metamorphose until they have reached a certain degree of development. In *Nereis*, the overriding importance of the target in reacting differentially to the same hormone concentration is particularly well demonstrated.

What determines this differential reactivity of the blastema after amputation of various numbers of segments? One favoured answer is that there is an axial gradient of growth potential in the intact animal, high anteriorly and declining posteriorly. But this explanation really only restates the problem in a different way, for at present the nature of this axial gradient is obscure. Such gradients of influence have been recognized in the embryonic development of many animals, including vertebrates, and in the regeneration of coelenterates and flatworms (pp. 35, 43). The hormonal control of growth and regeneration in polychaetes thus abuts onto major problems in developmental biology.

REPRODUCTION AND EPITOKAL METAMORPHOSIS IN POLYCHAETES

Polychaete families show considerable variations in their life-cycles. In monotelic species which breed only once, maturation and its control can be viewed as a unique progression. Polytelic species, on the contrary, may breed several times in a year during a prolonged breeding cycle, or

individuals may breed several years in succession. In polytelic species, therefore, the events associated with reproductive activity must be cyclic. Somatic growth makes considerable demands upon the resources available to an animal, and the metabolic needs of gametogenesis, particularly in the female, may also be large. Consequently, somatic growth and reproduction are often antagonistic, occurring alternately as in many crustaceans (p. 233), or separated completely, the reproductive period not beginning until somatic growth has ended. In the polychaetes, however, this temporal separation of somatic growth and sexual maturation is not as generally applicable as at one time seemed likely. It holds for the monotelic nereids, where the hormonal control of growth and reproduction is concerned with the suppression of sexual reproduction until the initial period of somatic growth is completed. But in some polytelic cirratulids, for example, growth is not incompatible with sexual maturation and continues until the coelomic oocytes are maturing.[2141]

In many polychaetes, the form of the body changes at the approach of sexual maturity. The change in form may be slight, for example the modification of the nephromixia to produce ciliated ducts specialized for gamete release, but it can be so extensive in nereids that a highly specialized reproductive individual is produced. The sexual form, the epitoke, is produced from the sexually unripe animal, the atoke, either by transformation or budding. In nereid species, the epitoke is produced by transformation and is called a heteronereis. The heteronereis has specialized muscles, those of the pre-reproductive period breaking down; it develops membranous frills on its parapodia, and special oar-shaped chaetae; its eyes increase considerably in size, and are much more sensitive to light (Fig. 4.8). These physical changes are associated with behavioural differences: the heteronereis is a more active swimming form than the atoke.[55]

In the syllids, the gonads are confined to the posterior segments, and these segments may break off as a free-swimming unit, often with a newly developed head, but lacking jaws and pharynx. This process of stolonization in the syllids is akin to the asexual reproduction of some oligochaetes, but is linked with sexuality and swarming; it also bears some resemblance to the formation of the nereid epitoke (Fig. 4.9).

In the monotelic nereids, reproduction is thus extreme in that after a period of somatic growth there is a profound metamorphosis and after spawning the individuals do not survive. In the stolon producing syllids the situation is similar, and the stolons die after spawning. Thus in these monotelic polychaetes, the whole metabolism can be diverted to reproductive activity after somatic growth has finished. In polytelic species, the somatic changes which accompany sexual maturation are reversible, and the somatic tissues must continue to function during and after reproduction, so extensive metamorphic changes are not found in these polychaetes.

In view of this diversity of life cycles, it is possible that a variety of hormonal mechanisms will control the two separate, but linked, processes of polychaete reproduction: gametogenesis and epitokal metamorphosis.

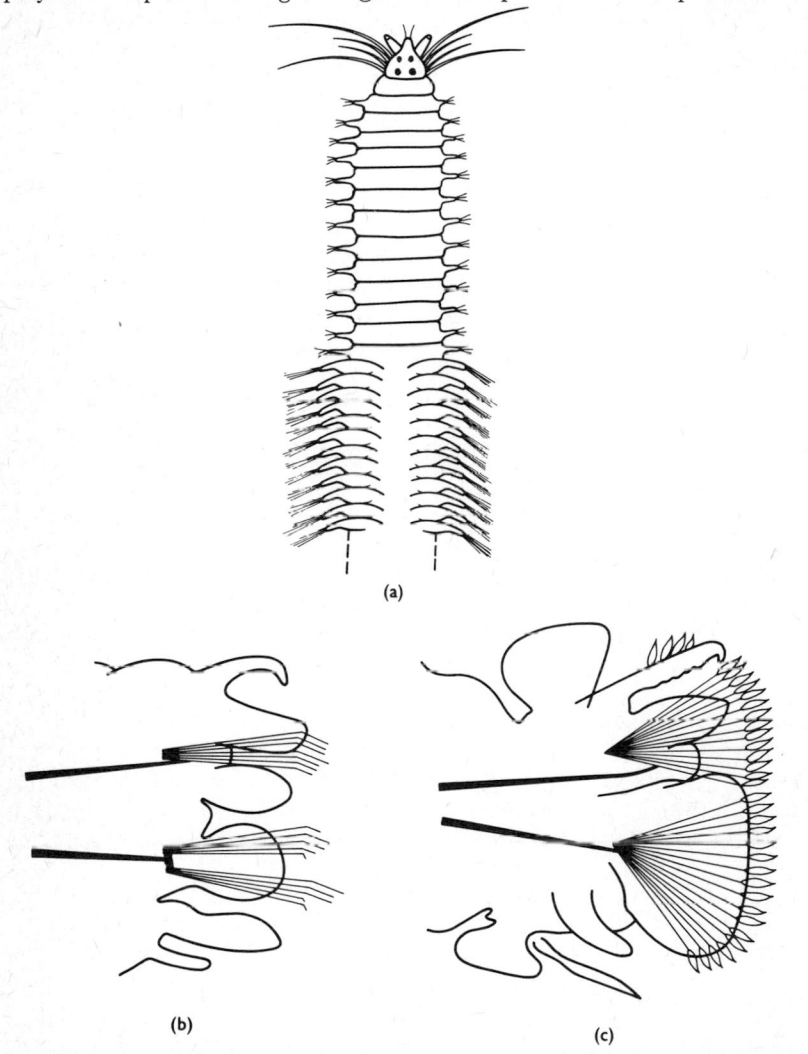

Fig. 4.8   The heteronereis. (a) Dorsal view of the anterior part of a nereid worm showing the structure of the parapodia in the heteronereis form (compare Fig. 4.1). (b) Cross-section of parapodium of a pre-reproductive worm. (c) Cross-section of parapodium in the reproductive worm. Note the greater number of oar-shaped chaetae and increased area of the heteronereis parapodium.

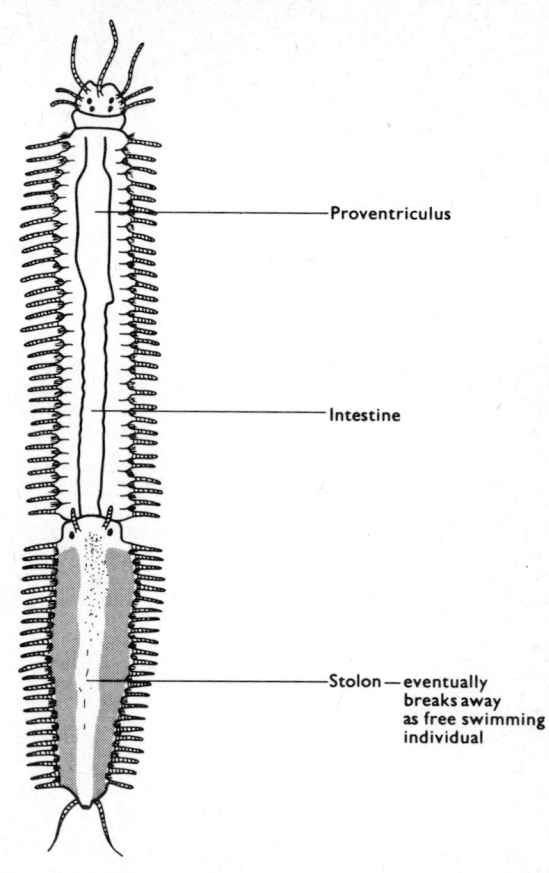

Proventriculus

Intestine

Stolon—eventually breaks away as free swimming individual

Fig. 4.9   Formation of a stolon in a syllid polychaete. The posterior half of the worm, which contains the gonads, becomes modified and breaks away to form a free-swimming individual.

Endocrine control of gametogenesis

In most polychaetes, vitellogenesis and the greater part of spermatogenesis take place after the gametocytes have been shed into the coelom from the germinal epithelium (Fig. 4.10). More rarely, the oocytes remain in contact with the germinal epithelium and form discrete gonads attached to the peritoneal walls, from which eggs are released only during the final stages of maturation. Occasionally, as in *Tomopteris*, abortive oocytes may form nurse cells, but it is more usual for coelomic cells to perform this function.

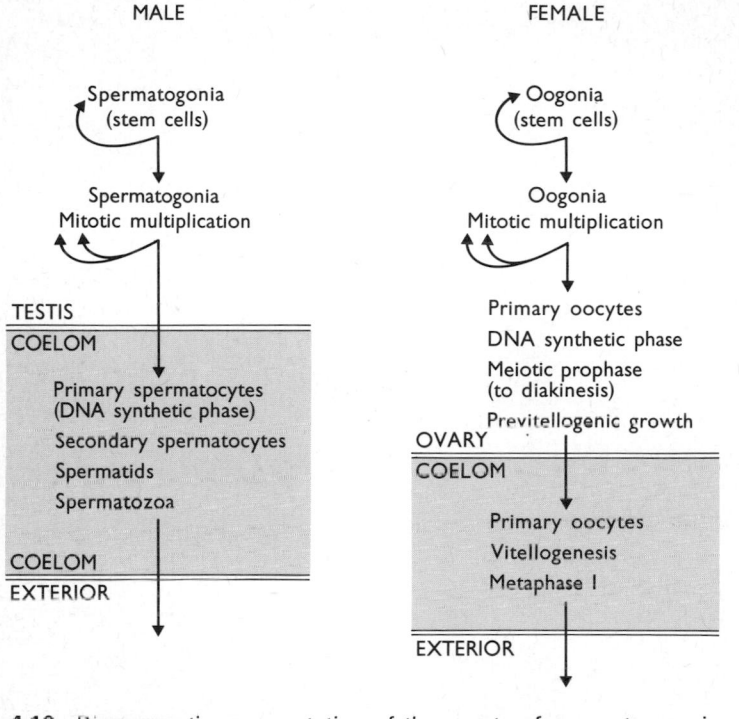

Fig. 4.10 Diagrammatic representation of the events of spermatogenesis and oogenesis in polychaetes. (After Clark and Olive[58a])

Early studies on gamete development in polychaetes concentrated upon oocyte development in the coelom, because estimates of sexual maturity could easily be made by examination of the contents of the coelomic fluid. These studies also concentrated upon the monotelic nereid species where a single temporary phase of gametocyte proliferation was all that was necessary to ensure reproductive success. In polytelic species, however, the activity of the germinal epithelium must be cyclic if batches of mature gametes are to be produced synchronously, and this implies a direct regulation of germinal proliferation. Regulation of gametogenesis must therefore be expected to occur at two levels: the proliferation of the germinal epithelium, and the maturation of the gametocytes in the coelom.

Most investigations of the control of maturation of the gametocytes in the coelom have centred upon nereid species. *Nereis diversicolor* has small gametocytes in the coelom by the 30-segment stage, i.e., only a few weeks after metamorphosis from trochophore larva to young worm. The oocytes

are shed into the coelom when they are 15–20 μm in diameter. Their subsequent growth, to about 70 μm, is slow, but this is followed by a rapid increase in diameter to about 140 μm and then once more by slow growth to the final diameter of 200 μm[56] (Fig. 4.11). Protein and lipid yolk is synthesized early in vitellogenesis and acid mucopolysaccharides are formed later.[77b, 224d]

Removal of the brain has no effect upon the proliferation of the oocytes from the germinal epithelium nor upon their initial slow growth. But once the oocytes have reached a diameter of about 30 μm, brain removal precipitates the rapid growth which normally begins at a diameter of 70 μm.[56] Even when this rapid growth has begun normally, brain removal accelerates the process. But the eggs produced by this precocious and accelerated growth are smaller than usual, and the development of cytoplasmic inclusions, including yolk granules, is abnormal. The brain clearly plays an important part in the regulation of synthetic processes in the oocyte, and since in the absence of the brain the ribosome–ergoplasmic reticulum

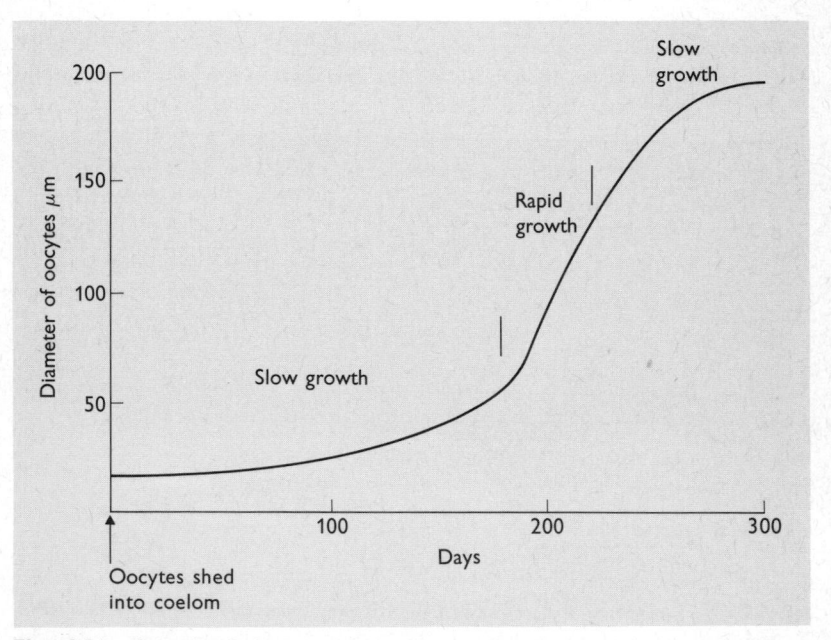

Fig. 4.11 Growth of the oocytes in *Nereis diversicolor*. This occurs in three phases : (1) a period of slow growth from the time the oocytes are shed into the coelom until they reach a diameter of 70 μm ; (2) a phase of rapid growth from a diameter of 70 μm to about 140 μm ; (3) a further period of slow growth from 140 μm in diameter to the final diameter of about 200 μm. (After Clark[56])

system of the oocytes undergoes a marked hypertrophy, it is likely that the brain hormone regulates the synthetic activities of the Golgi apparatus, and controls RNA and protein synthesis in the early stages, and muco-polysaccharide synthesis in the later stages, of oogenesis.[58a, 77a] The precocious eggs formed after brain removal can be fertilized, but fail to develop beyond the 8- or 16-cell stage. Finally, if the brain is removed from females in which the oocytes have completed their rapid growth phase, the operation is without effect.[56]

When a brain from an immature *Nereis* is implanted into a decerebrate individual, precocious oocyte growth is prevented. Moreover, oocyte development is also inhibited if a brain from an immature individual is implanted into an otherwise normal mature *Nereis*. But brains from mature individuals have no effect when implanted into either decerebrate or normal mature worms.

In *Nereis*, therefore, oocyte development is normally inhibited by a hormone, or hormones, produced by the brain, and vitellogenesis proceeds rapidly when this inhibition is lifted (Fig. 4.12). But this 'inhibitory' control of oocyte development is not simple: when the inhibition is *suddenly* removed, by brain removal for example, although the oocytes grow rapidly, they do not lay down protein yolk in the normal manner. Does the brain, therefore, also secrete a hormone that promotes vitellogenesis?

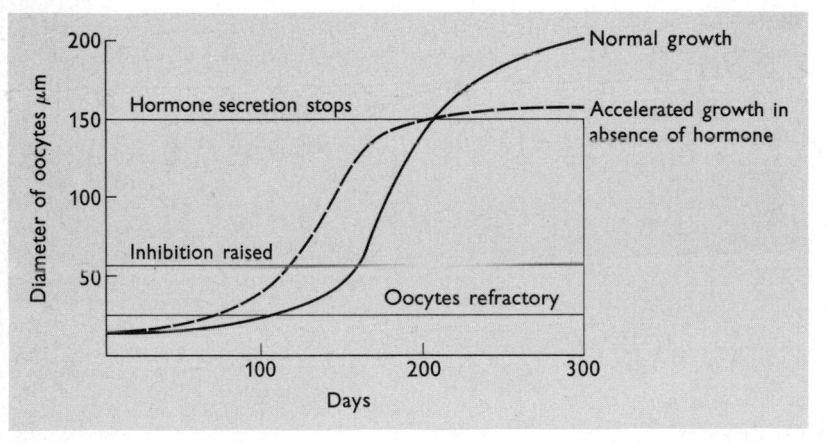

Fig. 4.12 The hormonal control of oocyte development in nereids. A decreasing concentration of a single hormone controls all phases of oocyte development. Initially, a high concentration of the hormone inhibits oocyte development. As the hormone concentration falls, inhibition is lifted and the oocytes grow. Complete absence of the hormone, achieved by removal of the brain, causes accelerated but abnormal growth of the oocytes. Normal progressive growth of the oocytes depends upon a progressive decrease in the amount of hormone. (After Clark[56])

The evidence is that the brain does *not* secrete a vitellogenic hormone, at least in the sense of a hormone which is distinct from the oocyte-inhibiting hormone.[125] When the head from a young *Platynereis dumerilii* is implanted into a single parapodium of a headless individual, oocyte development in that parapodium is completely inhibited. But implantation of a similar head into a headless fragment of 30 segments is followed by oocyte growth *and* vitellogenesis, whereas in headless segments without any implanted brain, abnormal vitellogenesis accompanies the rapid oocyte growth. It must be concluded, therefore, that when the brain hormone concentration is high, as in a single parapodium containing an implanted brain, oocyte development is completely inhibited; when the hormone is absent, as in decerebrate individuals, oocyte growth is rapid but vitellogenesis is abnormal; but at an intermediate hormone concentration, obtained by diluting the hormone by implanting a small brain into headless fragments of 30 segments, oocyte growth and vitellogenesis occur more or less normally. Thus a reduced concentration of the 'inhibitory' hormone from the brain has a *positive* effect upon vitellogenesis. In nereids, therefore, normal oocyte development, including vitellogenesis, results from the *progressive* decrease in the amount of a single hormone produced by the brain, and not by the action of two antagonistic hormones.[125]

In some nereids, the developing coelomic oocytes play a role in regulating the endocrine activity of the brain. In *Perinereis cultrifera*, the implantation of submature oocytes into the coelom of young males or females induces complete maturation and metamorphosis, indicating that the secretion of the inhibitory brain hormone in the hosts is reduced.[224a, 224c] A negative feedback system between gametes and brain thus controls the decline in brain hormone secretion in *Perinereis cultrifera*. This seems not to be the situation in *Platynereis dumerilii*, where environmental factors are more important (p. 78).

Spermatogenesis in *Nereis diversicolor* is also controlled by the gradual withdrawal of an inhibitory brain hormone. A method of assaying the endocrine activity of nereid brains has been developed, using an *in vitro* organ culture technique: a 'diversicolor unit' has been defined as the minimum quantity of an aqueous brain homogenate which when introduced into the culture medium inhibits the onset of spermatogenesis.[81a] This technique has confirmed that the hormonal content of the brain decreases as the oocytes increase in size. It has further demonstrated that each stage in oogenesis which is characterized by different metabolic events takes place at a new and lower level of hormone activity (Fig. 4.13).[81a]

The *in vitro* bioassay has revealed that the rate of decrease of inhibitory hormone in the brain is related to the life span of the worm. In *Nereis diversicolor*, which lives for 1 or 2 years, the decrease is linear when plotted on a logarithmic scale, whereas in *Perinereis cultrifera*, which lives for 3

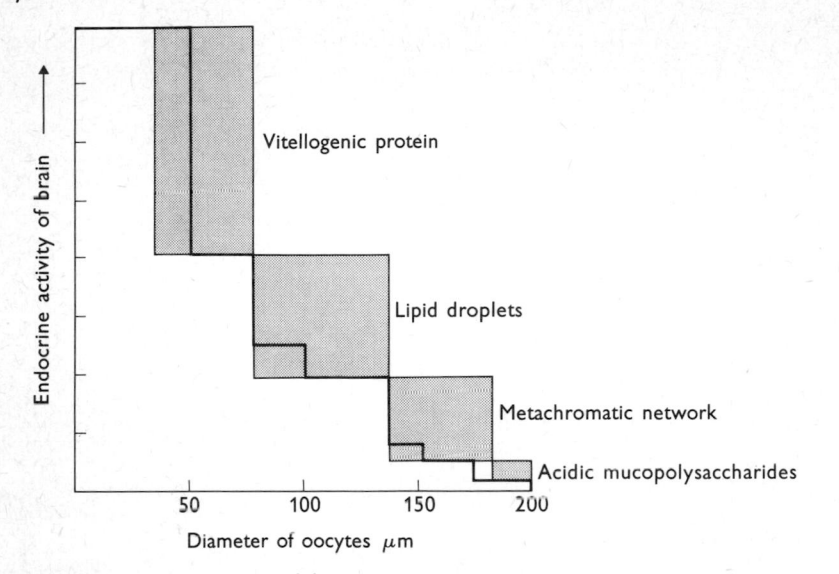

Fig. 4.13 The synthetic activity of the oocytes of *Perinereis cultrifera* related to oocyte size and the endocrine activity of the brain. (After Durchon and Porchet[81a])

years, the decrease is at first slow until the oocytes reach a diameter of 130 μm, and subsequently, during the final phase of oocyte development, the decrease is more rapid. There is also a difference in the endocrine activity of the brains of male and female *Perinereis*, which can be related to the different rates of gametogenesis in the two sexes. Spermatogenesis is completed more rapidly than oogenesis, and so females begin their sexual development soon after the beginning of the second year, but in males the onset of sexual development is delayed until the end of the second year. In parallel with these changes, the endocrine activity of the female brain begins to decline at the beginning of the second year, but in the male the decline does not start until the end of the second year.[81a]

It is not surprising, in the light of these results, that the effects of decerebration in different nereid species depend upon the life span. In *Perinereis cultrifera*, the oocytes are refractory to the effects of decerebration until the second winter of the life cycle: thereafter, the oocytes *degenerate* after decapitation if they are less than 120 μm in diameter.[77c] Worms with an oocyte diameter of about 120 μm can be considered pubertal, and their brains are weakly inhibitory; when such brains are implanted into young animals, they induce the precocious development of inclusions within the oocytes.[224b] In *Nereis grubei*, however, oocytes less than 50 μm in diameter are resorbed after decerebration of the worms,[244a] but this species lives

for only 2 years, and oocytes of this size are probably equivalent to the larger oocytes in *Perinereis cultrifera*. In *Nereis diversicolor*, an even shorter lived species, the oocytes never degenerate after decerebration. These results also illuminate the very important fact that the brain hormone does not merely inhibit oocyte development prior to vitellogenesis, but is necessary for the *maintenance* of normal development of pre-vitellogenic oocytes.

An excellent illustration of the principle that the precise effects of decerebration in nereids depend upon the life-cycle is provided by *Nereis limnicola*—a sibling species to *Nereis diversicolor*, but viviparous and living in fresh water. *Nereis limnicola* is a self-fertilizing hermaphrodite, and the developing larvae are retained within the coelom until the 20–30 pairs of parapodia stage, when they are released by rupture of the body wall. The brain hormone exerts an inhibitory influence over gamete development in *Nereis limnicola*, as in other nereids, but the normal mature oocytes resemble the abnormal oocytes obtained after decerebration in *Nereis diversicolor*. When the brain is removed from *Nereis limnicola*, accelerated oocyte growth occurs but these oocytes develop normally after fertilization and normal larval stages are found in the coelom.[8d] The development of viviparity in *Nereis limnicola* has transferred to the coelom the function of providing nutrients for the growth of the embryos and the oocytes are thus relieved of the intense metabolism of vitellogenesis—and its endocrine control—which is such a feature of other species.

Little is known of the control of germinal proliferation in monotelic polychaetes, but it is likely that in the nereids at least, a precise control is unnecessary. Immature oocytes of some nereid species degenerate when the brain hormone titre is low (see above) and as this titre is zero just before spawning, only mature oocytes will survive. In other monotelic families, such as the Glyceridae and Eunicidae, the germinal epithelium ceases to proliferate during gamete maturation and may even degenerate.[126a, 246a]

Preliminary studies on a few species in families other than the Nereidae suggest that, with the exception of the Syllidae, the nereid-type of control of gametogenesis is absent. Syllids possess a mechanism, comparable with that of the nereids, for the control of gamete maturation and stolonization. But the brain is unequivocally *not* the source of the hormone. Instead, removal of the muscular proventriculus precipitates the production of reproductive individuals—the stolons—and the gametes contained therein undergo accelerated development. Implantation of a proventriculus, but not other parts of the pharynx, into animals without their proventriculus prevents the precocious development. The proventriculus is thus clearly the source of an inhibitory hormone[291c] although what elements produce the hormone is unknown.

In the polytelic polynoid, *Harmothoë imbricata*, decerebration during

sexual development *inhibits* gametogenesis: vitellogenesis is arrested and the larger oocytes may degenerate; in males, spermatogonial proliferation is blocked.[65a] This strongly suggests a gonadotrophic influence from the brain, but so far it has proved impossible to reverse the effects by implanting brains into decerebrate individuals. In the Pacific palolo worm, the monotelic *Eunice viridis*, decerebration prevents coelomic gametocyte development.[126a]

A different kind of cerebral control over gametogenesis is exercised in the Arenicolidae. In the lugworm, *Arenicola marina*, the development of the gametocytes when first shed into the coelom from the germinal epithelium is *not* controlled by a brain hormone. In the female, oogonial division and the prophase of the first maturation division take place in the ovary. Nuclear activity is then arrested as the oocytes are released into the coelom and vitellogenesis takes place. Decerebration of an immature *Arenicola* has no effect upon vitellogenesis, but decerebration of a mature female delays spawning indefinitely and the oocytes remain in the suspended prophase of the first maturation division. In the male, sperm development up to the morula stage will occur in the absence of the brain, but the spermatocytes do not separate from the morulae to become mature spermatozoa. Spawning in decerebrate animals can be induced by the injection of homogenates of the brain.[151, 151a] The brain thus releases a maturation-stimulating hormone which causes the entry of oocytes into metaphase I of meiosis, and the separation of mature spermatozoa from the morulae. The hormone can be extracted from the brain only shortly before the breeding season; after spawning the hormone content of the brain declines. The hormonal control of oocyte maturation in *Arenicola*, associated with spawning, is particularly interesting since a quite similar mechanism has been shown to be present in starfish (p. 49).

In *Arenicola*, as in other polytelic polychaetes, the activity of the germinal epithelium is cyclic. In the cirratulid *Cirratulus cirratus* and the ampharetid *Melinna cristata* a new generation of gametocytes begins to develop shortly after the previous one has been spawned. In *Cirratulus*, the breeding of individuals in the population is unsynchronized but each female breeds once every year, and for a period of 2–3 months before spawning gametocyte release into the coelom ceases, ensuring that the coelomic oocytes become uniformly mature. The cessation of germinal proliferation is caused by the inhibition of oocyte release from the gonads, which is mediated by the coelomic oocytes.[214l] The production of oocytes is thus controlled by a negative feedback system. In *Melinna*, the breeding season of individuals in a population is closely synchronized and the renewed proliferation of gametocytes immediately after spawning is thus also synchronized. This involves both the renewed output of oocytes from the ovaries and renewed oogonial cell division.[151e] It is likely that the inhibition

of gametocyte production is lifted at spawning and the next generation of coelomic oocytes begins to develop immediately[58a] (Fig. 4.14).

A similar control over spermatogonial renewal in *Arenicola* is exercised by the accumulating coelomic spermatocytes by a negative feedback system.[151b, 214j, 214k] If an adult worm is stripped of spermatozoa, the gonads respond with an outburst of mitotic activity. Such an internal regulation of spermatocyte production would ensure the development of a uniform population of mature gametocytes in ripe animals. But this cannot be the only mechanism operating in *Arenicola*, since spawning is not followed immediately by germinal proliferation; the testis remains

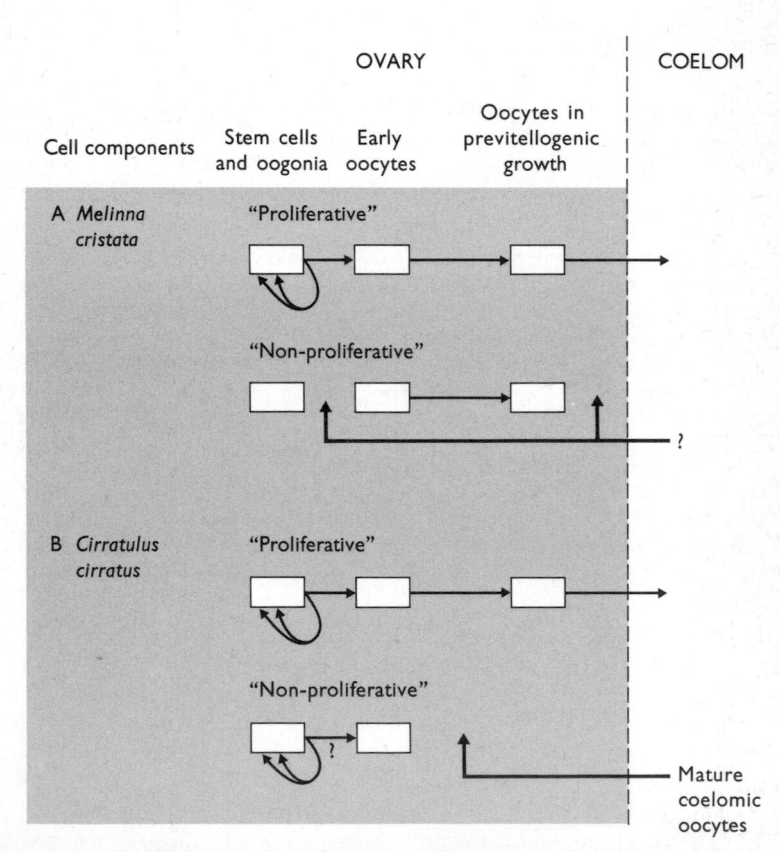

Fig. 4.14 Models of ovarian oogenesis in *Melinna cristata* and *Cirratulus cirratus*. Bold arrows indicate the stages which are regulated in the transition from the proliferative to the non-proliferative states. (After Clark and Olive[58a])

dormant for two months before proliferation is renewed.[214j] Although the mitotic response to the stripping of spermatozoa does not occur in the absence of the brain,[151b] there is no evidence for a separate trophic hormone which increases the frequency of mitosis. It is possible that the maturation hormone released shortly before spawning has an additional effect on the testis to cause the arrest of cell development.[58a]

In all these examples of a negative feedback control by developing gametocytes over germinal proliferation acting *via* the endocrine activity of the brain (p. 69), the gametocytes are floating freely in the coelomic fluid, and so must exert their effects by chemical means. It must be concluded, therefore, that the gametocytes are the source of a hormone or hormones.

The hormonal mechanisms controlling reproduction in polychaetes can be grouped into two broad categories. In one, illustrated by *Arenicola* and *Cirratulus*, a hormone has only a single, tissue-specific, effect. In the other, illustrated by the nereids, the hormone has a profound effect upon the whole metabolism of the animal, and its gradual withdrawal stimulates the development of the germ cells and somatic reproductive changes. In this case the withdrawal of the brain hormone switches somatic growth to sexual development, and the situation recalls that in *Hydra* (p. 37). The results of these experiments in polychaetes underline the importance of the target in its reaction to the same hormone concentration, and form a notable addition to our understanding of endocrine mechanisms in all animals.

The control of epitoky

Removal of the brain from a young *Nereis* is followed by premature epitokal metamorphosis, as well as by precocious and accelerated oocyte growth. Implantation of a brain from a young individual into a decerebrate mature *Nereis* not only prevents epitoky but also prolongs the life of the worm. The brain of a young *Nereis* thus produces a hormone which inhibits epitoky.

However, as with the control of gametogenesis, the brain hormone is not purely inhibitory, but plays an active part in promoting epitoky once the initial inhibition has been lifted. Thus if the brain of *Platynereis dumerilii* is removed from an individual whose oocytes are larger than 100 μm in diameter, epitokal metamorphosis is completed in 5–7 days, rather than the normal time of 8–15 days. But if the oocytes are less than 80 μm in diameter at the time of brain removal, then epitokal metamorphosis is abnormal and the animal dies within 12 days. Brains taken from very young females with no visible oocytes and implanted into headless individuals with oocytes 100–120 μm in diameter will retard

further oocyte development in the hosts, prevent further epitokal meta-morphosis, but induce segment regeneration. If these implanted brains are then *removed* after two weeks, metamorphosis continues to completion. A brain from a small, very young *Platynereis* inhibits metamorphosis completely when implanted into the parapodium of an older individual, but when implanted into a decerebrate fragment of 30 segments, metamorphosis is initiated and proceeds normally. Thus, exactly like the hormone from the brain which controls oocyte growth and vitellogenesis, the brain hormone controlling epitokal metamorphosis is inhibitory in high concentration but stimulatory in lower titre. Normal metamorphosis must therefore depend upon the gradual decrease in hormone production by the brain.[125]

As the nereid type of hormonal regulation of gametogenesis is not found in other polychaete families, it is unlikely that the somatic changes which accompany sexual maturation will be controlled by an inhibitory hormone in these families. In *Cirratulus cirratus*, removal of the brain from immature animals does not result in the precocious hypertrophy of the nephromixia, nor does brain removal in *Harmothoë imbricata* cause the elongation of the ventral nephridial papillae which normally accompanies sexual maturation.[58a] Clearly, further endocrinological studies on non-nereid polychaete families are necessary.

The relationship between gametogenesis and epitoky

The endocrine control of gametogenesis and epitokal metamorphosis is so similar that it seems very likely that only one hormone from the brain controls both processes. Preliminary purification of nereid brain extracts shows that the water-soluble fraction inhibits both gametogenesis and epitoky,[81] but whether this is because only one hormone is involved, or that at least two hormones with similar chemical properties separate together in the purification process, must await further and more detailed analysis.

That brain removal from a young *Platynereis* causes accelerated but abnormal gametogenesis and epitoky, whereas brain removal from older individuals allows both to proceed normally, is often taken to mean that the two processes have similar threshold responses, and must therefore be reacting to a single hormone. But the possibility cannot be eliminated that two hormones are involved, closely synchronized both in production and action.

In *Perinereis cultrifera*, homogenates of the coelomic contents of females with oocytes 200–250 μm in diameter injected into females with oocytes 30–60 μm in diameter sometimes accelerate metamorphosis (in 6 individuals out of 27 in one experiment) without any concomitant effect

upon oocyte growth.[80] This could mean that in the injected worms, the production of gametogenic hormone is unaffected by the homogenate, whereas the production of metamorphic hormone is prevented, so allowing metamorphosis to proceed. Such a differential effect would imply the existence of two distinct hormones. But in similar experiments with *Platynereis dumerilii*, none of the injected worms are affected in any way by the homogenate.[126] In both nereid species, when gametogenesis is prevented by x-irradiation or starvation, epitokal metamorphosis occurs more or less normally.

The situation in the viviparous *Nereis limnicola* (p. 71) is interesting because sexual and somatic maturation do not occur synchronously: the somatic changes follow sexual maturation and are associated with the presence of larvae within the coelom.[8d] This might suggest that separate inhibitory hormones control sexual and somatic maturation in this species. However, it is equally likely that a single inhibitory hormone is present in *Nereis limnicola* and that somatic maturation is inhibited by a very low titre of the hormone—a concentration insufficient to inhibit sexual maturation; it is also possible that the somatic tissues are not competent to respond until after the gametes have matured and larval development has begun.

At present, the most economic hypothesis is that a single brain hormone controls both gametogenesis and epitoky in nereids. The two processes must be intimately linked; but in insects the equally important association of vitellogenesis and accessory gland development, both conclusively shown to be controlled by the same hormone (p. 170), provides a precedent for a single hormone affecting two separate but related processes.

Reproduction, growth and regeneration

In most polychaetes, somatic growth is at first rapid, with a high rate of segment proliferation continuing until about half the usual segment number is achieved. Then both segment proliferation and enlargement slow down considerably, and eventually stop altogether. Vitellogenesis and the final maturation of both oocytes and spermatocytes occur during the latter period of somatic growth. If segments are amputated during this time regeneration will not occur. It might therefore be supposed that at an appropriate time, nutrients are switched from somatic growth to gametogenesis and that this switch is manipulated by hormones. The epitokal transformation must also use considerable energy. It is possible that all polychaetes at one time metamorphosed in a manner similar to the nereids, but have since become neotenous, that is, reproduction is associated with a juvenile body form, with a consequent saving in energy. On this interpretation the nereids must have retained the original trans-

formation because the advantages of a different reproductive habitat outweighed in this group the disadvantages of epitoky. However, if there is a complete lack of unity in the endocrine control of gametogenesis and epitoky in polychaetes, it is equally likely that epitoky has arisen independently in several polychaete families.

The brain of a mature nereid will not induce regeneration when implanted into a decerebrate young individual. Thus production of growth and regeneration hormone by the brain must cease as the individuals mature. But a brain from a young nereid will induce segment proliferation in decerebrate mature worms, and regeneration in amputated mature worms. So the tissues of mature nereids retain the potentiality for growth and regeneration, but the potential is not realized because of the normal absence of the appropriate hormone. Moreover, the implanted young brain also inhibits gametogenesis and epitoky in its mature host; both processes go on normally if the implanted brain is subsequently removed (p. 74). Is the growth and regeneration hormone therefore identical with the gametogenesis and metamorphosis hormone(s)?

In *Nereis diversicolor*, the secretion by the brain of growth-promoting hormone and of oocyte-inhibiting hormone ceases at about the same time, suggesting that only one hormone is involved in both processes. But when segments are amputated early in the second, decreasing growth phase, oocyte development is not inhibited although segment regeneration occurs. It is suggested that this result indicates the existence of distinct growth-promoting and oocyte-inhibiting hormones: the oocytes develop in a normally decreasing inhibitory hormone concentration, although the growth hormone titre is increased to facilitate segment regeneration. But in *Platynereis dumerilii*, segment regeneration *does* retard oocyte development, a result consistent with a single hormone hypothesis. It is possible, of course, that in *Nereis diversicolor* the oocytes are inhibited by a growth hormone concentration higher than that achieved during segment regeneration in older individuals, such a concentration being present in young actively growing worms. So the difference between *Nereis* and *Platynereis could* be due to variety in the oocyte threshold of response to a single hormone, rather than to the existence of different hormonal mechanisms in the two species. But the problem will not be definitely resolved without further experiments and the use of much more highly purified extracts of brain hormones.

REPRODUCTION AND ENVIRONMENT

The spawning of polychaetes is often related to environmental events, and there is usually some degree of co-ordination in spawning between the members of a population. In certain species, this has acquired the

strictest periodicity. For example, in the palolo worm *Eunice viridis,* which lives in the reefs of the Southern Pacific, as the day of the last lunar quarter of the October–November moon dawns, the posterior half of the body breaks off and swims to the surface to spawn. The anterior end of the worm regenerates its missing portion and spawns again the following year. The times of swarming are so predictable that they are included in the local Fijian calendar. Of course, in many other so-called palolo worms, the periodicity is nothing like so precise as in *Eunice.*

In the vast majority of polychaetes, spawning is not so well co-ordinated as in *Eunice* and other species. But even so, some degree of co-ordination is necessary. It is theoretically possible for environmental events to synchronize three distinct phases of sexual development in polychaetes: the onset of germinal epithelium proliferation and early gametogenesis; the initiation of sexual and somatic maturation; and the induction of spawning and gamete release.

Although negative feedback from developing gametocytes may synchronize the renewed phase of germinal epithelium proliferation in polytelic species (p. 72), this cannot account for the initial synchronization in virgin worms. Some evidence exists for an environmental control of early gametogenesis. For example, if immature *Glycera dibranchiata* are maintained above ambient temperature, premature gonadal development is induced, and sexual maturity in the serpulid *Hydroides dianthus* can be accelerated by changes in food quality and the light regime.[189c] Environmental events probably do affect early gonadal development, but the processes involved are little understood.

In nereids, the decline in brain hormone production and the consequent sexual and somatic development is influenced by environmental events. In *Platynereis dumerilii,* the breeding season extends from March to October in the Mediterranean and there is a monthly periodicity of surface swarming, the largest numbers appearing at the time of the new moon and the smallest at the full moon. *Platynereis* swarms on the night following the completion of metamorphosis, so the time at which metamorphosis *begins* will determine the time of swarming. Since metamorphosis takes about 1 to 2 weeks to complete, it must be initiated at about the time of a full moon.[124]

It might be expected then that increasing photoperiod could be the trigger for the initiation of metamorphosis. When *Platynereis* is kept in constant light, the periodicity of swarming is eventually abolished. Exposed to 12 hours light and 12 hours dark in every 24 hours, followed by 6 days constant light at the end of every month, *Platynereis* will spawn 16–20 days after the end of the period of constant light. It appears, then, that increased photoperiod initiates metamorphosis, presumably by causing the cerebral neurosecretory cells to stop producing hormone (p. 74).[124]

But *Platynereis* will spawn even though the previous critical full moon was obscured by cloud. This together with the maintenance of swarming periodicity in constant illumination for a period of up to 3 months, suggests that an endogenous cycle becomes imprinted in the worms, perhaps by exposure to exogenous lunar cycles in early life.

Blinded worms can be synchronized to an artificial photoperiod in precisely the same way as normal worms. The eyes therefore are not the only pathway which transmits information about the environmental photoperiod to the cerebral neurosecretory cells. Exactly the same phenomenon is seen in insects and vertebrates (Chapter 14). It is possible that the neurosecretory cells themselves, or perhaps some nervous centre in close proximity, react directly to environmental photoperiod.[124]

Another factor involved in the co-ordination of swarming in *Platynereis* is the release of a pheromone (p. 287) by the female. The pheromone is released into the water during spawning, and co-ordinates the final swarming behaviour and gamete release in the whole population.[21b]

Very little is known of the role of the environment in initiating the decline in production of the brain hormone in nereids other than *Platynereis*. In many species, metamorphosis takes much longer than the two weeks for *Platynereis* and so it is unlikely that synchronized breeding will result solely from environmental factors affecting brain hormone production. Interactions between individual worms must play an important part. The syllid *Autolytus edwardsii*, for example, shows increased swimming frequency when exposed to sudden changes in light intensity in the laboratory; in its natural environment, the swarming of stolons takes place shortly after sunrise and sunset. Individuals are also positively phototaxic and so they come into close association at the surface of the sea. Here the females release a pheromone which induces rapid swimming and mating behaviour in the male. There are thus three sequential factors involved in mating in *Autolytus*: a sudden change in light intensity which results in surface swimming; the pheromonal stimulation of rapid swimming in the male; and finally contact between males and females leading to synchronized mating and semen release.

Even less is known of the factors controlling the onset of reproductive activity in non-swarming polychaetes. The synchronized spawning of populations of *Arenicola* may be induced by the first really low air temperature during an autumnal low tide, which stimulates the release of the maturation hormone.[150a] Even so, attributing such control to only one environmental factor is probably an over-simplification.

GROWTH AND REPRODUCTION IN OLIGOCHAETES

Although endocrine mechanisms in the oligochaetes have been far less intensively studied than those in the polychaetes, there is sufficient evidence to show that neurosecretory cells in the brain of oligochaetes are involved in the control of somatic and reproductive growth and some aspects of metabolism. The evidence may be sparse, and is sometimes conflicting, but it does indicate the lines that future investigations can take.

Endocrine control of growth

Histological changes in the neurosecretory cells of the brain of *Eisenia foetida* provided the first evidence that growth in earthworms might be influenced by neurosecretory hormones.[126f] This has since been confirmed experimentally: removal of the brain from a young *Eisenia* in the rapid prepubertal growth phase inhibits growth. The inhibition is not permanent for the brain regenerates and, as it does so, growth recommences, the amount of growth being proportional to the degree of brain regeneration.[43a] The implantation of an extra brain into a normal animal causes an acceleration of growth. The obvious conclusion to be drawn from these results— that the brain secretes a hormone which stimulates growth—should be accepted with caution, since the effects upon growth of the implantation of additional brains depends both upon the number of brains implanted, and upon the stage of growth of the host. Thus when several additional brains are implanted into prepubertal worms, growth is *inhibited*; the implantation of an extra brain into pubertal worms (in which the clitellum is developed) accelerates growth, yet the implantation of two additional brains into these pubertal animals inhibits growth.[43b] A possible explanation for these apparently conflicting results is that in normal animals, both the hormone titre and the reactivity, or competence to respond, of the tissues vary during growth. It is suggested that in prepubertal animals, the hormone concentration is low and the reactivity of the tissues has a correspondingly low threshold so that growth results; in pubertal worms, although the concentration of hormone may be greater than in prepubertal animals, the activation threshold of the tissues is relatively higher, so that growth will result only when the hormone titre is raised by the implantation of an extra brain.[43b] This hypothesis, however, does not explain why additional brains, over the minimum that will accelerate growth, will inhibit growth in both prepubertal and pubertal worms. It might rather be expected that a plateau rate of growth stimulation would be achieved, and additional brains would have no further effect. Another possibility, which has not been tested, is that the brain of *Eisenia* produces a growth-inhibiting hormone in addition to the growth-stimulating one. If the

inhibitory hormone was normally present in low concentration compared with the stimulatory hormone, its effects would not be apparent when small numbers of brains were implanted. But with larger numbers of brains, the concentration of the inhibitor could be such as to counteract the effects of the stimulatory hormone. Whatever the explanation, the curious results of brain implantation in *Eisenia* would repay further study.

Diapause and Regeneration

Many earthworms enter a period of diapause during the summer, when reproductive activity ceases and the reproductive organs regress. *Eophila dollfusi*, for example, diapauses for six weeks between April and July, and during this period material accumulates in the neurosecretory cells of the brain and ventral ganglia.[102] Removal of the brain or electrocoagulation of the neurosecretory cells in the posterior part of the brain in non-diapausing worms induces the onset of diapause.[102a] This suggests that diapause results from the withdrawal of a brain hormone.

Associated with diapause in some earthworms is the ability to regenerate caudal segments. In *Eophila*, removal of the brain followed by amputation of caudal segments induces diapause during which segment regeneration occurs. Amputated worms with intact brains heal the wound but do not regenerate their lost segments. This might suggest that the brain produces a hormone which *inhibits* regeneration. It is possible, however, that the diapause induced by brain removal, during which reproductive activity ceases, is a state favourable for regeneration and brain removal itself has no endocrine effect upon regeneration. Moreover, regeneration could be stimulated by neurosecretory hormones from the ventral ganglia.[44d]

Regeneration is not associated with diapause in all earthworms: *Eisenia foetida* can regenerate amputated caudal segments at all times. Further, in *Lumbricus terrestris* the brain must remain intact for 48 hours after caudal amputation if regeneration is to occur, although its removal thereafter has no effect.[151c] These results argue for a stimulatory effect of brain neurosecretion on regeneration. It is clear that in all earthworms which have so far been examined, the brain plays a role, either direct or indirect, in the control of regeneration. More detailed experimental work on a much greater variety of species is required to clarify the processes involved.

Control of Reproduction

In *Eisenia foetida*, removal of the brain immediately arrests gonadal development, and leads to the disappearance of structures, such as the clitellum, associated with copulation and egg laying. Histological cycles

in the various types of neurosecretory cell within the brain can be correlated with normal reproductive development. Reimplantation of the brain into decerebrate worms will stimulate the production of spermatids from spermatogonia. Removal of the sub-oesophageal ganglion also prevents egg laying, but only after an interval of time, and the operation probably interferes with the passage of neurosecretion from the brain.[127, 128] How the production of stimulatory hormones for growth and reproduction is correlated in normal earthworms is unknown, but the situation is clearly quite different from that which obtains in the polychaetes (p. 76).

Control of Osmoregulation and Metabolism

Individuals of *Lumbricus terrestris* placed in distilled water increase in weight for about two hours, but thereafter the weight is gradually restored to normal; the worms are clearly regulating the uptake of water. When the brain is removed, animals placed in distilled water increase in weight rapidly for about 6 hours, and continue to do so, although more slowly, thereafter. When brains are implanted into such decerebrate individuals, or brain homogenates are injected, the normal response is regained.[158m] This suggests that the cerebral neurosecretory cells of *Lumbricus* produce a hormone which controls water balance in the animal, although whether this is a diuretic hormone, or controls water balance in some more subtle way, for example by altering the ionic balance between tissues and body fluids, is not known.

The brain of *Lumbricus terrestris* also produces a hyperglycaemic hormone; brain removal results in the blood glucose level falling to zero within 24 hours, whereas the injection of brain homogenates into decerebrate animals restores the blood glucose level to normal.[185b]

Neurosecretory cells in the brain of oligochaetes thus appear to control a variety of developmental and physiological events. Whether this indicates a corresponding variety in the number of hormones produced by the neurosecretory cells, or whether a much smaller number of hormones (even one) have diverse effects, is likely to prove an even more acute problem to solve in this group than in, for example, the insects.

REPRODUCTION IN LEECHES

The Hirudinea are the third major group of annelids. Leeches, like earthworms, are monoecious (hermaphrodite), and, although gametogenesis occurs simultaneously in both ovaries and testes, oogenesis lags slightly behind spermatogenesis.

The leech testis shows a cycle of activity correlated with the seasons.

Spermatocytes are budded as single cells from the thin walls of the testis and are released into the fluid filled testicular lumen. Here mitotic and meiotic divisions produce clusters of spermatozoa attached by their heads to a central cytoplasmic syncytium. In *Hirudo medicinalis* many mature sperm clusters are present during August, and are completely absent from November to April.[120]

It was first suggested that this testicular development was hormonally controlled when it was found that certain neurosecretory cells (the α-cells) in the brain decreased both in number and in their content of neurosecretion during the latter part of the year, reaching a minimum in December, but then increased to a maximum from April to June.[120]

The possibility of such a causal relationship between the cyclic production of neurosecretion and testicular activity needs to be verified by experiments involving the removal and reimplantation of the brain. The brain of the leech is very difficult to remove through the outer body wall, but fortunately the animals can be turned inside-out, and the brain removed through the gut wall. Naturally, the effects of the operation are compared with those in control animals, also turned inside-out, but from which the brain is not removed.[120]

In such decerebrate animals, the total number of gamete clusters becomes very much less than in the controls. Homogenates of brains taken from leeches in early February injected into decerebrate individuals markedly increase the total number of gamete clusters.[120] It can be concluded, therefore, that the leech brain produces a gonadotrophic hormone which has a positive effect upon spermatogenesis (Fig. 4.15). This situa-

Treatment	Average numbers of gametes (per field of view)	
	First half of development	Second half of development
Untreated	306	393
Brain removed	307	69
Brain removed *plus* injection of brain homogenate	296	237

Fig. 4.15 Control of spermatogenesis by the brain in *Hirudo*

tion is therefore similar to that in the oligochaetes, and differs from that in the polychaetes, a conclusion which is in line with the supposedly close relationship between oligochaetes and leeches.

5

Endocrine Mechanisms in the Mollusca

As has been shown in Chapter 2, neurosecretory cells have been identified histologically within the nervous systems of coelenterates, nematodes, annelids, etc., and subsequent experiments have suggested that these cells are the source of hormones controlling development. Because of the nature of the neurosecretory systems in these animals, it cannot be proved unequivocally that such cells actually produce the hormones. But the combination of histology and experiment provides good circumstantial evidence for such a view.

In the Mollusca, the problem is more acute. Many neurones within the ganglia of the nervous system stain with dyes, such as paraldehyde-fuchsin and chrome haematoxylin-phloxine, which have been used to characterize neurosecretory cells in other animal groups.[246, 100] But since these staining reagents are not specific for neurosecretory material, histological criteria alone provide insufficient evidence that the neurones are neurosecretory. The stained inclusions within such neurones sometimes show cyclic changes associated with other events in the animal, but this does not provide additional evidence for a neurosecretory function, since cyclic changes in neuronal inclusions, such as lysosomes or glycogen granules, are not improbable. Even if some of these neurones are neurosecretory cells, it is impossible to decide whether the cyclic changes in their histology indicate a neurosecretory control over some developmental process in the animal, or whether the neurosecretory cells *and* the developmental process are both reacting to some other factor. The neurosecretory cells could even be reacting to the process itself in order to control some quite different event within the animal.

The way to resolve this difficulty, of course, is to relate such histological observations to experiments in which the neurones suspected of being neurosecretory are removed and subsequently reimplanted and the effects of the operations noted. But in fact, few attempts have been made to determine experimentally the hormonal functions of suspected neurosecretory cells in the molluscs. To a large extent, this lack of experimentation is due to the inherent difficulties of removing small numbers of cells buried deeply within nerve ganglia. Total removal of any ganglion often results in gross malfunction of the operated individual followed by death. In only a few instances have histological studies been accompanied by surgical removal of ganglia. This account of molluscan endocrinology will be confined to these examples, since they demonstrate that experimentation rarely produces results consistent with hypotheses established by previous histological observations on supposed neurosecretory cells. Such contradictions should be borne in mind if reference is ever made to the wealth of literature on the histology of molluscan neurosecretion.

HORMONES AND REPRODUCTION IN GASTROPODS

The opisthobranch and pulmonate gastropods are usually hermaphrodite, and the control of gamete development and sex reversal in these groups has attracted considerable attention. In the pond snail, *Lymnaea stagnalis*, the male component of the ovotestis shows a clear periodicity, with a peak of spermatogenesis during April and May. But the proportions of early oocytes, fully grown oocytes surrounded by follicle cells, and degenerating oocytes remains constant throughout the year. Clearly, as egg masses are laid, there is continual development and replacement of new oocytes. However, periodicity in spermatogenesis and its absence in oogenesis suggests that the development of the two components of the ovotestis is controlled by different mechanisms. What may these mechanisms be?

The cerebral ganglia of *Lymnaea* each contain three groups of neurosecretory cells (Fig. 2.6),[190] the histology of which changes with the seasons. The cells of the mediodorsal and laterodorsal groups stain in a similar manner, and contain very little neurosecretory material in winter.[158] But in spring, neurosecretion accumulates in both the axons and the perikarya of the cells, the rate of accumulation slowing down during summer and autumn (Fig. 5.1). These changes are said to indicate maximum production, transport and release of neurosecretion during April and May by the cells of the mediodorsal and laterodorsal groups. But such an interpretation should be viewed with caution, since it is difficult to relate static

histological pictures to an essentially dynamic process (see p. 171 and Fig. 7.30).

The caudodorsal groups of neurosecretory cells in the cerebral ganglia (Fig. 2.6) stain differently from those in the other groups, which might

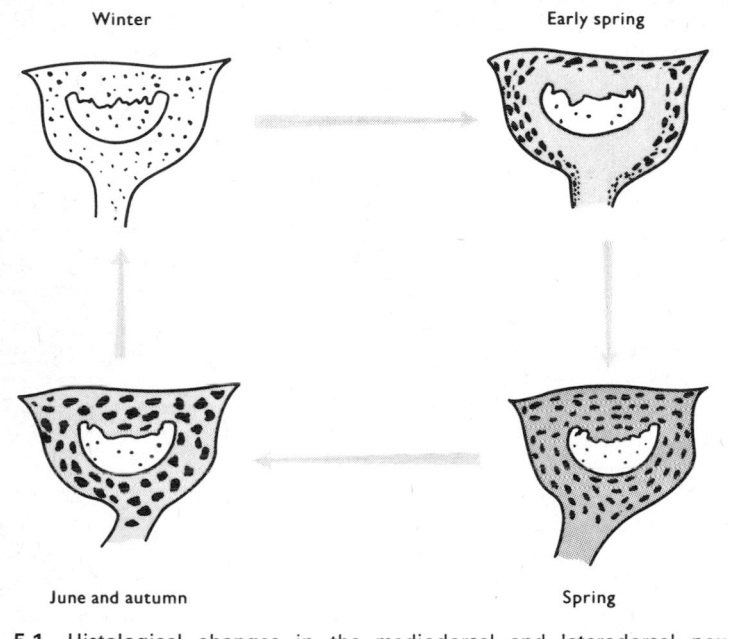

Fig. 5.1 Histological changes in the mediodorsal and laterodorsal neuro-secretory cells of the pond snail, *Lymnaea stagnalis*. During winter the cells contain only a small amount of neurosecretory material, consisting of small spherical granules dispersed throughout the cytoplasm. In early spring, the quantity of neurosecretory material begins to increase: flakes of material occur near the peripheries of the cells with very small granules near the nuclei. In late spring, the flakes of material increase in number and fill each cell. During summer and autumn the flakes seem to become more compact, occupying less of the cytoplasm of each cell. These histological changes may represent a cycle of synthesis and release of neurosecretory hormone during early spring (but see Fig. 7.30). (After Joosse[158])

indicate the production by these cells of a different neurosecretory hormone. The cells go through a seasonal cycle of histological change similar to that in the mediodorsal and laterodorsal cells (Fig. 5.1), and they are also considered, in consequence, to be inactive in winter, and highly active in spring.[158]

If the histology of the mediodorsal and the laterodorsal neurosecretory cell groups in the cerebral ganglia has been interpreted correctly, these

cells are maximally active during the peak of spermatogenesis. In young animals, the caudodorsal neurosecretory cells do not show this presumed activity, and no egg masses are produced in such animals. Consequently, it is possible that the mediodorsal and the laterodorsal neurosecretory cells control spermatogenesis, and the caudodorsal cells (with different staining properties) control oviposition and perhaps oogenesis also. How does this possibility stand the test of direct experimentation?

When the cerebral ganglia of _Lymnaea_ are completely removed, feeding activity, lung ventilation, copulation and egg-laying all stop, and the snail moves about slowly and only for short distances. The operation clearly has very drastic effects upon the life of the animal. However, when the neurosecretory cells alone are destroyed by cautery, oocyte vitellogenesis proceeds more or less normally although the growth of the shell and other parts of the body are retarded. But when the dorsal bodies are removed, vitellogenesis is inhibited (Fig. 5.2); reimplantation of the bodies

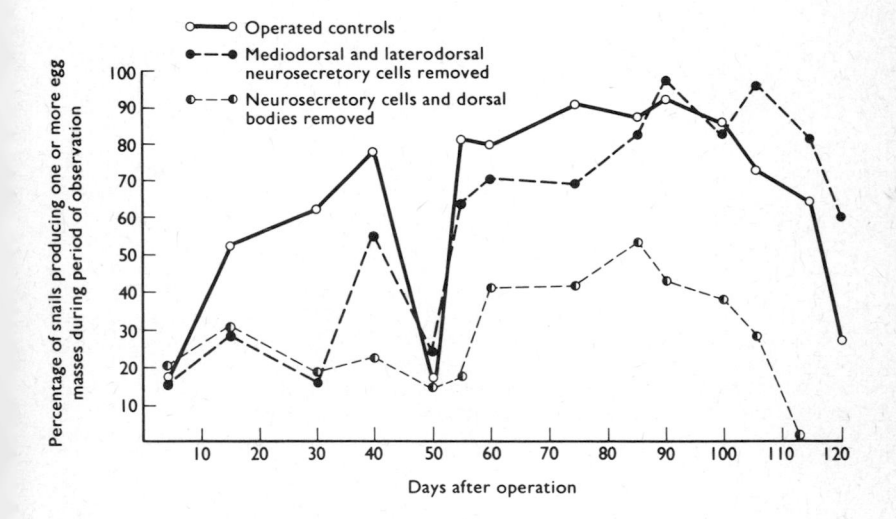

Fig. 5.2   Effects of removal of the medio-dorsal and the latero-dorsal neurosecretory cells and the dorsal bodies on egg mass production in the pond snail, _Lymnaea stagnalis_. Removal of the neurosecretory cells results in a lowered production of egg masses during the first half of the experimental period, but during the second half production of egg masses is not significantly different from that in the control animals. When the neurosecretory cells _and_ the dorsal bodies are removed, the rate of production of egg masses remains low throughout the experimental period. These results suggest that the dorsal bodies, rather than the neurosecretory cells, control the production of egg masses in _Lymnaea_. (After Joosse[158])

restores yolk deposition to normal[158a, 158b] (Fig. 5.3). Thus the histologically defined activity cycles within the cerebral neurosecretory cells cannot be *directly* related to vitellogenesis in *Lymnaea*, although if they are concerned in the control of somatic growth, an indirect correlation with yolk production is possible.

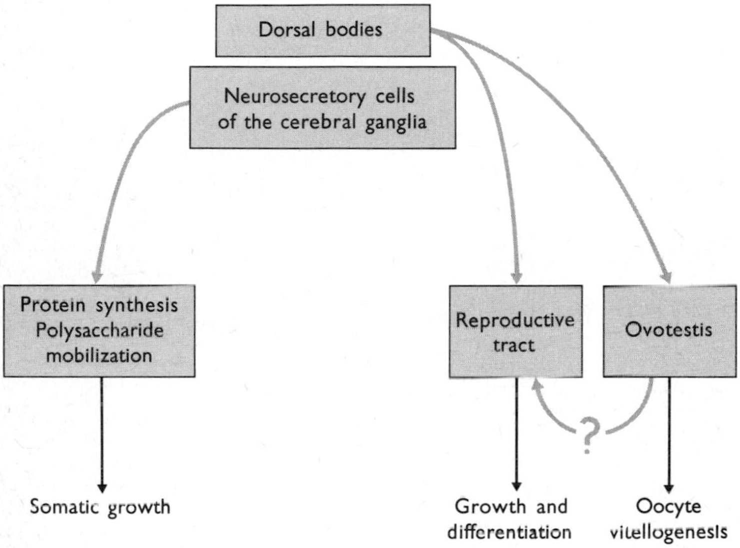

Fig. 5.3 Endocrine control of growth and reproduction in *Lymnaea stagnalis*. A hormone from the dorsal bodies controls oocyte vitellogenesis and development of the reproductive tract, the latter perhaps also affected by hormones from the ovotestis ; somatic growth is controlled by a hormone (or hormones) from particular neurosecretory cells in the cerebral ganglia.

In *Lymnaea*, early oocytes with accompanying follicle cells migrate from their origin in the germinal band in each lobe of the ovotestis to a region of the lobe which is in contact with the adjacent digestive gland. One side of the oocyte is thus in close proximity with the digestive gland and it is likely that this arrangement is necessary for the transfer of materials into the oocyte during vitellogenesis[158c] (Fig. 5.4).

It is now firmly established in *Lymnaea* that the dorsal bodies and neurosecretory cells in the cerebral ganglia control oocyte vitellogenesis and somatic growth respectively. The role of the dorsal bodies recalls that of the optic glands in cephalopods (p. 97). It is probable that similar

mechanisms will be found to operate in other gastropods. But the differentiation of male and female gametes from the hermaphrodite ovotestis has not been so thoroughly investigated experimentally, although the incubation of gonads with various suspected endocrine centres has provided valuable information.

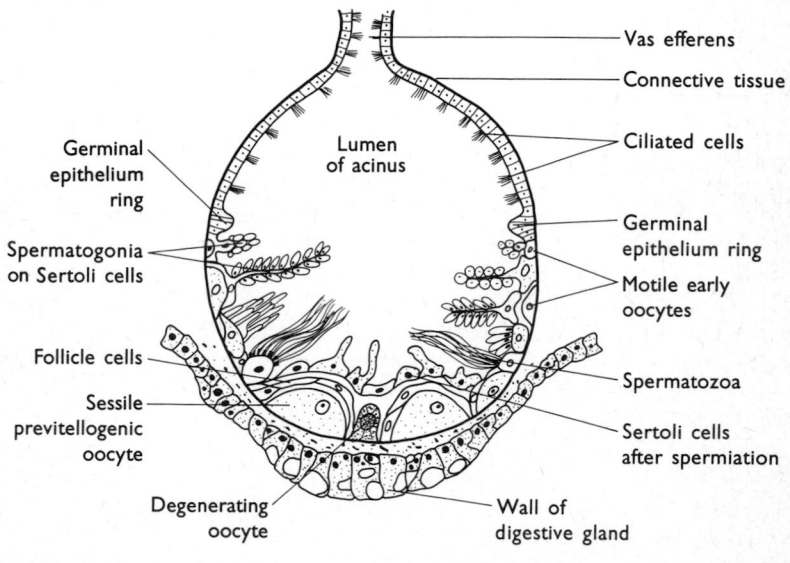

Fig. 5.4 Diagrammatic section through an acinus (blind sac) of the ovotestis of *Lymnaea stagnalis*. The germinal epithelium ring gives rise to both oocytes and spermatocytes with their associated follicle and Sertoli cells. The spermatogenic zone forms a broad band adjacent to the germinal epithelium ring, whereas the young oocytes migrate to a region close to the digestive gland where they become sessile and vitellogenic. After spermiation, the Sertoli cells possibly become phagocytic. (After Joosse and Reitz[158d])

The gonads of many gastropods can be cultured for relatively long periods in a suitable medium. When cultured alone, the gonads of *Helix aspersa, Calyptraea sinensis* and *Patella vulgata* are transformed into ovaries, even when the gonad is in the male phase at the beginning of incubation. [51c, 51d, 116a, 118k, 254a] When the gonads are cultured with cerebral ganglia obtained from individuals in the male phase, or in a medium containing haemolymph from such individuals, then spermatogenesis instead of oogenesis ensues.[51d, 116a, 118l, 254a] Gonads incubated with cerebral ganglia from female phase individuals show accelerated oogenesis with many oogonial mitoses[51e] and eventually vitellogenesis. This suggests,

by comparison with the experimental work in *Lymnaea*, that the factor originates not from the cerebral ganglia themselves, but from the associated dorsal bodies which were not removed in these incubations.

Three conclusions emerge from the results of these incubation experiments: that the hermaphrodite gonads of gastropods in isolation autodifferentiate into ovaries; that the production of increased numbers of oocytes, together with vitellogenesis, depends upon a hormone produced very probably by the dorsal bodies; and that the autodifferentiation of gonads into ovaries is prevented, and development switched to testis formation, by the presence of a hormone produced by the cerebral ganglia or associated structures at particular times. These conclusions are of considerable interest, because the situation in gastropods can now be compared with that in the crustaceans and many vertebrates, among others, where testis formation requires the presence of an additional (masculinizing) factor to prevent autodifferentiation of the gonad into an ovary, even when hermaphroditism does not normally occur.

In the gastropod molluscs, protandric hermaphroditism is a common condition, and the control mechanisms which must operate can now be described in general terms. The ovotestis (hermaphrodite gland) is initially transformed into a testis by the production of a particular hormone by the cerebral ganglia. When this factor is withdrawn, transformation to an ovary occurs and the oocytes so produced grow and deposit yolk under the influence of a hormone, or hormones, from the dorsal bodies. This general hypothesis would clearly need modification to account for the variety of conditions encountered in many gastropods.

The clear periodicity in spermatogenesis in *Lymnaea* and many other gastropods probably has a different control system. It has been reported that the transformation from the male to the female phase can be accelerated by the removal of the optic tentacles, and that the injection of tentacle extracts into individuals can suppress premature spermatogenesis or induce the spermatogenic 'repose' which occurs normally in many protandric hermaphrodites.[51d, 51e, 118f] Moreover, it has been claimed that when gonads are incubated with cerebral ganglia from male phase individuals *together with* optic tentacles, spermatogonial mitoses are inhibited.[51d, 51e] This suggests that the gastropod optic tentacles contain endocrine constituents which control spermatogenesis and perhaps the phase change from male to female in protandric hermaphrodites. However, in the snail, *Helix aspersa*, and the slug, *Arion*, the optic tentacles seem to have the opposite effect: spermatogenesis is reduced and oogenesis increased after removal of the tentacles, whereas spermatogenesis increases and oocyte production returns to normal after the subsequent injection of tentacle extracts.[180, 220]

The controversial role of the optic tentacles in gastropods is further confused by dispute about the presence of possible endocrine cells within the tentacles. Neurosecretory cells of various types described in the optic tentacles of pulmonate gastropods[180, 181] are now thought to be glandular cells secreting materials largely to the exterior.[227i, 246] It is perhaps significant that in *Ariolimax columbianus* tentacle removal increases the galactogen content of the albumen gland,[206a] and in *Helix aspersa* the operation leads to modifications of the digestive gland.[229g] These effects are not necessarily the result of any endocrine mechanism: the tentacles are important in daylength perception and for orientation in feeding, and their removal could thus have significant metabolic results. Moreover, the tentacles have a great capacity for regeneration, and this could also affect the general body metabolism. When tentacular extracts are injected, the recipients will receive large amounts of mucous cell products, which may have significant metabolic effects. Any alterations in the general body metabolism would inevitably affect gonad development. It must therefore be concluded that an endocrine status for the gastropod optic tentacles is unproved, and many more precisely controlled experiments are necessary before the role (if any) of the tentacles in the control of reproduction can be established.[158a]

The development of the reproductive tracts in gastropod molluscs seems also to be hormonally controlled. In hermaphrodite gastropods, the proximal parts of the reproductive tract are also hermaphrodite, but distally the tract separates into male and female parts, each with their own accessory glands. In terrestrial pulmonates, when the juvenile, undifferentiated, tracts are implanted into a host in the male phase, the distal male part of the implant, together with its glands, is induced to develop; implantation of the juvenile tract into a female phase host evokes the development of the female parts of the implant.[185a, 229d] This suggests that two hormones control the development and differentiation of the reproductive tract: one is produced during the early stages of maturation and influences the development of the male parts of the tract; the other is produced later and induces the female parts to develop. When gonads in the male or female phase are implanted into juvenile hosts, it is claimed that the reproductive tract of the host develops its male or female parts according to the phase of the implanted gonad.[185a] This implies that the gonad is the source of each of the two hormones,[1] which are produced sequentially as the gonad transforms. However, in the slug *Ariolimax californicus*, the cerebral ganglia are said to be the source of the maturation hormones,[118g] and in *Calyptraea sinensis* and *Crepidula*, the optic tentacles have been implicated in the control of the development of the male tract,[254b] in a manner similar to that postulated for the primary gonadal transformation. In *Calyptraea*,

when a fully developed (male phase) penis is cultured *in vitro* with nerve ganglia from a female phase individual, the penis regresses to the form characteristic of the female phase; culturing the regressed (female phase) penis with ganglia from a male individual induces the full development of the male form.[254a] In *Lymnaea stagnalis*, growth and differentiation of the complete reproductive tract is inhibited when the dorsal bodies of juveniles are destroyed by cautery; the tract grows rapidly on the subsequent implantation of dorsal bodies.[158b]

The gonads of gastropod molluscs are so intimately associated with the digestive gland that castration is lethal. Direct experimentation to prove or disprove the conjecture that the gonads are the source of hormones controlling reproductive tract development is therefore impossible. On balance, it seems likely that the cerebral ganglia and/or the dorsal bodies are the more probable source of the hormones. If this indeed proves to be so, then gonadal and reproductive tract accessory gland development would both be independently controlled by hormones produced outside the reproductive system—a situation not dissimilar to that which has been described in insects (p. 170). However, it would also be wise to consider that in a group as diverse as the gastropod molluscs, hormones which have the same ultimate function could well originate in a variety of different sources.

Gonadal differentiation and transformation, together with the development of the reproductive tract, are clearly controlled by hormones. In addition, the final stages of reproductive development in the female—ovulation and spawning—are also hormonally controlled, and in one species the source of the hormone is very firmly established.

In the sea-hare, *Aplysia californica*, sea water extracts of the neurosecretory cells ('bag cells') associated with the parietovisceral ganglion (Fig. 2.7) will induce egg-laying when injected into another individual.[178c, 178d, 256a] Extracts of the distal parts of the pleurovisceral nerves and of the parietovisceral ganglia from which the bag cells have been removed will also induce egg-laying when injected into other individuals—because these regions contain the neurohaemal areas in which the bag cell axons terminate[264a] (Fig. 2.7).

In *Aplysia*, the eggs are laid about 1 hour after the injection of bag cell extract. This is not the result of any delayed reaction to the hormone, since eggs appear in the hermaphrodite duct next to the ovotestis within 1 minute after injection of the extract.[61c] The remainder of the time within the genital tract is occupied with packaging the eggs into a string of gelatinous capsules. It is therefore suggested that the primary function of the bag cell hormone is to cause the contraction of muscle cells around the ovotestis, thereby forcing the ripe oocytes into the hermaphrodite duct. It is also possible that the hormone may additionally cause the dissolution

of the junctions between the ripe oocytes and their follicle cells,[61c] recalling the spawning mechanism in starfish (p. 49).

Electrophoretic separation of *Aplysia* bag cell extracts reveals the presence of two specific proteins not present in other nervous tissue; one of these proteins is present also in the sheaths of the parietovisceral ganglia and the distal parts of the pleurovisceral nerves.[256a, 264a] Both the egg-laying hormone and the bag cell specific protein are denatured by heat and destroyed by the enzyme pronase. Because the distribution and chemical properties of the hormone and the specific protein are so similar, it is likely that the protein is the hormone itself, or at least that the protein, like the neurophysin of vertebrates, is a carrier for another active molecule.

During the process of maturation in *Aplysia*, the number of bag cells increase about three times, while the amount of specific protein within the cells increases nine times. The hormone is present throughout the year in mature animals, but extracts of the bag cells have maximum effects upon spawning during the summer months. This could indicate the presence of a spawning inhibitor in the ovotestis, as in the starfish ovary (p. 51), although since the ovotestis of *Aplysia* has a seasonal rhythm, the varying effectiveness of bag cell extracts may merely reflect the cyclic maturation of oocytes in the ovotestis.

Under normal conditions, the immediate cause of spawning seems to be stimuli received during copulation, since the egg string is released either during or a little time after mating. The bag cells are electrically silent in isolated individuals, and do not respond to stimulation of peripheral nerves. But when the pleurovisceral connectives are experimentally stimulated, the bag cells respond with repetitive spike activity lasting for as long as 55 minutes.[179] All the bag cells in one cluster respond synchronously, and those in the opposite cluster may also respond. The cells together contain about five times the threshold amount of hormone required for normal egg-laying, and it is suggested that the electrophysiological properties of the cells are such that more or less predetermined amounts of hormone are released after stimulation.[179] It can be assumed that stimuli received during copulation are transmitted *via* the cerebral and pleural ganglia along the pleurovisceral connectives to activate the bag cells to release their hormone. In this way, the release of oocytes from the ovotestis will be synchronized with sperm transfer from the partner, and the fertilization of the eggs will be ensured.

The nervous control of hormone release from the bag cells of *Aplysia* contrasts with that of the control of release of water balance hormone in *Lymnaea stagnalis*, where ultrastructural studies suggest that nervous mechanisms are not involved (p. 101). This difference is not surprising when the functions of the two hormones are taken into account: in *Aplysia* to regulate and synchronize oocyte release with fertilization, and in *Lymnaea*

to fulfil the longer term need for the regulation of water balance. However, it is of interest that a rapid ovulation control mechanism in *Lymnaea* also has been suggested.[158a]

HORMONES AND REPRODUCTION IN LAMELLIBRANCHS

In the edible mussel, *Mytilus edulis*, material accumulates in the neurosecretory cells of the cerebropleural and visceral ganglia during the period of gamete maturation. Discharge of the gametes is preceded by the disappearance of neurosecretion from the cells.[194] Removal of the cerebropleural ganglion has no effect upon gamete maturation but apparently accelerates gamete discharge. Removal of the visceral ganglion retards oviposition.[195]

If it is assumed that ganglion removal has no traumatic effect upon *Mytilus*, the relationship between neurosecretory cell histology and the results of extirpation are equivocal, to say the least. If the sudden disappearance of neurosecretion means the release of a hormone which causes gamete discharge, then the acceleratory effect of removal of the cerebropleural ganglion is inexplicable. It is likely that experiments involving reimplantation of the nerve ganglia, or the injection of homogenates, would produce results which would clarify the situation enormously.

In the zebra mussel, *Dreissena polymorpha*, there is an even closer parallel between the histology of neurosecretory cells in the cerebropleural and visceral ganglia and the reproductive cycle. In *Dreissena*, both neurosecretory and reproductive cycles begin at the end of summer. The histology of the neurosecretory cells suggests a slight discharge of material at the end of autumn or beginning of winter, and simultaneously there is a spurt in oocyte growth. In the spring and summer, maximum discharge of neurosecretion is correlated with intensive oocyte growth and spawning.[3] Neurosecretion is said to be discharged when the amount within the cells diminishes.

But when the cerebropleural ganglion is removed in March—at the onset of oocyte growth—the oocytes still grow to maturity. Spawning can also take place some weeks after removal of the cerebropleural ganglion. These experimental results contradict completely the relationship between neurosecretion and oocyte development and spawning, established from interpretations of neurosecretory cell histology. Removal of the visceral ganglion results in rapid death.[3]

In summary, the equivocal and sometimes contradictory results of histological and experimental investigations of possible endocrine mechan-

isms controlling reproductive processes in the lamellibranchs allow no definite conclusions to be drawn at the present time.

HORMONES AND REPRODUCTION IN CEPHALOPODS

The cephalopod molluscs include the squids and octopi, which are perhaps the most highly developed of all invertebrate animals. They have a large and complicated brain (Fig. 2.9) and their eyes are functionally equivalent to those of the vertebrates. Both octopi and squids are remarkably intelligent creatures, and are very popular animals for use in experiments upon learning and behaviour. In fact, the best account of the endocrine control of sexual maturation in *Octopus* arose as a secondary consequence of such experiments.

In the immature female *Octopus*, the optic glands (Fig. 2.9) are small and pale, but increase in size some ten times and become bright orange as the ovaries enlarge.[268] M. J. Wells, investigating the effects upon learning of the removal of different parts of the brain of *Octopus*, or of cutting connections between its various parts, discovered that a certain proportion of his experimental females became precociously mature, and these all had enlarged optic glands. Further investigations showed that three kinds of operation upon the brain causes enlargement of the optic glands: (a) removal of the subpedunculate lobes (Fig. 5.5); (b) cutting the nerves between these lobes and the optic glands; and (c) blinding the animals by cutting the optic nerves, or by cutting the optic stalks distal to the optic gland and removing the optic lobes (Fig. 5.5). Ovarian enlargement is more rapid after operations (a) and (b) than after (c). Lesions elsewhere in the brain have no effect upon maturation, and do not increase the size of the optic glands: these can be considered control experiments for the three operations which do cause precocious maturation. Moreover, when the optic glands are removed, no matter what other lesions are made in the brain, precocious maturation does not occur. Finally, when the lesions that cause enlargement of the optic gland are made unilaterally, the gland on the other side is unaffected.[268]

What conclusions can be drawn from these results? First, it seems clear that a hormone from the optic glands controls ovarian enlargement: this is caused largely by yolk deposition in the existing oocytes, and so the hormone is vitellogenic and presumably acts upon the follicle cells around the oocytes. Secondly, the effects of unilateral nerve sections suggest that optic gland enlargement is not the result of any intermediary endocrine control: the contralateral optic gland would also enlarge if this were so. Consequently, the optic glands must be inhibited in the immature female by a nervous centre in the subpedunculate lobes of the brain;

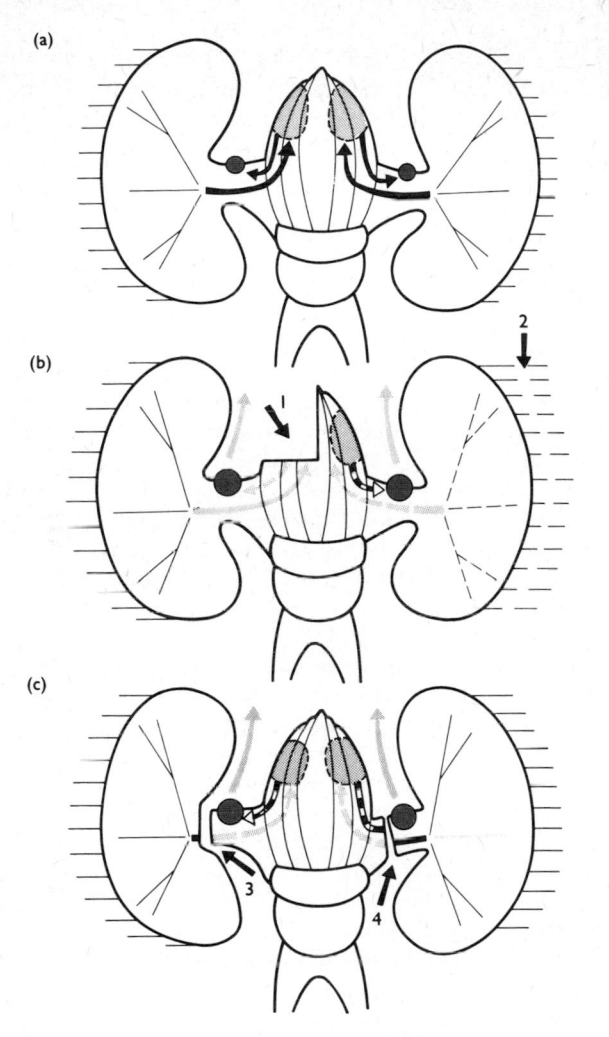

Fig. 5.5 Hormonal control of sexual maturity in *Octopus*. (a) In an immature, unoperated animal, the activity of the optic glands is suppressed by inhibitory nerves. It is likely that changes in photoperiod govern the inhibitory centres in the brain and hence the secretion of the optic glands. (b) Operations (shown by black arrows) which cause enlargement—and secretion—of the optic glands and result in development of the gonads: (1) removal of the source of the inhibitory nerve supply; (2) section of the optic nerves, causing blindness. (c) Operations (shown by black arrows) which cause enlargement of the optic glands and gonads, and which eliminate the possibility of stimulatory innervation of the glands; (3) removal of the optic lobes; (4) section of the optic tract. Compare Fig. 2.9. (After Wells and Wells[268])

these centres are themselves controlled by nervous impulses coming from the eyes. This hypothesis would explain the more rapid ovarian enlargement that occurs after removal of the subpedunculate lobes or section of the optic tract proximal to the optic glands, when compared with the effects of section of the optic nerves or optic tract *distal* to the glands (see Figs. 2.9 and 5.5).

There is also ultrastructural evidence that the nerve tract to the optic gland contains two types of fibres: one synapses with other axons within the gland (axo-axonal), and the other synapses with gland cells (axo-glandular). The axo-axonal synapses are absent from active optic glands and it is suggested that the axo-glandular synapses are the endpoints of *stimulatory* impulses to the glands, this stimulation being inhibited by the fibres with axo-axonal synapses in immature animals.[95d] During normal maturation, therefore, the optic glands may be activated not merely by the lifting of a nervous inhibition, but by active stimulation in addition. This perhaps would ensure coordinated and gradual activation of the optic glands. However, since isolated optic glands cultured in a suitable medium increase in size within a few days and show active hormone production,[81b, 227b] as they do *in situ* when both inhibitory and stimulatory fibres are severed (p. 95), the stimulatory mechanism can clearly be superseded, at least under experimental conditions.

When immature ovaries of *Sepia officinalis* are incubated with optic glands, large numbers of mitoses appear within the germinal epithelium, and oocytes and follicle cells proliferate.[227b] The enlarging oocytes also lay down some yolk platelets (primary vitellogenesis) but the full development of the eggs is not achieved under these culture conditions.

In the cephalopods, therefore, the hormone from the optic glands appears to control oogenesis at two stages: the initial oogonial and follicle cell proliferation, and the secondary vitellogenesis that brings the eggs to their full size. When radioactively labelled leucine is injected into the blood of *Octopus vulgaris*, the amino acid is incorporated into protein in the ovaries within a few hours (Fig. 5.6).[214c] Removal of the optic glands prevents the accumulation of labelled protein in the ovaries, and when labelled blood protein is withdrawn from one individual and injected into another, the protein is not taken up by the ovaries. Moreover, in ovariectomized individuals there is no large accumulation of labelled protein in the blood or the liver. All this evidence points to the ovary—and most probably the follicle cells—as being the site of synthesis of the vitellogenic proteins, and hence the target for the optic gland hormone. The cephalopods thus differ from invertebrates such as insects and crustaceans, and also from the vertebrates, in which vitellogenic protein synthesis occurs at sites outside the ovaries, and the proteins are carried to the oocytes by the blood (Fig. 5.6, compare Fig. 7.20).

During sexual maturation in *Octopus*, the growth rate is reduced and may even become negative; the bases of the arms become unusually thin, and digestive enzymes in the liver and salivary glands decline.[214c] The intense ovarian protein synthesis induced by the optic gland hormone is thus accompanied by reduced somatic protein synthesis. The hormone could thus function as a switch to control the antagonistic processes of somatic and reproductive growth.

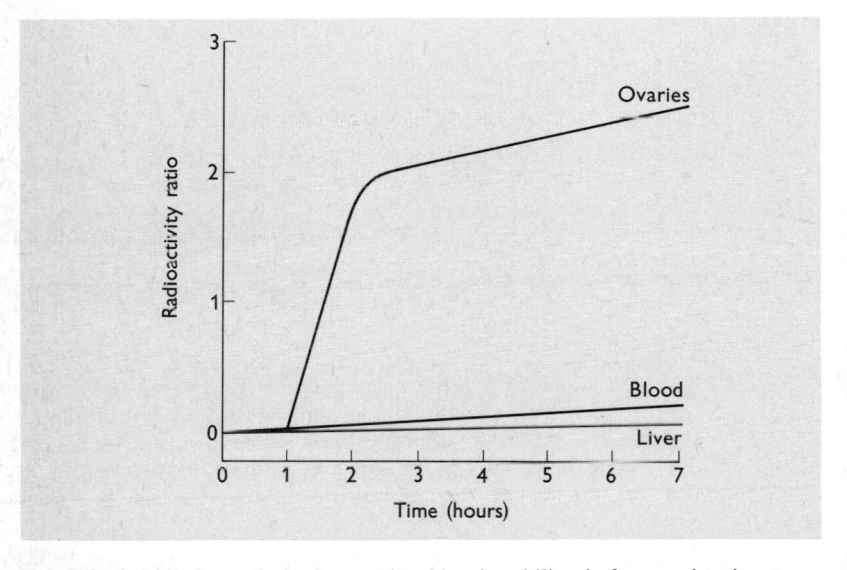

Fig. 5.6 Labelled protein in the ovaries, blood and 'liver' of precociously matur-ing *Octopus vulgaris* at different times after the injection of radioactive leucine. The radioactivity ratio is the incorporation of the amino acid into unit amounts of protein relative to the amount injected into each animal, allowing correction to be made for animals of different weights. The early and marked incorporation of leucine into ovarian protein suggests that the ovaries synthesize the protein themselves, and that it is not produced in the 'liver' and transported to the ovaries. Compare Fig. 7.20. (After O'Dor and Wells[214c])

A similar control over testis development by the optic glands in the male *Octopus* can be demonstrated, except that the testes apparently react to a lower titre of hormone. Furthermore, the males normally mature much earlier than the females, and when the optic glands are removed, the testes can actually regress. This suggests that the development of the testes needs a continuous supply of optic gland hormone; it is likely that the same applies to ovarian development.

The secretory activity of the optic glands is thus inhibited in the immature octopus by nervous means. There is no evidence for the intervention of neurosecretory hormones in this inhibition. Nor is it clear how the nervous inhibition is lifted during normal maturation. Wells suggests that special mechanisms for delaying maturation are found in the arthropods and vertebrates as well as in the cephalopods, and that such mechanisms are important only in these highly developed animals because the central nervous system must develop its full potentialities before the gonads become mature. On this view, the control of endocrine activities by the central nervous system is not surprising. But the cephalopods differ from the arthropods and vertebrates in their *direct* nervous control over the endocrine organs responsible for maturation. It may be that the development of the full potentiality of the central nervous system is only one aspect of the general antagonism between somatic and reproductive processes found in many animals already discussed in Chapter 4.

ENDOCRINOLOGY OF REPRODUCTION IN *LYMNAEA* AND *OCTOPUS*

Major experimental investigations have now been made into the endocrine control of growth and reproduction in the gastropod *Lymnaea stagnalis* and the cephalopod *Octopus vulgaris*, together with supporting observations on related species in the two groups. Although it may be unwise to draw general conclusions from the results of experiments upon two such widely disparate animals, it is instructive to compare what is known about the endocrinology of these species.

In *Lymnaea*, the dorsal bodies are closely associated with the cerebral ganglia and produce a hormone which controls vitellogenesis and the growth and differentiation of the female reproductive tract and associated glands. In *Octopus*, the optic glands are similarly in close juxtaposition with the brain and are also associated with oogonial proliferation and vitellogenesis, and with the development of the female reproductive tract. The gastropod dorsal bodies can thus be considered functionally equivalent to the cephalopod optic glands. In *Lymnaea*, particular neurosecretory cells in the cerebral ganglia control protein synthesis, the mobilization of stored polysaccharides and general body growth.[104a, 158c] In *Octopus*, the optic gland hormone may be involved in the reduction of somatic protein synthesis and growth during maturation, but it is possible that the subpedunculate lobe–pharyngo-ophthalmic vein neurosecretory system (p. 20) may be involved in somatic growth and maturation in this species, since the system contains many inclusions in immature animals which

disappear in the mature animal.[95d] If this proves to be so, then the degree of parallelism between the endocrine control of growth and reproduction in *Lymnaea* and *Octopus* may indicate a pattern common to both the gastropod and cephalopod Classes, and perhaps even to all molluscs.

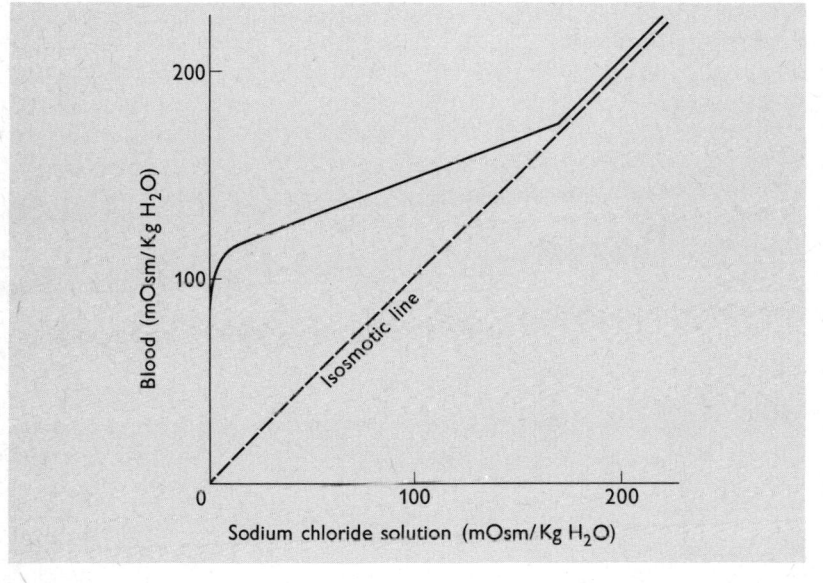

Fig. 5.7 Osmolality of the blood of *Lymnaea stagnalis* exposed for 24 hours at 20°C to different concentrations of sodium chloride solution. The snails maintain an osmotic differential between the blood and the external medium up to about 0·1 M NaCl; at higher salinities, the blood is slightly hypertonic. (After Wendelaar Bonga[268d])

OTHER HORMONALLY CONTROLLED PROCESSES IN MOLLUSCS

Lymnaea stagnalis is a freshwater animal, and therefore has an internal environment which is hyperosmotic to its surroundings. Like all freshwater animals, *Lymnaea* has to cope with potential water influx and salt loss.

When placed in deionized water and in salt solutions of different concentrations, *Lymnaea* maintains an osmotic differential between its blood and the external medium up to about 0·1 M NaCl (Fig. 5.7). Thereafter, the blood is only slightly hypertonic to the medium.[268d]

The ability to osmoregulate suggests the possibility of a controlling mechanism: that this is hormonal in *Lymnaea* has been amply proved by extirpations and reimplantations of parts of the nervous system. Of all the ganglia (Fig. 2.6), only when the pleural ganglia are removed is there any remarkable change in the water balance of the snail: the body swells and increases in weight due to the accumulation of water.[191, 126e] Removal of the right pleural ganglion causes greater water accumulation than removal of the left, but rather less than extirpation of both ganglia. Reimplantation of the right pleural ganglion prevents any water accumulation, and the single injection of an homogenate of the right pleural ganglion causes a rapid reduction in the weight of the animal after the pleural ganglia had been removed some time previously. These results suggest that the pleural ganglia of *Lymnaea* produce a diuretic hormone.

A large number of a particular type of neurosecretory cell—the dark green cell—is found in each pleural ganglion in *Lymnaea*. When snails are kept in deionized water, both histological and ultrastructural changes in the cells, and in their associated neurohaemal areas, indicate increased synthesis and release of secretions, whereas in snails kept in saline solutions, the production and release of secretions by the dark green neurosecretory cells are suppressed.[268a, 268c, 268d] These observations, taken in conjunction with the experimental results of extirpation and reimplantation of the pleural ganglia, support the view that the cells produce a diuretic hormone. Pleural ganglia implanted into host snails, and thus with all their peripheral nervous connections severed, react ultrastructurally in the same way as pleural ganglia *in situ*, when the hosts are kept in solutions of different osmolarities: this implies that the neurosecretory cells in the ganglia are reacting directly to the effects of the different environments and not to an intervening neuronal message.[229a]

In a related species, *Lymnaea limosa*, cauterization of the pleural ganglia induces water accumulation and reduces urine flow,[44b] but in *Lymnaea limosa* cauterization of the parietal ganglia (Fig. 2.6) reduces the turnover of sodium ions. In *Lymnaea stagnalis*, the parietal ganglia and the visceral ganglion contain neurosecretory cells (yellow cells) which are different in kind from the dark green cells of the pleural ganglia,[268a, 268b, 268c] but which show signs of synthesis and release in animals kept in deionized water, and their suppression in animals kept in saline solutions.[268d] Of particular interest is the fact that neurohaemal areas containing axon terminals apparently from the yellow cells are found not only centrally, but also in the pallial and anal nerves and in the region of the kidney and ureter. It is therefore possible that in *Lymnaea*, the ionic balance may be controlled hormonally, although the evidence for such a control is not as well documented as that for the control of diuresis. But it would not be surprising to find in *Lymnaea* endocrine mechanisms which control and

coordinate both water and ionic balance, as has been found, for example, in crustaceans (p. 222) and vertebrates.

In the prosobranch gastropod, *Littorina littorea*, which lives in a marine environment, there is also evidence that hormones are involved in osmoregulation. In seawater concentrations down to about 25% of normal, *Littorina* is an osmoconformer, its body fluids being isosmotic with those of its surroundings. *Littorina* begins to osmoregulate at concentrations below 25% seawater, but when the left cerebral, left pleural or right pleural ganglia are extirpated, the ability to osmoregulate is significantly reduced. The concentration of sodium ions in the blood also falls following extirpation of these ganglia from animals kept in 25% seawater. The reimplantation of individual ganglia does not restore the animal's ability to osmoregulate, and it is suggested that neurosecretory cells in the three ganglia in some way interact upon each other to produce their effects.[291a]

In the cephalopod molluscs, the existence of the neurosecretory system of the vena cava (Fig. 2.10) is well established morphologically from both histological and ultrastructural studies.[1i, 1j, 17a, 201b] The extensive neurohaemal areas in the anterior vena cava are found beneath the muscular coat adjacent to the inner surface of the vessel.

When the isolated heart of *Eledone cirrosa*, perfused with sea water, is treated with a sea water extract of the vena cava, both the frequency and amplitude of beating are markedly increased and remain high for between 10–20 minutes.[17a] Extracts of other blood vessels have no such effect, although they frequently produce transient inhibitory and/or excitatory effects. Extracts from regions of the vena cava which contain the densest areas of neurosecretory terminals have the greatest cardio-excitatory effects, and extracts of the inner layers of the anterior vena cava have ten times the activity of the outer layers—which approximates to the ratio of nervous tissue in the two layers of the wall. There is little doubt that the neurosecretory system of the vena cava is the source of the cardio-excitatory principle(s).

The cardio-excitatory hormone is not species specific, although extracts of the anterior vena cava of octopods must be diluted 10–20 times to produce similar effects in decapods, and those of decapods concentrated by a similar amount to produce effects in octopods.[18] The nerve endings are denser in the octopod species and hence presumably contain more neurosecretion.

The cardio-excitatory hormone is quite different in its effects from 5-hydroxytryptamine, adrenaline or noradrenaline and its action is not blocked by the 5-hydroxytryptamine inhibiting drug, 2-bromo lysergic acid diethylamide. Extracts of the cardio-excitatory hormone prepared in 50% acetone are much more potent than sea water extracts, although the activity of the latter is increased by heating, suggesting that perhaps

more hormone is released after rather harsh treatments.[17a] The enzymes trypsin and pepsin have no effect upon the activity of vena cava extracts, although treatment of the extracts with pronase at $37°C$ for 2–3 hours destroys 80–90% of the cardio-excitatory activity, suggesting the presence of some peptide links. Extracts have been purified some 95 times, and the active principle has a molecular weight of about 1300.[18]

What is the function of the cardio-excitatory hormone in the normal animal? It was suggested that the hormone could be released in response to 'stress', but when the neurosecretory nerves in anaesthetized *Eledone* were electrically stimulated, in only one out of eleven experiments was a cardio-excitatory effect produced.[17a] However, when the vena cava and associated neurosecretory nerves are dissected out into oxygenated sea water and the nerves stimulated electrically, the bathing medium is found to contain the hormone when assayed on other isolated hearts.[18] The *in vivo* experiments were unsuccessful either because of the practical difficulties of maintaining the animals at a level of anaesthesia which ensures a relatively 'normal' physiological state but prevents violent movement,[17a] or else the hormone had already been released as a result of experimental handling and further release of the hormone would produce no greater effect. The latter explanation is the more likely, since when animals are caught and taken from a sea water tank, their blood contains the cardio-excitatory principle; but blood from tentacles cut off very quickly from individuals contains no cardio-excitatory activity.[17m] Thus the hormone is most probably released as the result of stimuli which overall may be called stressful. The position of the neurohaemal areas, together with the direction of blood flow, suggests that when released the hormone could affect both the systemic and branchial hearts. The advantages of an increased flow of blood through the gills and the rest of the body in 'fight or flight' conditions are obvious.

6

Endocrine Mechanisms in the Insecta—I

Those invertebrate animals, the endocrinology of which has already been discussed, all possess dispersed neurosecretory cells, are without neurohaemal organs (or else the glands are very poorly developed), and with the exception of the cephalopod molluscs are devoid of epithelial endocrine glands. Consequently, only rarely can the sources of hormones be located precisely.

But in the insects, some neurosecretory cells are grouped conspicuously within the brain and are associated with particularly well developed neurohaemal organs. At least two pairs of epithelial endocrine glands are present (Chapter 2). Insects are therefore much more favourable animals for the investigation of endocrine mechanisms than any of the less highly organized invertebrates. The great economic importance of insects, together with the relative ease with which many species can be reared in the laboratory, are added inducements for such investigations. As a result, more is known about the endocrinology of insects than any other invertebrate group.

Because of the nature of the insect endocrine system, experiments can be more sophisticated and produce more definite conclusions than any described previously. An insight into the mechanisms of hormone action has been gained, and some insect hormones have been isolated, purified and identified chemically (Chapter 12). A full discussion of insect endocrinology would occupy more space than this book can provide: the accounts which follow have been considerably condensed.

GROWTH, MOULTING AND DIFFERENTIATION

In insects, growth and differentiation are profoundly influenced by the cuticle which must be detached before the epidermal cells can grow and divide, and shed before the individual can increase in size. This moulting of the cuticle usually occurs a fixed number of times in any species. The interval between moults is called a stadium, and the form that the insect takes in any stadium is called an instar.

The overall change in form between the first larval instar and the adult is called metamorphosis. This may be slight, as in the primitive wingless, or apterygote, insects like the silverfish, involving merely the appearance of scales, coxal styles and external genitalia (Fig. 6.1). It may

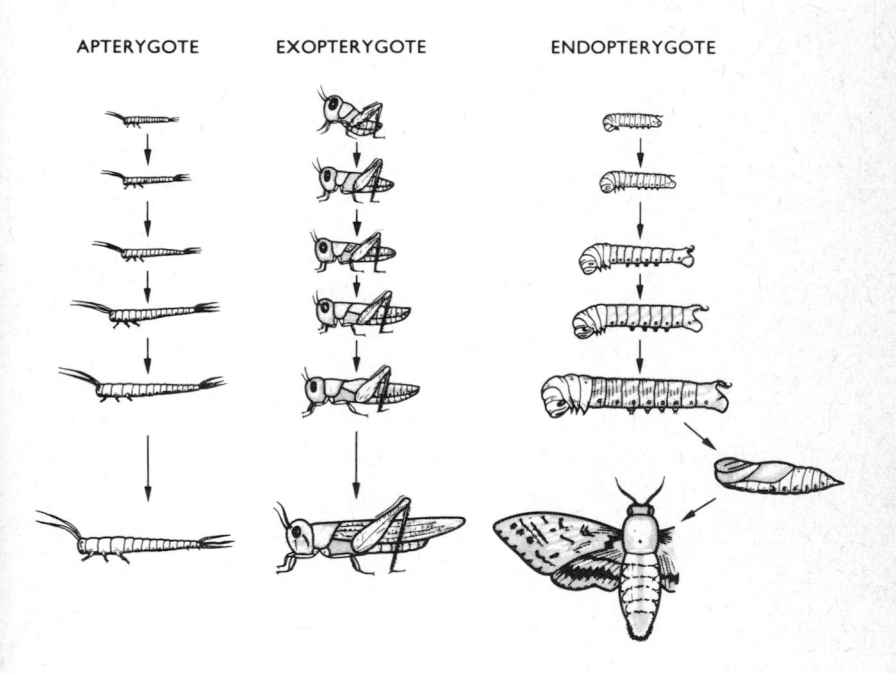

Fig. 6.1 Metamorphosis in insects. Metamorphosis is the overall change in form between the first larval instar and the adult. Change in form only becomes apparent during the periodic moults. In the primitive apterygote insects metamorphosis is slight, consisting in the progressive appearance of scales, coxal styles and external genitalia. The exopterygote insects show greater changes: wing pads appear in the later larval instars, developing into functional wings at the last moult; the external genitalia also develop progressively. In the endopterygote insects, the larval instars are similar and metamorphosis takes the form of a sudden transformation of the last larval instar to pupa and hence to adult.

be more pronounced as in the exopterygote insects (cockroaches, grass-hoppers, bugs etc.), where the appearance of wing pads in the later larval instars and their transformation into functional wings at the last moult may be more obvious signs of progressive differentiation than the development of genitalia (Fig. 6.1). Finally, metamorphosis may take the form of a sudden transformation of the larval stage into a form—the pupa—quite unlike it in appearance with the subsequent emergence of the adult (Fig. 6.1). This complete or indirect metamorphosis occurs in the endopterygote insects—beetles, ants and bees, butterflies and moths, flies etc.

In endopterygote larvae, bodily form, mouthparts and sense organs (particularly in the absence of compound eyes) are all different from those in the adult, indicating the quite separate ways of life of the larva and adult. The pupal stage is that in which many of the purely larval structures are destroyed and replaced by those of the adult. But the precursors of the adult structures are present in the larvae in the form of imaginal buds, nests of cells separate from the functioning larval tissues. In some endopterygotes, wings, legs and mouthparts are formed from these imaginal buds which undergo some development even in the larval stages. But in other endopterygotes, all the adult body, including the epidermis, is developed from imaginal buds and cells. In the exopterygotes and the apterygotes, on the other hand, the daughter cells which give rise to larval tissues at an early moult are the direct precursors of those which eventually form the adult tissues.

A comprehensive theory of the endocrine control of moulting, growth and differentiation in insects must explain these different metamorphoses.

Endocrine control of development

Current views on the endocrine control of insect development can be summarized as follows. A hormone produced in the neurosecretory cells of the brain is carried by the neurosecretory cell axons to the corpora cardiaca, from where it is released into the blood. This hormone then stimulates the thoracic glands (or their equivalent) to produce a hormone which causes the epithelial cell to begin the processes which lead to moulting. At the same time, at all moults except the last, the corpora allata secrete a hormone which ensures the development of another larval instar. During the last moult, the corpora allata are inactive; and the adult is produced (Figs. 2.12, 6.2, 6.3).

These hormones are named according to their function. The brain hormone, because of its effect upon the thoracic glands is called *thoracotrophic hormone*; that from the thoracic glands, *ecdysone* because it initiates moulting (ecdysis); and that from the corpora allata,

juvenile hormone (*or neotenin*), because it ensures the retention of juvenile characters.

Upon what evidence are these statements based? In insects the classical extirpation: reimplantation experiments (Chapter 1) are not always possible, but the equivalent of gland removal can be obtained by ligaturing off that part of the body containing the suspected endocrine organ.

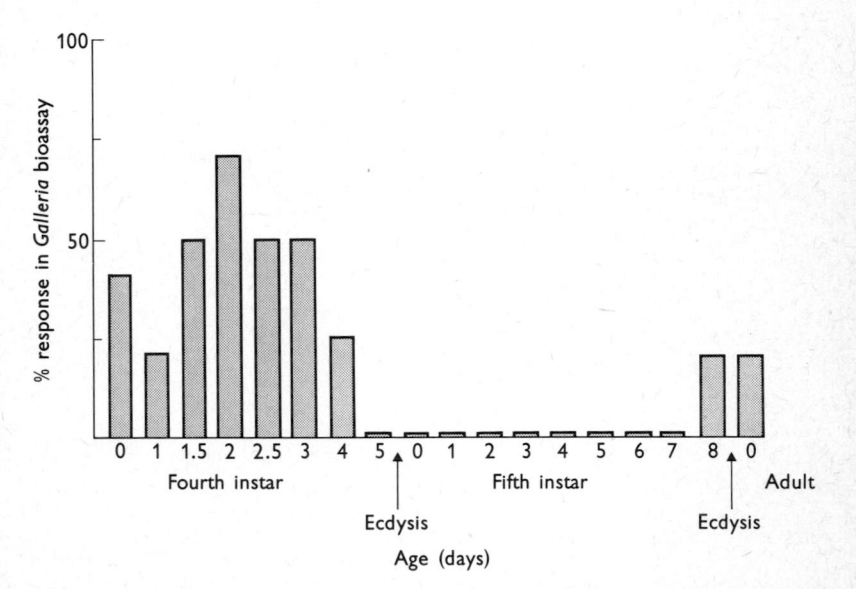

Fig. 6.2 The titre of juvenile hormone in the blood of fourth and fifth instar larvae of the migratory locust. Juvenile hormone is present in the blood during the fourth larval instar but is absent during the fifth. Small quantities of juvenile hormone are present in the blood when the fifth larval instar moults to the adult; the reason for this is unknown. The juvenile hormone concentrations were measured by bioassay (p. 276). (After Johnson and Hill[156a])

Thoracotrophic hormone

In many species of insect, the cerebral neurosecretory cells have been removed or destroyed by cautery or by x-rays.[271, 261, 133, 112] When the operation is performed early enough in the instar, the insect does not moult. Instead, it often goes into a state of suspended development for what may be a very long time, similar in many respects to the developmental arrest, or diapause, which intervenes normally in many life histories (see p. 177*ff*). When neurosecretory cells, or even whole brains,

are reimplanted into such operated animals, their suspended development is lifted and they subsequently moult. The cerebral neurosecretory cells are therefore essential for the moulting process.

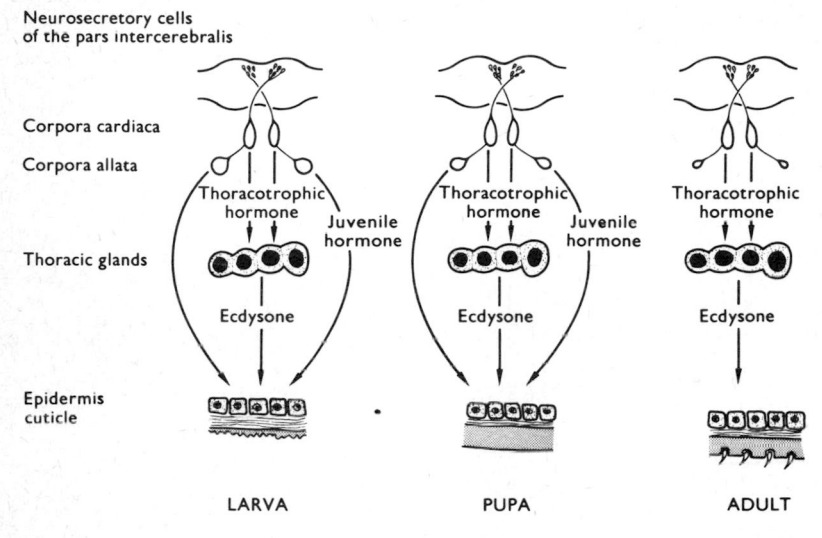

LARVA   PUPA   ADULT

Fig. 6.3   Endocrine control of insect metamorphosis. Three hormones are involved: *thoracotrophic hormone* from the cerebral neurosecretory cells is released into the blood from the corpora cardiaca and stimulates the thoracic glands (or their equivalents) to produce *ecdysone*. Ecdysone induces the epithelial cells to begin the processes which lead to moulting. In all larval instars except the last, the corpora allata secrete *juvenile hormone* which ensures the development of another larval form. In the exopterygote (hemimetabolous) insects, progressively less juvenile hormone is secreted during each instar and results in progressive differentiation towards the adult form. In the endopterygote (holometabolous) insects, a consistently high level of juvenile hormone is secreted in each larval instar which maintains the more or less constant larval form. Juvenile hormone concentration is diminished in the last larval instar and the pupation moult results. Before the final moult in both exopterygote and endopterygote insects, the corpora allata are relatively inactive, little or no juvenile hormone is secreted, and the adult form is produced.

When the cerebral neurosecretory cells are removed or destroyed, neurosecretion disappears from the corpora cardiaca.[133] Ligaturing or cutting the neurosecretory nerves between the brain and the corpora cardiaca similarly depletes the amount of neurosecretion in the glands, but in addition neurosecretion accumulates on the *brain* side of the

ligature or cut.[234, 261] This can only mean that neurosecretion is transported from the cells in the brain to the corpora cardiaca. In fact, in a very few insects, the movement of material along the living nerves has been seen.[261]

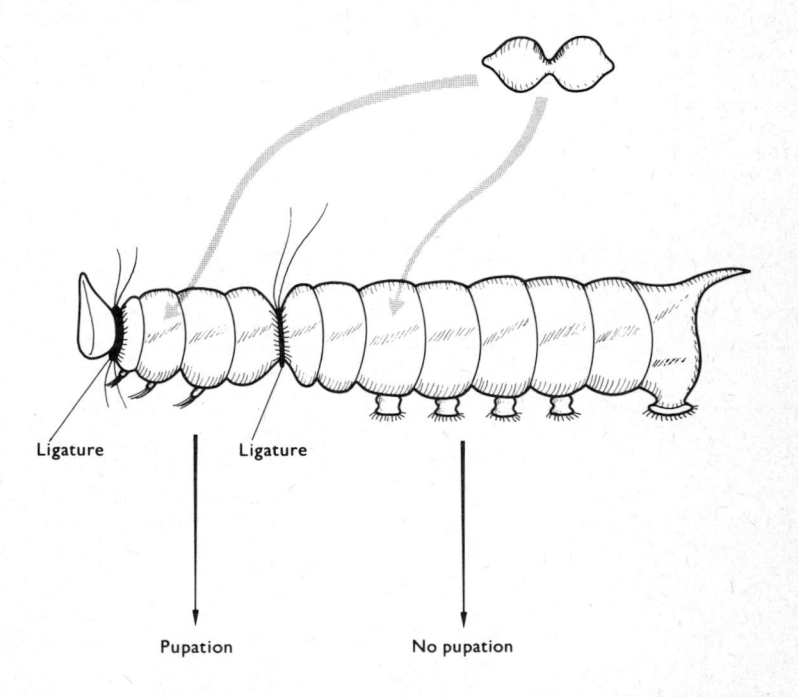

Ligature   Ligature

Pupation   No pupation

Fig. 6.4 An experiment to show the activation of the thoracic glands by the brain. A ligature behind the head isolates the source of the larva's own thoracotrophic hormone. When a brain is implanted into the abdomen, separated by a ligature from the thorax, no pupation occurs. But when the brain is implanted into the thorax, this part will pupate due to the production of ecdysone by the thoracic glands. By varying the level of the ligature, the precise source of ecdysone can be found.

When whole brains are implanted immediately into individuals whose own neurosecretory cells have been removed, moulting is delayed compared with normal animals. Moulting is similarly delayed, but not prevented, when corpora cardiaca are removed. The reason for this is that the neurosecretory nerves, severed in the two kinds of operation, have to develop a new neurohaemal organ before release of the hormone into the haemolymph approaches normal.[257]

Although these experiments show the necessity of an intact cerebral neurosecretory system for normal moulting, the thoracotrophic action of the brain hormone is not yet proved. This can be demonstrated as follows.

The insect has its neurosecretory cells removed and is ligatured between the thorax and the abdomen. When a brain is implanted into the abdomen, moulting does not occur: brain hormone alone does not activate the epithelial cells. But when the brain is implanted into the thorax, this subsequently moults. By varying the level of the ligature, it can be shown that only that part of the body containing thoracic glands will moult after brain implantation (Fig. 6.4).[286, 275]

Ecdysone

The previous experiment also provides evidence that the thoracic glands produce the moulting hormone (but see p. 276). In many insects, definite proof by removal and reimplantation that the thoracic glands are necessary for moulting is difficult because the glands are delicate structures, almost impossible to remove completely. But the operation has been performed in a small number of insects, and has confirmed their importance for the moulting process.[256] Recently, pure ecdysone has been synthesized and is now available in large quantities. The injection of the hormone into insects ligatured to separate off their own thoracic glands is always followed by moulting.

Juvenile hormone

In many insects, when the corpora allata are removed in early instars, moulting takes place as usual, but instead of another larva being produced, a diminutive adult emerges (Fig. 6.5). When the glands are removed in the last larval instar, normal moulting to the adult follows. But when several corpora allata from earlier instars are implanted into the last larval instar, a supernumerary larva is produced instead of an adult. This is usually larger than any normal larva, and may subsequently moult to a giant adult. These experiments confirm that the corpora allata produce a hormone in all larval instars except the last which prevents the emergence of the adult instar.

The overall control of development by hormones thus rests upon a good experimental foundation. But how is this control modified to permit the progressive differentiation towards the adult form in exopterygote insects, and the complete metamorphosis of the endopterygotes?

The classical experiments of Wigglesworth carried out upon the blood-sucking bug, *Rhodnius prolixus*, have provided an answer to this question. *Rhodnius* feeds only once in each instar, and the insect moults at a fixed

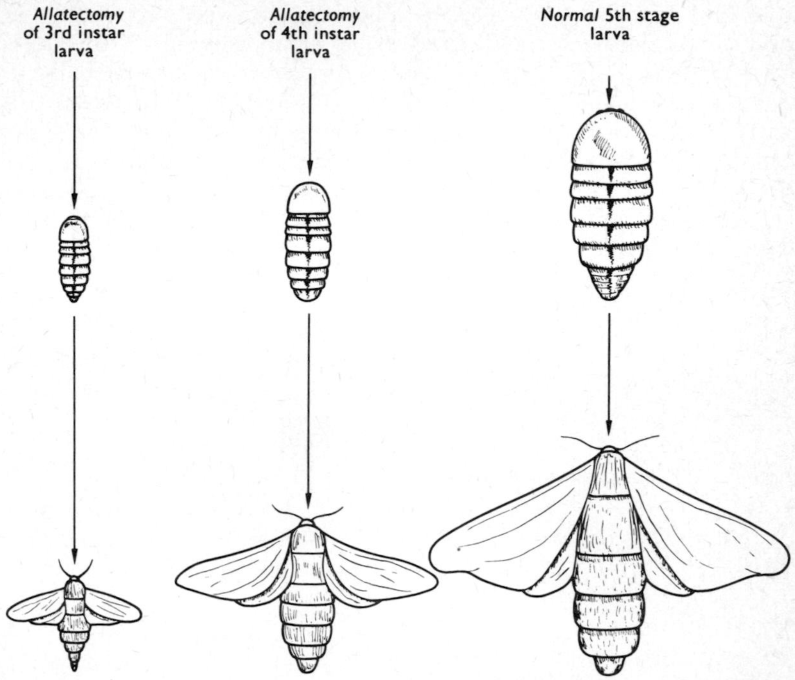

Allatectomy
of 3rd instar
larva

Allatectomy
of 4th instar
larva

Normal 5th stage
larva

Fig. 6.5 The effects of removal of the corpora allata at different times during the life of the silkworm, *Bombyx mori*. Allatectomy during the third larval instar results in a diminutive pupa, and a correspondingly small adult. The same operation performed in the fourth larval instar produces a larger pupa and adult, although still smaller than normal. (Redrawn from Wigglesworth[277])

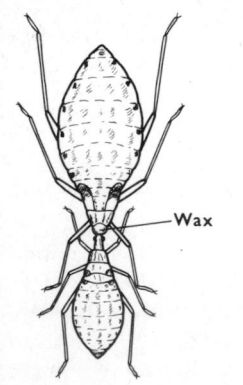

—Wax

Fig. 6.6 Parabiosis between two larval stages of *Rhodnius prolixus*. One or both of the larvae are fed, parts of their heads amputated, and the individuals joined with glass capillary tubing sealed into place with paraffin wax. The elongated head of *Rhodnius* allows amputation at different levels, so that the corpus allatum can be left in place or removed as desired. (After Wigglesworth[269])

time after its meal. Two or more insects can be joined together by fine capillary tubing so that their haemocoeles are in continuity[269] (Fig. 6.6); their epidermal cells grow and make contact within the insides of the tubes. Insects joined together in this way are said to be in *parabiosis*. They are like Siamese twins, with separate sets of organs, but with a common blood system. Clearly, both members of a parabiotic pair will be influenced similarly by blood borne factors. Thus the parabiotic insects moult synchronously, even though only one member may have fed before parabiosis to initiate the moulting.

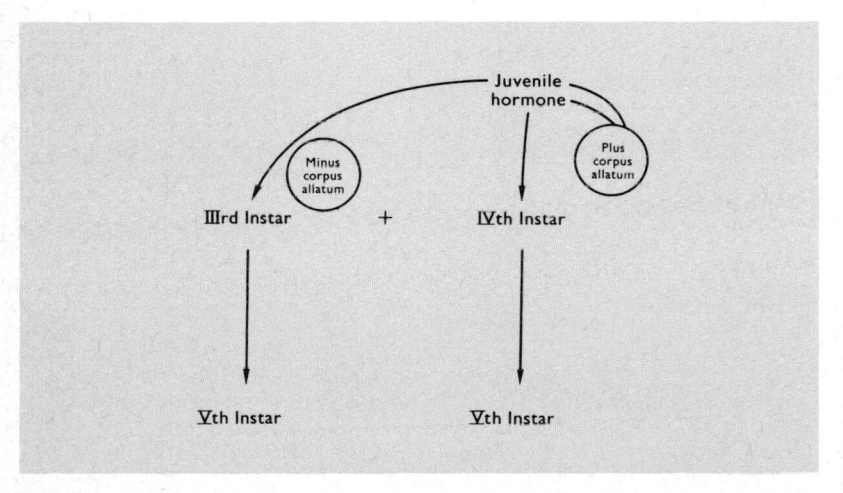

Fig. 6.7 Individuals of the third and fourth larval instars of *Rhodnius* joined in parabiosis as in Fig. 6.6. The third instar is decapitated *behind* the corpus allatum, the fourth instar *in front of* the gland. Both individuals therefore develop under the influence of the fourth instar corpus allatum. The fourth instar moults normally to a fifth instar larva, but the third instar does not moult to a normal fourth. Instead, it develops more the characteristics of a fifth instar larva. Compare Fig. 6.8.

When a third instar *Rhodnius*, minus its corpus allatum, is joined in parabiosis to a fourth instar, which retains its corpus allatum, the fourth instar differentiates during the subsequent moult to a normal fifth instar. However, the third instar member of the parabiotic pair does not moult to a fourth instar, but under the influence of the corpus allatum of its fourth instar companion, moults to a form more like the fifth instar than the fourth (Fig. 6.7). When the contrary situation is tested—the third instar retains its corpus allatum and the parabiotic fourth instar is allatec-

tomized—then the fourth instar moults to *another* fourth instar, instead of a fifth as it would do with its own corpus allatum[269] (Fig. 6.8). These results suggest that the third instar corpus allatum is more potent in suppressing adult differentiation than is the fourth instar gland. Similar experiments with parabiotic pairs of earlier instars lead to the conclusion that the potency of the corpus allatum in *Rhodnius* falls progressively in successive instars, until in the last larval instar the gland is inactive and the adult is produced. Progressive differentiation at successive moults in the hemimetabolous insects is therefore controlled by the corpus allatum.[277, 279, 284]

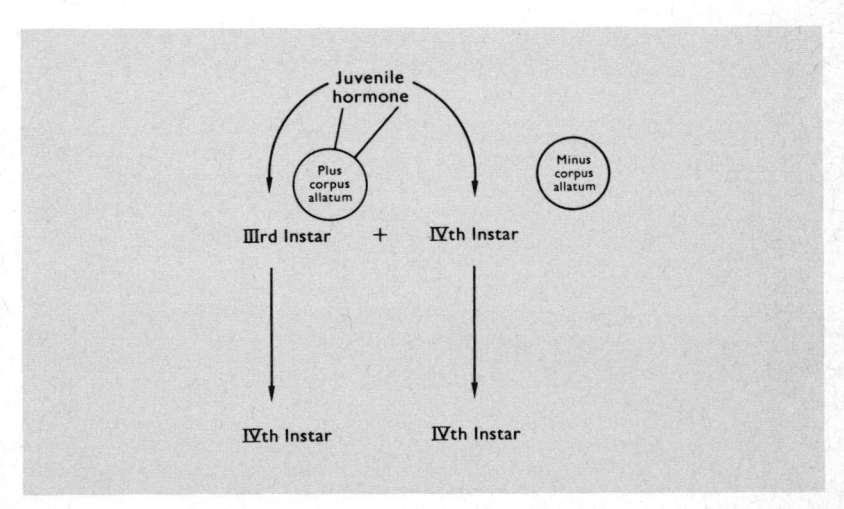

Fig. 6.8 The reverse situation of that in Fig. 6.7 : the third instar retains its corpus allatum, the fourth instar is without its gland. The third instar moults normally to a fourth instar. But under the influence of the corpus allatum of the third instar larva, the fourth instar moults to a form more like another fourth instar, rather than the fifth instar to which it would otherwise have given rise. Compare Fig. 6.7.

The results of the experiments described above could be explained in terms of a decreasing concentration of juvenile hormone in the blood— the corpus allatum becomes less 'active' in successive instars. But another explanation is possible, as the next experiment shows. Two fourth instar *Rhodnius* are joined in parabiosis, one member of the pair seven days after feeding, the other one day after feeding. The physiologically younger larva moults to a form intermediate between the fourth and

fifth instars (Fig. 6.9). So when juvenile hormone is introduced *earlier than normal* into the moulting process, differentiation towards the adult form is delayed. Thus the control over metamorphosis in hemimetabolous insects exercised by the corpus allatum can be effected in one of two ways: either by the gland becoming less active in successive instars, or by the gland secreting its hormones progressively later in successive instars. In practice, there is little difference between these two possible situations.

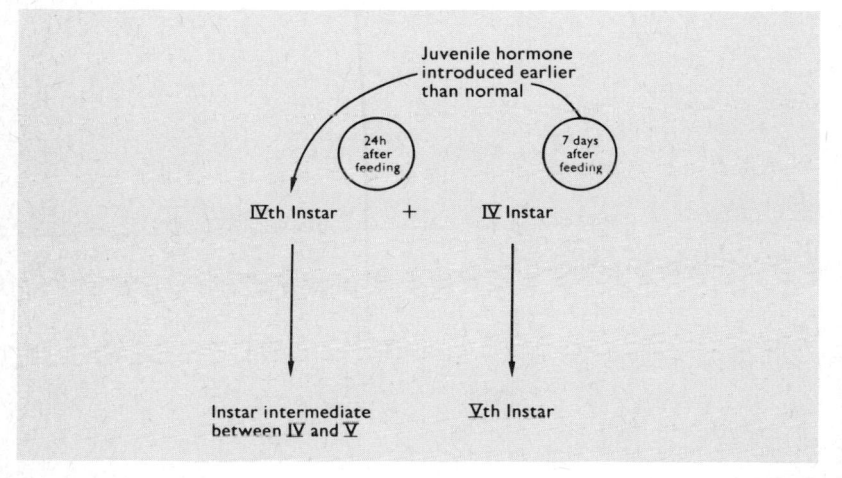

Fig. 6.9 Two fourth instar larvae of *Rhodnius* joined in parabiosis, one 7 days after feeding the other 1 day after feeding. The physiologically older individual moults normally to a fifth instar form, but the physiologically younger Individual moults to a form intermediate between the fourth and fifth instars. In this individual juvenile hormone has been introduced earlier than normal in the moulting process, and the resulting larva is more juvenile than it would otherwise be. (After Wigglesworth[277])

Does the corpus allatum control metamorphosis in holometabolous insects in a similar way? Arguing from the situation in the Hemimetabola, it might be expected that the larval corpora allata secrete juvenile hormone at a *continuously* high level in the larval instars, thus preventing metamorphosis, that the glands are less active in the last larval instar to induce pupation, and that the pupal corpora allata are inactive, producing the adult after the pupal-adult moult. Gilbert and Schneiderman have shown that this is indeed so. Having developed a bioassay for juvenile hormone (p. 276), they extracted and tested the hormone from various larval stages, pupae and adults of the giant silkmoth, *Hyalophora cecropia*. Its concentration remains high in the early larval stages, falling markedly

in the last instar larva and the hormone is not detectable in the later pupal stages[110] (Fig. 6.10). Moreover, the injection of purified juvenile

Stage	Juvenile hormone activity per gram fresh weight
Unfertilized eggs	8·15
7 day embryos	7·09
1st instar larvae	7·47
5th instar larvae	0·57
Diapausing pupae (male)	1·47
Diapausing pupae (female)	1·11
2-day old developing adult (male)	0
17-day old developing adult (male)	0
20-day old developing adult (male)	13·70
22-day old newly emerged adult male	201·64
Adult male, 7 days after emergence	418·00
Adult female, 7 days after emergence	11·63

Fig. 6.10   Juvenile hormone activity during the life history of *Hyalophora cecropia*. Note the very low activity in the last (5th) instar larva, and the high activity in mature males. (Data from Gilbert and Schneiderman[110])

hormone extract into pupae often results in the formation of a *second* pupal stage after the moult[108, 288]; the injection of ecdysone into last instar larvae sometimes produces a supernumerary larval stage, proving the presence of some juvenile hormone at this time, and if the last instar larvae are allatectomized early enough in the stadium (i.e. before the corpora allata begin to secrete) an adult, or a monster intermediate between a pupa and an adult, is produced.

Moulting in adult insects

Many apterygote insects moult frequently as adults, whereas the pterygotes normally do not with the exception of the preimago of the mayfly, which moults a fine, transparent cuticle shortly after emergence. What is the cause of this difference? Simply, the thoracic glands of the pterygote insects degenerate and disappear during or shortly after the last moult, while those of the apterygote insects do not. Adult pterygotes can be induced to moult by exposing them to moulting hormone by parabiosis or implants of thoracic gland[271, 275] (Fig. 6.11). The reason for the dissolution of the thoracic glands is a little obscure: they break down during a moult which follows allatectomy, whereas if corpora allata are implanted into last instar larvae to produce supernumerary larval instars, the thoracic glands persist. In other words, the structural integrity of the thoracic glands is maintained during a moult in which juvenile

hormone is present, but breaks down on undergoing a moult in the absence of the hormone.

It has been suggested that an additional factor (perhaps hormonal) is necessary for breakdown of the thoracic glands.[284, 278] But this would make the process unnecessarily complicated. In the holometabolous insects

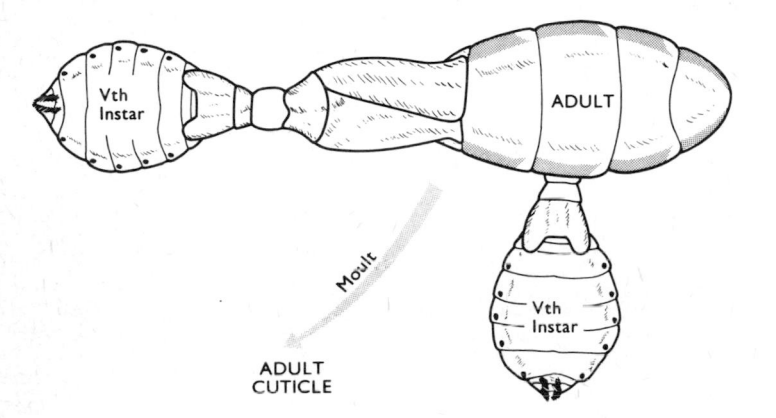

Fig. 6.11   Adult moulting in *Rhodnius prolixus*. Parabiosis between an adult and two fed fifth instar larvae causes synchronous moulting in all three individuals: the adult moults into a second adult instar. Compare Fig. 6.12. (After Wigglesworth[271])

particularly, many purely larval tissues disappear during metamorphosis. The thoracic glands are larval tissues *par excellence*. There is, therefore, no need to invoke a special dissolution factor for the thoracic glands, additional to the factors which are normally involved in metamorphosis.

Mode of action of developmental hormones in insects

In the presence of ecdysone alone, the insect moults and differentiates fully to the adult form. When juvenile hormone is also present, adult development is prevented to a greater or less degree (p. 110). Ecdysone has consequently been called a *moulting and metamorphosis* or a *growth and differentiation* hormone, and juvenile hormone an *inhibitory* or *status quo* hormone. But the functions of the hormones implied by these names need to be examined very carefully.

There is abundant evidence that in the absence of ecdysone, its target tissues neither grow nor develop and go into a developmental arrest very similar to natural diapause (p. 107). In *Rhodnius*, exposure of the epidermal cells to ecdysone causes enlargement of the nucleolus, the

appearance of RNA in the cytoplasm, and an increase in the numbers, size and branching of the mitochondria[279, 280] (Fig. 6.13). These changes in the cells are clearly those associated with the restoration of protein synthesis.

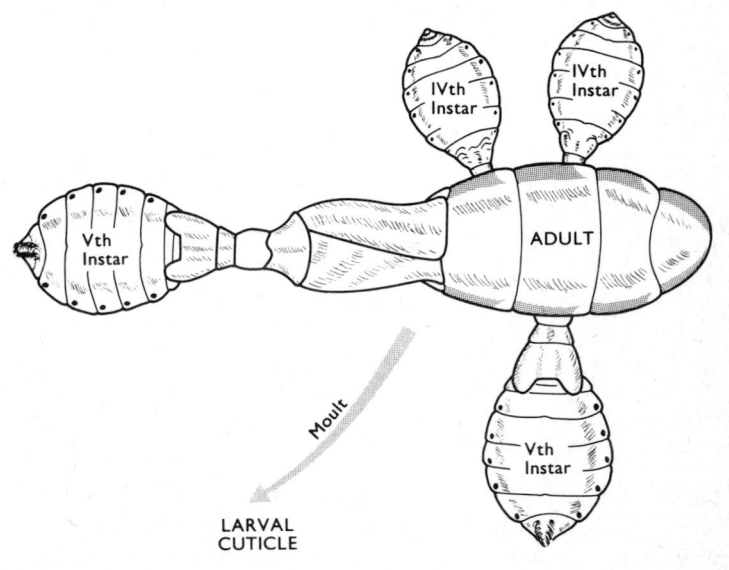

Fig. 6.12 As in Fig. 6.11, except that two fourth instar larvae, decapitated in front of the corpus allatum, are additionally joined in parabiosis to the adult. Juvenile hormone supplied by these larvae influences the moult so that the adult regresses to a more juvenile form. Legs have been omitted from individuals, and the adult wings cut short. (After Wigglesworth[271])

The modern view of cellular protein synthesis is that the genetic material in the nucleus—deoxyribonucleic acid (DNA)—controls the manufacture of many specific kinds of ribonucleic acids (RNAs), each RNA built on a template of DNA consisting perhaps of a temporarily unwound part of the DNA double helix. The messenger RNA (m-RNA) passes out of the nucleus and becomes associated with protein bodies called ribosomes, located along the endoplasmic reticulum of the cell where they have ready access to raw materials. Meanwhile, small molecular weight RNAs, called transfer RNAs, become bonded to free amino acids within the cell cytoplasm. The individual constitution of each transfer RNA decides which amino acid will become attached. The combined transfer RNA and amino acid passes to the ribosome, and each

transfer RNA associates with a particular region of the m-RNA. Consequently, a sequence of amino acids is brought into close contact and a specific protein can be formed. The specificity of the protein is thus determined by the structure of the m-RNA on which it is formed, which in turn is dependent upon its original DNA template.

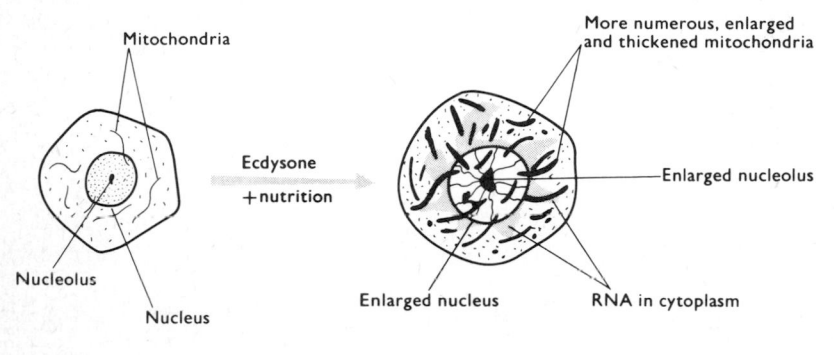

Fig. 6.13   The effects of ecdysone upon an epidermal cell in *Rhodnius prolixus*. The cell nucleus and nucleolus enlarge, RNA appears in the cytoplasm, and the mitochondria increase in size and number. These changes are associated with the restoration of protein synthesis by the cell. (After Wigglesworth[283])

Is it possible, therefore, that ecdysone can affect the nuclear DNA (the genes) in a cell so that specific proteins are produced? Karlson and his colleagues have investigated this possibility in considerable detail, working with one part of the many processes which characterize metamorphosis in the blowfly, the tanning of the puparium. Cuticular tanning is brought about by the cross bonding of protein molecules to form a stable, large molecular weight compound called sclerotin, through the intervention of a tanning quinone (Fig. 6.14). The tanning quinone in the blowfly puparium is derived from N-acetyl-dopamine. N-acetyl-dopamine comes ultimately from tyrosine; this is oxidized to dihydroxy-phenylalanine (DOPA) which is decarboxylated to dopamine. The dopamine is acetylated and oxidized to the tanning quinone: the two key enzymes in the process are dopa-decarboxylase and N-acetyl-dopamine phenoloxidase (Fig. 6.14).

Dopa-decarboxylase increases in the blowfly larva at the same time that ecdysone is produced, and also increases 7–10 hours after the injection of ecdysone into ligated larval abdomens; this increase can be inhibited by the simultaneous injection of actinomycin, mitomycin or puromycin, antibiotics which are known to inhibit RNA and protein biosynthesis. So it is concluded that ecdysone induces dopa-decarboxylase by the *de novo* synthesis of enzyme protein, and RNA is needed for

the process. The N-acetyl-dopamine phenoloxidase in the cuticle exists initially as a pro-enzyme; the phenoloxidase is produced as the result of the action of an activator enzyme (Fig. 6.14). Production of this activator enzyme is also controlled by ecdysone, possibly again by a process of *de novo* synthesis.[165, 166]

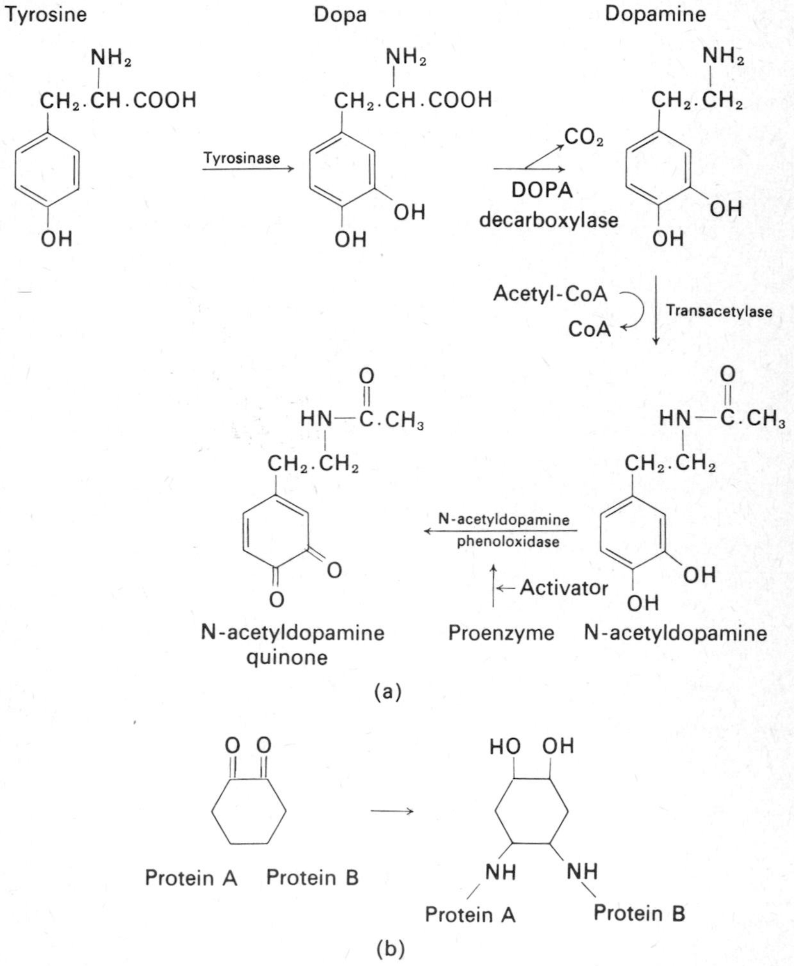

Fig. 6.14   (a) Synthesis of N-acetyl dopamine quinone in the puparium of the blowfly, *Calliphora erythrocephala*. (b) Production of sclerotin: two protein chains are linked by the quinone molecule and by repetition a complex quinone: protein macromolecule, called sclerotin, is formed.

When radioactively labelled ecdysone is injected into last instar blowfly larvae, about 45% of the radioactivity in the epidermal cells is recovered from the nuclei, 31% from the mitochondria and 25% from the microsomes, possibly indicating a primary effect upon the nucleus. After exposure to ecdysone, isolated epidermal cell nuclei will incorporate labelled uracil into m-RNA more rapidly than nuclei which have not been so exposed. The m-RNA produced after ecdysone induction will stimulate protein synthesis in an *in vitro* system, and the protein so produced has dopa-decarboxylase activity.[167] Thus in the normal *Calliphora* larva, ecdysone is said to induce dopa-decarboxylase in the epidermal cells by activating directly the nucleus to produce a specific m-RNA. In other insects, also, DNA-dependent RNA synthesis is said to be a primary result of ecdysone stimulation.

However, this relationship between the presence of ecdysone and the production of dopa-decarboxylase only holds true for that particular developmental stage of the larva. The enzyme is not formed when ecdysone acts during larval moulting. Moreover, no dopa-decarboxylase can be demonstrated in the blowfly pupa although ecdysone is present in high titre, and the enzyme increases just before adult emergence at a time when the hormone is not demonstrable.[160a] It is thus more likely that ecdysone does not have a direct action on specific gene loci, but instead acts as a stimulus which causes the cell to initiate a developmental programme appropriate to its stage. In the face fly, *Musca autumnalis*, the puparium is not tanned but is instead hardened with calcium salts; calcification is controlled by ecdysone.[95a] The hormone can again be considered to be activating its target cells in this species which then synthesize products according to their genetic constitution.

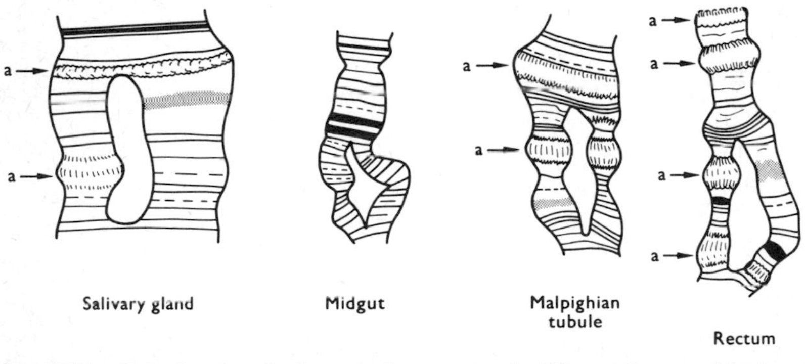

Salivary gland   Midgut   Malpighian tubule   Rectum

Fig. 6.15 Part of a giant (polytene) chromosome in different tissues of the larva of *Chironomus tentans*. The chromosome is recognised in each tissue by a small inversion. Different regions (a) are puffed in the different tissues. (After Beerman[10])

Supporting evidence for this point of view comes from experiments with the 'giant' or polytene chromosomes of Diptera. These polytene chromosomes consist of a large number of threads lying side by side. The chromosomes have a banded appearance, segments rich in DNA alternating with segments containing much less (Fig. 6.15). All the major landmarks of the chromosomes can be recognized in all tissues in which the chromosomes are reasonably well developed, although the actual form of the chromosomes may differ in different tissues[10] (Fig. 6.15). At the same developmental stage specific bands are inflated (puffed) in specific tissues. The puffs represent the main sites of RNA synthesis on the chromosomes. This chromosomal RNA differs significantly in its physical properties when extracted from different puffed segments, and the puffs consequently seem to elaborate specific m-RNAs. It is possible that the specific metabolism of different tissues is therefore controlled by m-RNAs generated in the tissue-specific puffs.[84] The non-puffing segments on the chromosomes would be suppressed genes, presumably the result of the processes of determination and differentiation in embryogenesis.

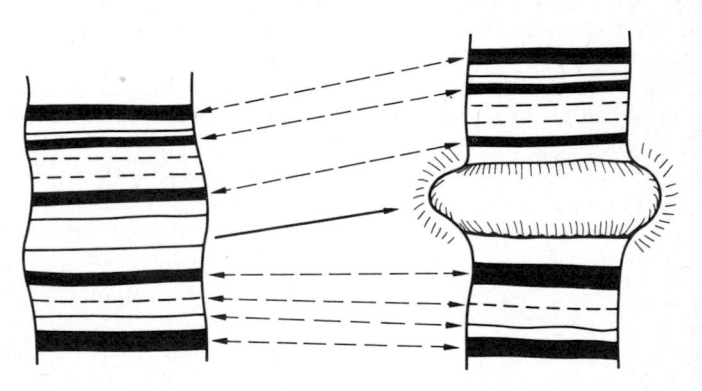

Fig. 6.16   A stage specific puff in part of a giant chromosome of *Chironomus tentans*. The same bands can be identified in both the mid-larval (on the left) and late larval (on the right) stages. One of the bands is puffed in the late larval, but not in the mid-larval, stage.

In addition to such tissue-specific puffing patterns, further series of puffs appear on the chromosomes which are related to the *developmental* stage of the insect (Fig. 6.16). Thus in the forelobe of the salivary gland of *Acricotopus lucidus*, a very large puff (Balbiani ring) is present in the mid-larval stage and retracts at the end of larval life.[206] Such stage specific puffs have now been identified and tabulated in a number of species. Are

these puffs therefore connected with the presence of hormones in particular developmental stages?

It is known that such a connexion does exist. The injection of pure ecdysone into mid-larval stages of the midge *Chironomus tentans* results in a puffing sequence in the salivary gland chromosomes characteristic of the *late* larval stage.[59] The first chromosomal puff appears only 15–30 minutes after injection, to be followed by a second 15–30 minutes later. Then 15–20 hours after the injection of ecdysone, a large number of further chromosomal puffs appear with still more forming 30–70 hours later. Among these later puffs are those characteristic of the prepupal period. If insufficient ecdysone is injected, the two very early puffs rapidly regress, and none of the subsequent puffs appear. The later puffing sequence is dependent upon the prior activity of the first two gene loci.

In the fruitfly, *Drosophila melanogaster*, the transplantation of salivary glands from younger to older larvae will induce chromosomal puffing, and such puffs are prevented from appearing when older salivary glands are transplanted to young larvae.[9] A ligature around the body which separates the salivary glands into two parts, and prohibits also the passage of endogenous hormones, prevents the appearance of chromosomal puffs in the salivary gland cells posterior to the ligature while anterior to the ligature the chromosomes puff normally. Chromosomal puffs which appear after the injection of ecdysone are therefore not experimental artifacts.

There is more direct evidence that chromosomal puffs are related to the metabolic activities of the cell. In *Chironomus tentans*, a specific type of secretion granule is produced by certain cells of the salivary glands, and there is a particular puff in one of the chromosomes of these cells. When *Chironomus tentans* is crossed with a sibling species, *Chironomus pallidivittatus*, which possesses neither puff nor secretion granule, the puff and granule are inherited in simple Mendelian fashion, and only those individuals with the puff produce the secretion.[11] However, in *Chironomus thummi*, there are only two tissue specific salivary gland puffs whereas many enzymes are produced by the glands. Since the enzymes are also present in other tissues of the body, and can be transported from the haemolymph borders of the salivary glands to their secretory canals, the relationship between the chromosomal puffs and enzyme manufacture is rather obscure. But it is possible that the salivary gland puffs in *Chironomus thummi* are in fact controlling the formation of packaging materials for the enzymes as they traverse the cells. Thus DNAase, which destroys DNA, and which is present in the form of granules in the salivary gland secretion, accumulates in the salivary glands before the onset of metamorphosis at a time when the other enzymes are

decreasing in amount (Fig. 6.17). The disappearance of the tissue specific puffs at this time could perhaps result in the loss of the membranes needed to isolate this and other enzymes, so that the gross disruption of the salivary glands characteristic of metamorphosis is facilitated. The salivary gland cells would be genetically programmed for death, and the sequence would be initiated at the appropriate time by the insect's hormones.[182, 183, 184, 185]

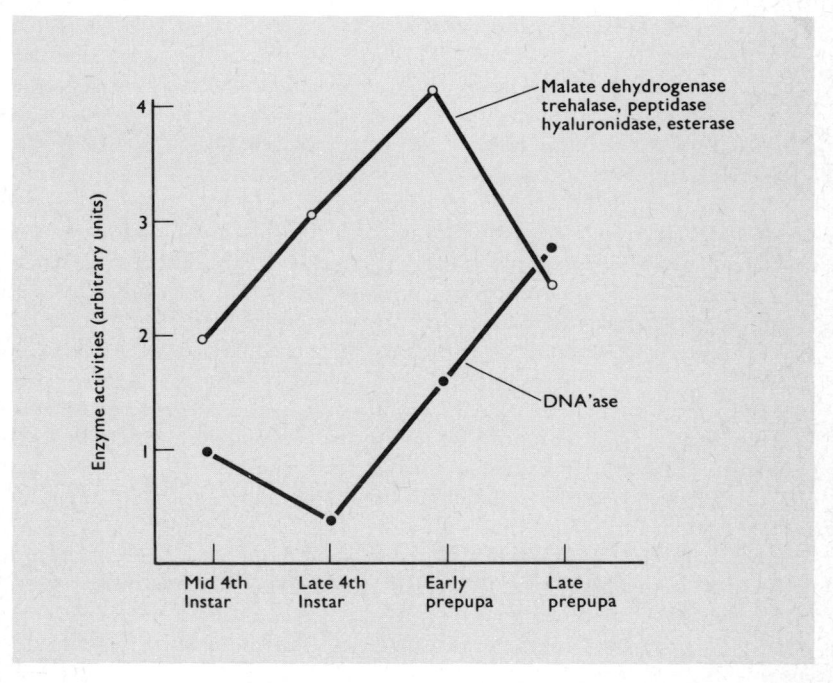

Fig. 6.17 Enzyme activities in the salivary glands of *Chironomus thummi* at different developmental stages. Only DNAase activity increases during the prepupal stages, associated with cellular dissolution. For further details, see text. (After Laufer and Nakase[184])

Changes in the puffing patterns of the giant chromosomes of Diptera due to the injection of ecdysone, or to variations in the titre of the endogenous hormone, might suggest the view that the hormone exerts its effect upon the genes. The very short time which elapses between ecdysone injection and the appearance of the first two chromosomal puffs could indicate that the hormone affects the genes *directly*, although the time is still more than sufficient for gene activation to have occurred as a

secondary consequence of a more primary effect of the hormone else-where.[126b] The antibiotic actinomycin blocks the DNA synthesis in cells; when this is injected together with ecdysone into *Chironomus* larvae, the first two puffs on the salivary gland chromosomes do not appear,[61] again suggesting that ecdysone is affecting the gene loci directly.

But is this really so? When the puffing patterns of the salivary chromosomes of *Chironomus* are examined throughout larval life, it is found that the early two puffs appear before *every* moult. It is only in the later sequences of puffs that differences appear between larva to larva moults and the larva to pupa moult.[60] Consequently, it must be concluded that the reaction of the early gene loci to ecdysone is not specifically associated with metamorphosis, but with moulting in general. The hormonal difference between the larval and metamorphosis moults lies in the presence of a high concentration of juvenile hormone in the former and a low or zero concentration in the latter. So it is not unreasonable to assume that the later sequences of gene activities differ between the two kinds of moult because of these differences in the concentration of juvenile hormone. Therefore the *de novo* enzyme syntheses associated with metamorphosis would be due not to the presence of ecdysone but as a major consequence of the absence or reduced concentration of juvenile hormone. This is to say that the genes in each cell follow a parti-cular programme of activities during a moult in the presence of juvenile hormone, and different programmes during moults in which juvenile hormone is present in reduced concentration or is absent. So the major hormone affecting differentiation in insects is the juvenile hormone and not ecdysone. But what then is the function of ecdysone?

In *Chironomus*, if the two early chromosomal puffs are prevented from appearing because actinomycin is also injected together with ecdysone, or if the puffs rapidly regress because ecdysone is injected in low con-centration, the later sequences of gene loci remain inactive and do not puff.[61] So the gene activities concerned with larval or metamorphic processes depend upon the prior activity of the genes which respond to ecdysone. The antibiotic puromycin will block cellular protein syn-thesis, not at the level of DNA synthesis like actinomycin, but at the cytoplasmic level. When puromycin is injected with ecdysone into *Chironomus* larvae, the two early puffs appear as usual, but all the later puffs are completely inhibited.[61] Consequently, the activity of these gene loci must depend upon some interaction between the early reacting genes and the cytoplasm. In practice, it can be said that ecdysone activates the cell through its effect upon the initial two gene loci. Moreover, since the subsequent gene activities depend upon interactions between the nucleus and the cytoplasm, juvenile hormone could exert its effects at either the level of DNA, or at some other point in the protein synthetic pathway.

Is there any evidence for this view of the importance of juvenile hormone as a programmer of gene activities, and of ecdysone as an activator of the cell? An adult *Rhodnius* can be induced to moult by joining it to moulting 5th stage larvae (Fig. 6.11) or by implanting active thoracic glands. But when juvenile hormone is also present, for example when the adult is joined to moulting 4th instar larvae (Fig. 6.12), the adult moults but its cuticle reverts to a larval form. Pupal epidermis from some insects implanted into larval hosts moults synchronously with the larvae and produces progressively more juvenile cuticle.[224, 270] Such reversal of metamorphosis does not occur in all insects, but that it does in some means that juvenile hormone must play a more positive part in determining the degree of differentiation than merely by opposing

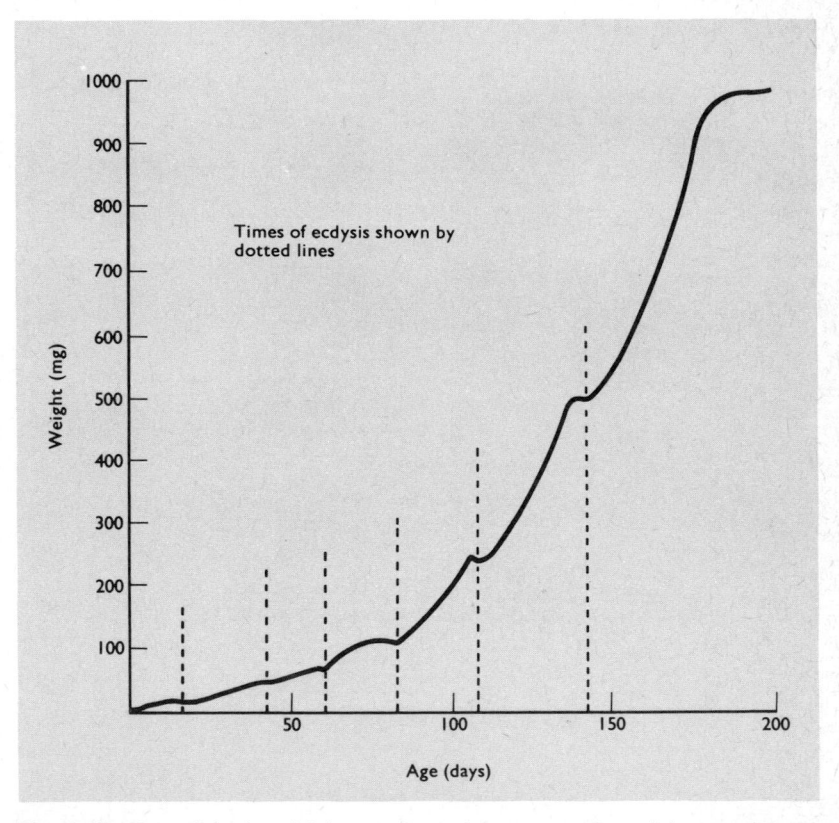

Fig. 6.18 Growth in the stick insect, *Carausius morosus*. Growth is a continuous process during each instar, the small discontinuities at each ecdysis being due to loss of the cast cuticle.

its natural expression. Juvenile hormone cannot be merely an inhibitory or status quo hormone.

Insect growth is often described as discontinuous, periods of increasing size corresponding with each moult and alternating with static periods

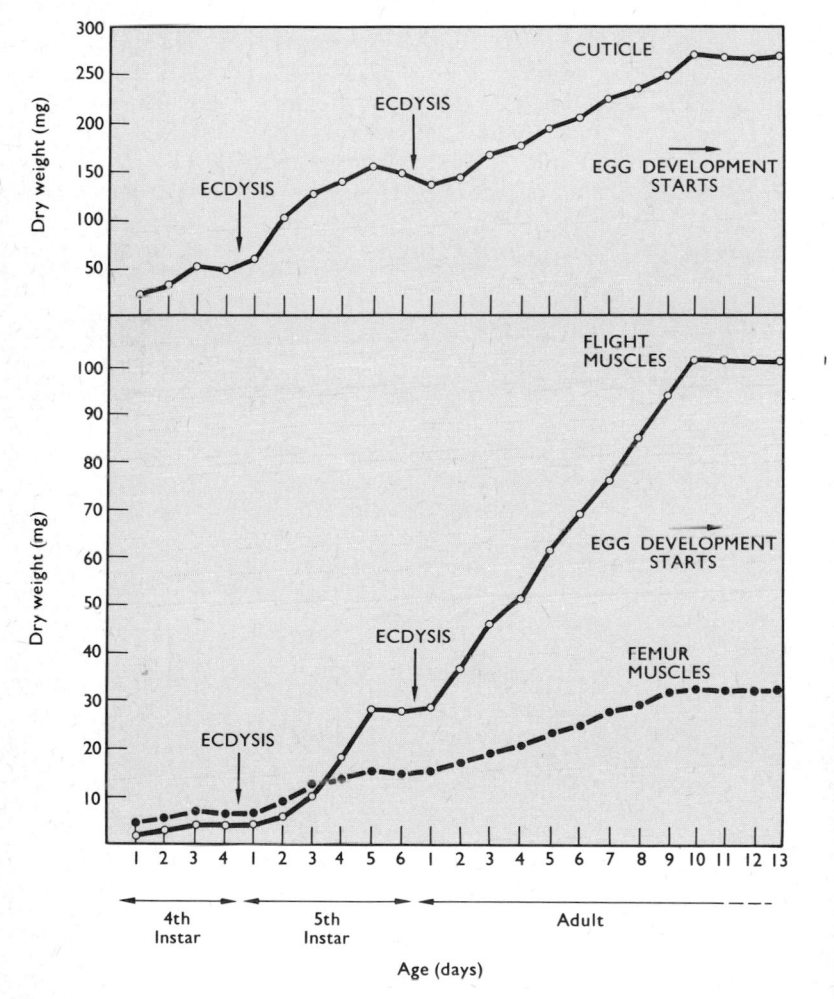

Fig. 6.19   Growth of the cuticle, flight muscles and femoral muscles during the fourth and fifth larval and the adult instars of the desert locust. The tissues increase in weight not only during the larval instars, but also in the adult after the final moult. Somatic growth in the adult female continues until just before oocyte development begins. (After Hill and Goldsworthy[146] ; Hill, Luntz and Steele[147])

between moults. But when growth is measured as increase in wet or dry weight, or in total nitrogen, no discontinuities appear in the growth curve except for small losses associated with ecdysis (Fig. 6.18). Although the production of ecdysone is cyclic, the growth increments continue throughout each larval instar, and even in the adult when the thoracic glands have disappeared.[147] Individual tissues in the adult show the same kind of growth increments so the weight increases are not associated solely with the storage of reserves (Fig. 6.19). Moreover, some insects can be induced to moult without any associated increase in size.[197] Neither is cell division causally related to the action of ecdysone. The imaginal buds of many holometabolous insects increase in size many times by cell division, and can undergo radical changes in form between larval moults. In larvae of *Locusta*, two waves of cell division pass forwards and backwards from the posterior and anterior ends of the body before the moult, related to but clearly not dependent upon the presence of ecdysone.[256] Cell division takes place during wound healing, even in adult insects where the prothoracic glands are no longer present. In *Rhodnius*, cell division in the epidermis and fat-body is induced by the mutual separation of the cells resulting from the stretching of the abdomen after feeding, and mitosis in the fat-body occurs in the absence of ecdysone[283] (Fig. 6.20). Greater or less mitotic activity in different parts of the body depends upon whether or not juvenile hormone is present.[271, 272]

Tissue	% increase in nuclei/unit area
Epidermis : stretched* by feeding +MH and JH	110
Epidermis : unstretched** +MH and JH	4
Fatbody : stretched*** +MH	89
Fatbody : unstretched** nutrition +MH	0
Fatbody : stretched**** nutrition alone	50

* 4th stage larva fed, decapitated after 24 hours and joined in parabiosis with 4th stage larva.
** 4th stage larva *unfed*, decapitated after 24 hours and joined in parabiosis with 4th stage larva.
*** Normal 4th stage larva after feeding.
**** 4th stage larva fed, decapitated 24 hours later.

Fig. 6.20 Mitosis in abdominal epidermal cells and fatbody cells during moulting in *Rhodnius prolixus*. In the epidermis, the increase in number of cells depends more upon stretching than upon the presence of hormones. Increase in cell number in the fatbody depends wholly upon stretching. (After Wigglesworth[283])

These observations suggest that individual cells have an inherent capacity for the kind of growth that they will undergo during any moult. This conclusion is underlined by the demonstration that every epidermal cell in *Rhodnius* has the capacity to divide and differentiate to form sensory plaques and dermal glands, and the expression of this potentiality is not dependent upon the presence of ecdysone or even juvenile hormone.[276, 277] Further, the overall pattern of pigmentation, areas of sensory plaques, hairs and even cuticular structure are fairly constant for any larval instar, and will vary predictably at each moult until the final adult pattern is attained. In *Rhodnius*, when pigmented epidermal cells migrate to heal a wound in the larval stages, the degree of displacement can be traced through each successive instar to the adult (Fig. 6.21): the

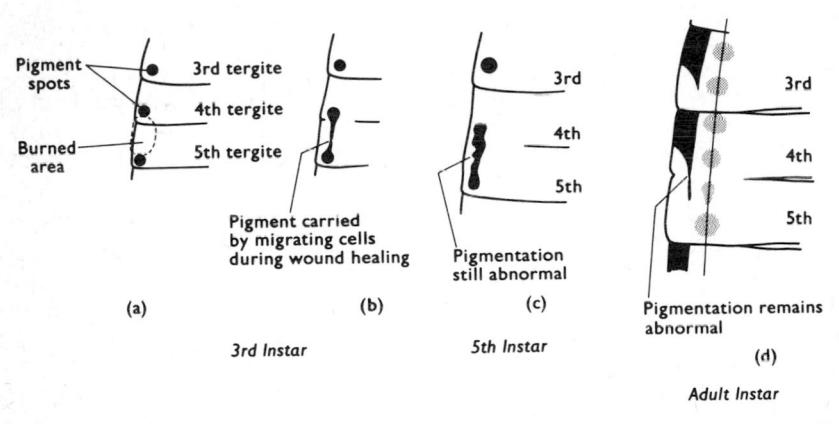

Fig. 6.21 Wound repair and cell lineage in *Rhodnius prolixus*. When a small part of the tergum of a third instar larva is burned (a), the area is healed by division and migration of the surrounding cells. If these cells are pigmented, the normal pigment pattern becomes distorted (b). This distortion is still evident in successive instars (c), and even in the adult instar (d), where the relationship between pigmented and unpigmented cells is different from that in the larvae. This experiment suggests the continuity of a cell line throughout development in this exopterygote insect. (After Wigglesworth[272])

cells retain the properties they had in their original site.[271, 272] So even in the larval stages, the epidermal cells possess the latent potentialities for particular adult structures. Thus the larval cells reveal characteristics which typify the larval body, and the same cells produce the characteristics of the unique body form of the adult of any particular species. In other words, the larval cells carry a dual pattern: a visible larval pattern, and a latent, invisible adult pattern.[271] The juvenile hormone can thus

maintain the larval pattern perhaps indirectly by participation in cyto-plasmic processes, or even perhaps by some action at the level of the genes.[277]

The imaginal buds (the wing discs, for example) of holometabolous insects can be cultured *in vitro* in a suitable medium. When ecdysone or ecdysterone is added to the medium, DNA and RNA synthesis is stimu-lated,[95c, 213f] specific protein synthesis is initiated,[214e, 229c, 291d] and finally the discs evaginate and deposit cuticle.[213e, 213g, 213h] Although the early stages of protein synthesis occur shortly after exposure to the hormone, a much longer 'contact period' is necessary before the discs continue their development independently.[81c, 81d, 214d] The hormone must also circulate *in vivo* for some time before it can exert its full effects.[81c, 81d, 284, 284b] But these results do not invalidate the general concept that the moulting hormone is no more than a messenger 'informing' the wing discs to follow their intrinsic developmental programme at the appro-priate time. If moulting hormone affects, for example, permeability pro-perties within the cell or its nucleus,[279, 283] it is not unreasonable to sup-pose that exposure to the hormone for a particular period of time is necessary before the precursors and metabolites for protein synthetic processes are fully established.

This interpretation of the events associated with metamorphosis recognizes the overriding importance of the individual cell in the deter-mination of its specific characters. Cell division and differentiation are the intrinsic properties of the cell, residing in its genes, the particular expression of which at any stage is controlled by the juvenile hormone. Although ecdysone induces the activation of its target cells, such as the epidermis, which is necessary before growth and differentiation can occur, other cells may be activated by quite different factors. Thus the fatbody cells of *Rhodnius* initiate protein synthesis when nutrients in the haemolymph become available after a meal.[283] Even so, their subsequent metabolism can be modified by circulating hormones, as when specific vitellogenic proteins are produced in the adult female (Chapter 7).

Some species of lepidopteran pupae, and adults, retract the epidermis from the cuticle when injected with ecdysone and mitomycin, but do not secrete a new cuticle or show any other signs of further development.[200] Mitomycin, like actinomycin, blocks DNA synthesis, and the subsequent lack of cellular activity after its administration is not surprising. But the separation of the epidermis from the cuticle, called apolysis, is normally a very early step in the moulting process and would suggest that the cells have been activated by ecdysone even though DNA synthesis is inhibited by mitomycin. This result would seem to contradict that in *Chironomus*, where the injection of ecdysone and actinomycin prevents the appearance of the early puffs on the salivary gland chromosomes

(p. 124). It is possible, of course, that in both the lepidopterans and the chironomids, ecdysone has a primary effect upon some cytoplasmic constituent(s) with a consequent effect upon the genes, and that apolysis is the result of this primary effect within the epidermal cell. But the actions of antibiotics upon cellular protein synthesis in different animals are not necessarily the same, nor are the effects of the same hormone upon *dissimilar* tissues in *different* animals. Moreover, the doses of ecdysone used in the experiments upon the moths were rather large, so that their effects could possibly be abnormal.

Insect growth and development thus depends upon the intrinsic properties of individual cells, together with cellular interactions which give gradients and polarized fields important in determining patterns and in coordinating growth.[191c] Ecdysone acts to 'inform' the cells that their next developmental cycle is to commence; whether this development is larval or adult will depend upon the concentration, or time of action within the cycle, of the juvenile hormone.

7

Endocrine Mechanisms in the Insecta—II

Hormones and Reproduction

THE INSECT REPRODUCTIVE SYSTEM

Insects generally are bisexual, and although parthenogenesis is anything but uncommon, frequently playing a very important part in normal life-cycles (e.g. those of many aphids), sexual reproduction is the usual rule.

Until recently, it was considered that the morphological differences between male and female insects were strictly genetically determined. The occasional appearance of intersexual mosaics provided good evidence for this view. It was thought that the development of secondary sexual characteristics, controlled by hormones produced by the gonads, so prominent a feature of sexual differentiation in the vertebrates, just did not apply in insects. But it is now known that in one insect at least (the glow-worm, *Lampyris noctiluca*) both the primary and secondary male characteristics are induced by an 'androgenic' hormone from paired endocrine glands associated with the testes[211] (Fig. 7.1). Transplantation of these glands into a female larva at a fairly early stage will induce masculinization. It is too early yet to say how widespread in insects this phenomenon may be, but even this one example shows the danger of generalizing about such aspects of the life of insects.

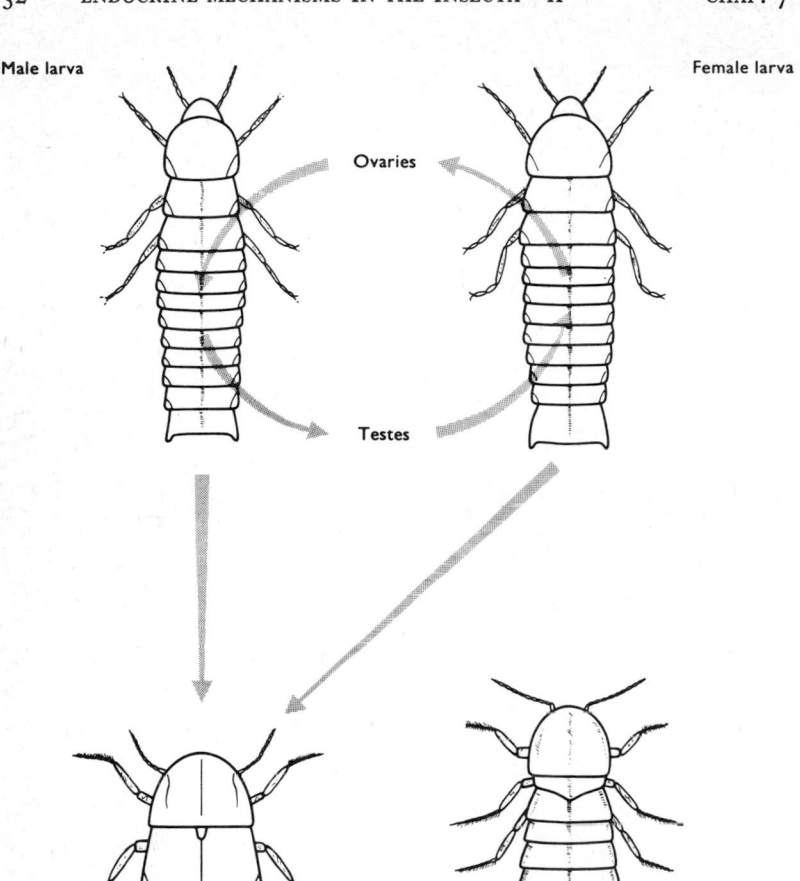

Male larva

Female larva

Ovaries

Testes

Adult male   (a)   Adult female

Fig. 7.1 Sex determination in *Lampyris noctiluca*. (a) The transplantation of testes together with androgenic tissue from a male to a female larva causes the latter to develop into a male. Ovary transplantation from female to male larvae is without effect. (b) *opposite* The table shows that the potency of testes to induce masculinization of females is reduced in the pupa and is negligible in the adult. Correspondingly, the competence of the female to react to the factor is lost after pupation. (After Naisse[211])

Testes from	Implanted into	Result
male larvae between moults 3 and 4	female larvae of same age	all males produced
	female larvae between moults 4 and 7	all males produced
	female pupae	females produced
male larvae between moults 4 and 5	female larvae between moults 3 and 4	all males produced
	female larvae of same age	all males produced
	female pupae	females produced
male pupae	female larvae between moults 3 and 4	80% males produced
male adults	females of all stages	all females produced

(b)

Fig. 7.1 Continued

The terminal segments of the abdomen in both male and female insects are deeply involved in the formation of the genitalia (Fig. 7.2). In the female, various kinds of ovipositor are developed from what must have been originally abdominal appendages; in the male, outgrowths are formed from the terminal segment which divide to form a protrusible phallus and associated clasping organs. In both sexes in addition, parts of the internal ducts and glands of the reproductive system are derived from embryonic ectodermal invaginations. The development of adult genitalia is a major characteristic of insect metamorphosis. In a limited sense, therefore, the combination of hormones which controls metamorphosis (p. 108) may also be considered to influence reproduction. But in general, reproductive endocrinology confines itself to those hormones which are active in the adult instar itself.

In most insects the ovaries are paired, each surrounded by a thin epithelial sheath containing connective tissue fibres and unstriped muscle cells in addition to the flattened epithelial cells. Each ovary is made up of a number of egg-tubes, or ovarioles, the proximal extensions of which are called the terminal filaments and attach the ovaries to the dorsal body wall (Fig. 7.3). There are usually 4, 6 or 8 ovarioles in each ovary, although numbers varying from one to several hundred are not uncommon. In queen termites, there may be several thousand ovarioles in each ovary.

Beneath the terminal filament in each ovariole is the germarium, in which the differentiation of oocytes from the primary germ layer takes place. Next in line along the ovariole is the vitellarium, which makes up the greater part of each ovariole and in which yolk deposition or vitellogenesis in the oocyte occurs. The ovariole joins the oviduct by a narrow neck or pedicel.

Two main types of ovariole are found in the insects, separated accord-ing to the way in which the oocyte is differentiated from the oogonium and the way in which it is nourished during the early growth period. In the ***panoistic ovariole***, the daughter cells of the oogonia metamorphose into oocytes (Fig. 7.6). These become associated with prefollicular cells

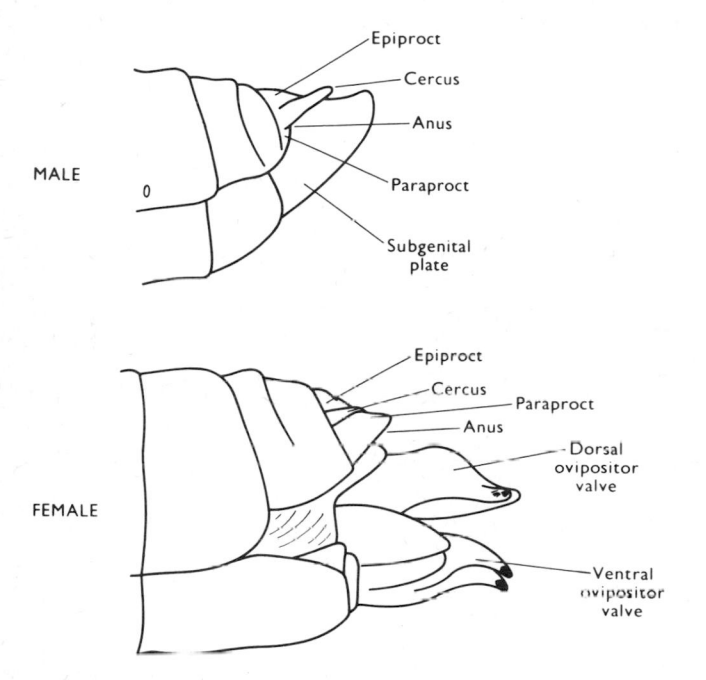

Fig. 7.2 External genitalia of the male and female desert locust, *Schistocerca gregaria*. The development of these specifically adult characteristics is controlled by the juvenile hormone during moulting.

of mesodermal origin, a little lower down the germarium. The pre-follicular cells become organized into a layer of follicle cells surrounding each oocyte and accompany the latter into the vitellarium. The follicle cells are concerned with all the stages of growth and yolk deposition in the oocyte, and after vitellogenesis is completed, the follicle cells secrete the shell, or chorion, around the egg. In the ***meroistic ovariole***, on the other hand, only one of the daughter cells from each oogonium meta-morphoses into an oocyte (Figs. 7.4, 7.5). The remaining daughter cells are connected with the oocyte by shorter or longer intercellular con-nections and nourish the oocyte during the early stages of its growth, or rather, supply it with material for biosynthesis; these associated cells

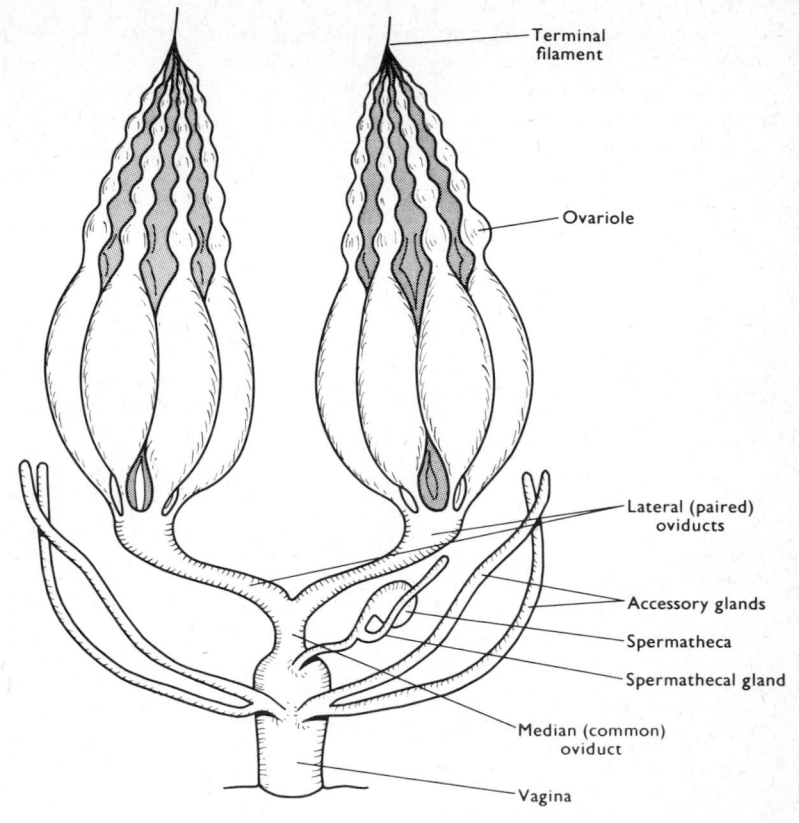

Terminal filament

Ovariole

Lateral (paired) oviducts

Accessory glands

Spermatheca

Spermathecal gland

Median (common) oviduct

Vagina

Fig. 7.3.   Ovaries of a generalized insect. Each ovary is composed of a number of egg tubes, or ovarioles, opening into a lateral oviduct. The paired oviducts lead into a common oviduct, from which a spermatheca for sperm storage opens. A spermathecal gland may be present, whose secretion may keep the sperm viable, or aid their movement down the spermathecal duct during fertilization. Accessory glands produce secretions to surround the eggs and lubricate their passage during oviposition.

are consequently called nurse cells. The meroistic ovariole may be further subdivided into two kinds. In one, the nurse cells remain in the germarium, and as the oocyte moves into the vitellarium its connections with these cells become extended to form nutritive cords (Fig. 7.5). In the other, the nurse cells remain in contact with the oocyte, moving with it into the vitellarium (Fig. 7.4). The first kind of ovariole is called *telotrophic* (feeding from afar) and the second, *polytrophic*. Prefollicular cells are associated with the oocyte in both kinds of meroistic ovariole: in the telotrophic ovariole, only the oocyte becomes invested

with follicle cells, which take over the nourishment of the oocyte when the nurse cells degenerate; in the polytrophic ovariole both nurse cells and oocyte are invested with follicle cells, although there is often a pronounced constriction between the two. The nurse cells degenerate at a fairly early stage in the development of the oocyte, but their remains can persist until a much later stage in vitellogenesis. The follicle cells in the meroistic ovaries finally produce the chorion of the egg in the same way as in the panoistic ovariole.

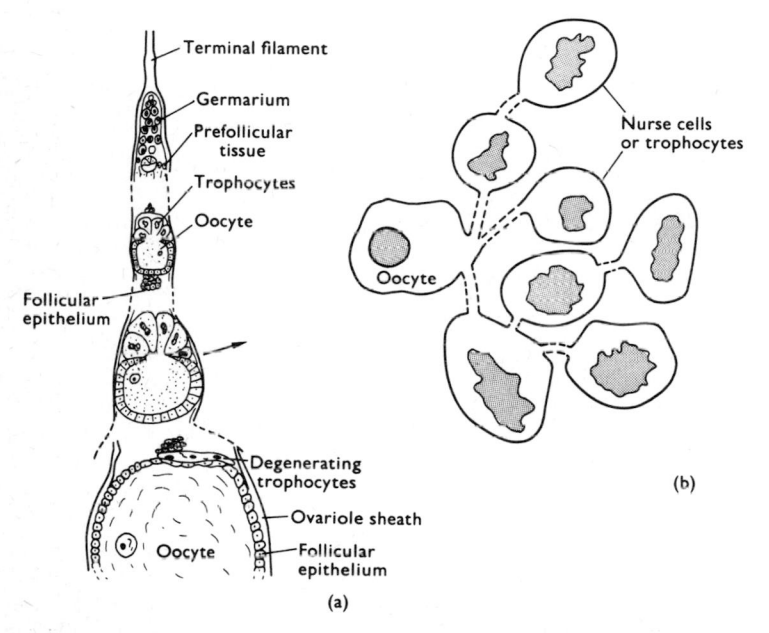

Fig. 7.4   (a) A polytrophic meroistic ovariole. Unlike the telotrophic type, the trophocytes remain in close association with the oocyte during its early passage along the ovariole. Trophocytes and oocyte together make up a follicular chamber. (b) This shows how the trophocytes are connected with each other and ultimately with the oocyte which they supply with materials during its early development. Oocytes and trophocytes are developed from a primary oogonium.

The essential difference between panoistic and meroistic ovarioles is that in the former the follicle cells are involved in the early growth of the oocyte, whereas in the latter this primary function has been taken over by the nurse cells, themselves very much modified oocytes. In both types of ovariole, the later development of the oocyte and the formation of the chorion are functions of the follicle cells. Since vitellogenesis in successive oocytes takes place in sequence down the length of the vitellarium,

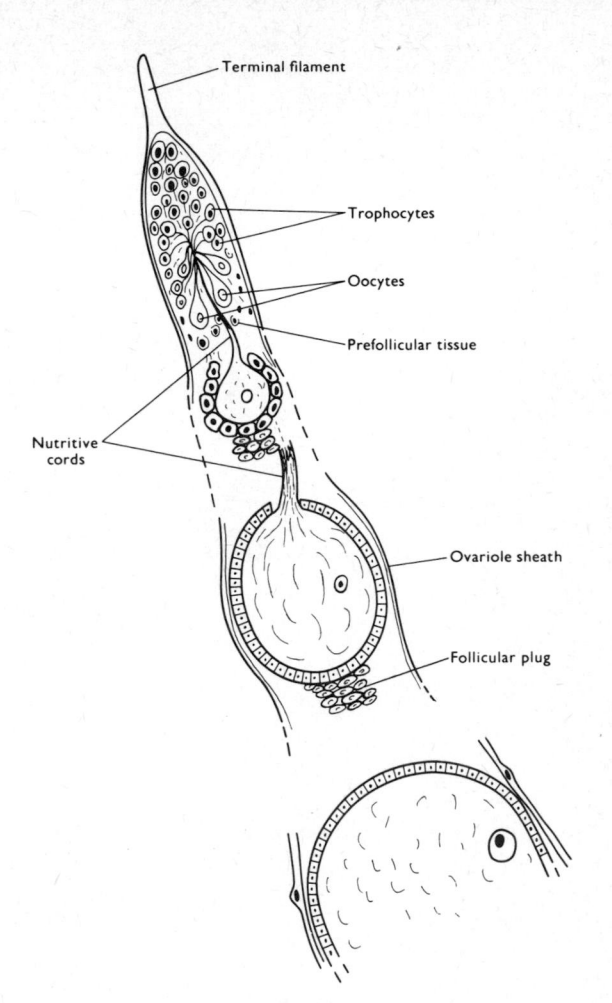

Fig. 7.5 A telotrophic meroistic ovariole. Nurse cells, or trophocytes, which have developed from primary oogonia, remain in the germarium but are attached to each oocyte by cytoplasmic filaments, the nutritive cords. The initial development of the oocytes depends upon the transmission of materials from the trophocytes along the nutritive cords.

various stages in oocyte development may be seen at one time in any kind of ovariole.

Panoistic ovarioles are more typical of the 'older' orders of insects—grasshoppers, termites, stone-flies and dragon flies, for example. The desert locust, *Schistocerca gregaria*, has panoistic ovarioles. Telotrophic

ovarioles are characteristic of the bugs, including *Rhodnius prolixus*, and some beetles. Most holometabolous insects, including the blowfly *Calliphora*, have polytrophic ovarioles.

The pedicel of each ovariole leads into the mesodermal paired oviduct on its own side of the body. The paired oviducts lead into the ectodermal common oviduct, which usually opens on the ventral surface of the ninth

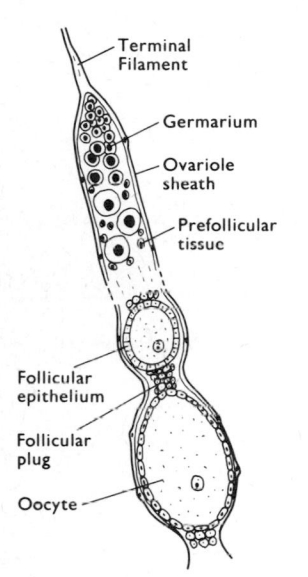

Terminal
Filament

Germarium

Ovariole
sheath

Prefollicular
tissue

Follicular
epithelium

Follicular
plug

Oocyte

Fig. 7.6 Section through a panoistic ovariole. The oocytes are each surrounded by follicle cells shortly after their emergence from the germarium.

abdominal segment. Associated with both the mesodermal paired oviducts and the ectodermal common oviduct are the accessory glands, which are variously concerned with secreting material to protect the eggs, or to cement the eggs to an appropriate substrate, or both, after laying (Fig. 7.7).

The testes are usually paired structures in the insects (Fig. 7.8), although many species have a single, fused median organ. Each testis is enclosed within a connective tissue sheath, which may be attached dorsally to the wall of the abdomen. The most distal part of the testis is the germarium where spermatogonia are produced. These are set free in the germarium where they become encapsulated by groups of somatic cells. The transformation of the cysts of spermatogonia to spermatocytes and spermatids, and the final metamorphosis to spermatozoa, may occur in different zones in the testis distally from the germarium. The process is not unlike oogenesis in the female; much modified, of course, because there is no vitellogenesis.

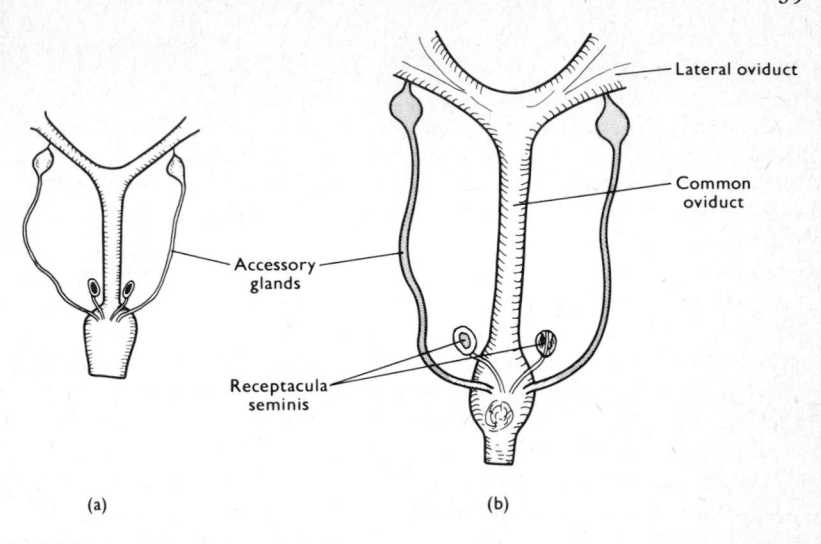

Fig. 7.7 Development of female accessory glands in the blowfly, *Calliphora erythrocephala*. (a) Reproductive system of a newly moulted female. (b) Reproductive system of a mature female.

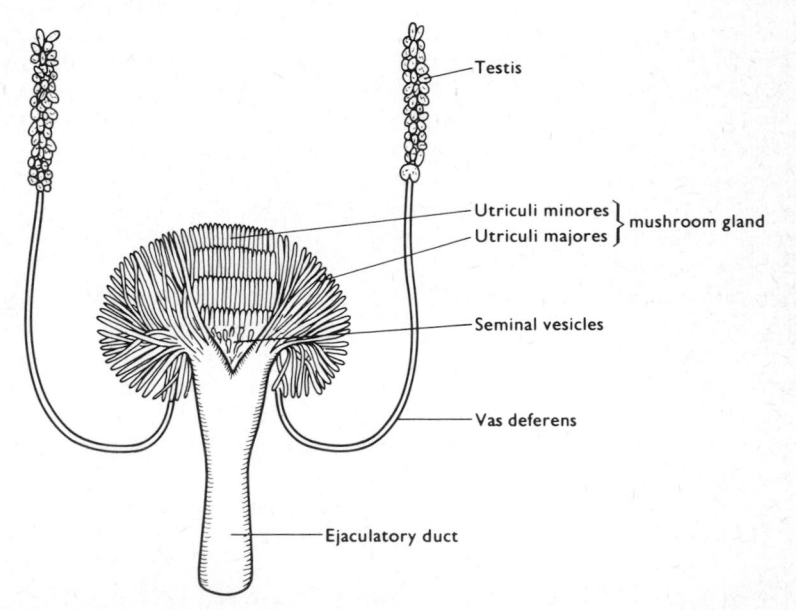

Fig. 7.8 Reproductive system of a male cockroach. The accessory glands are commonly called the mushroom gland.

The testes lead into the paired vasa deferentia, and the packets of spermatozoa are usually stored in dilatations of each vas deferens, the seminal vesicles. The paired vasa deferentia lead into a common ejaculatory duct, which opens into the phallos, and may have various structures developed within it to pump the sperm out of the system.[73] The ejaculatory duct and phallos are ectodermal in origin compared with the mesodermal vasa deferentia. Associated with both the mesodermal and ectodermal parts of the reproductive system are accessory glands, of a wide variety of form and structure in different insects, and producing a range of secretions which help in sperm transfer from the male to the female (Fig. 7.8).

All the winged insects copulate, sperm being passed from the male genitalia to the female reproductive opening during coupling. The only exception is found in the dragonflies, where the male has accessory reproductive mechanisms on the ventral side of the *anterior* abdominal segments, and sperm is transferred to these before coupling takes place. The terminal abdominal segments in the male bear clasping organs which hold the female's head during copulation, while she bends her abdomen forwards so that her reproductive opening comes into contact with the anterior accessory genitalia of the male. Sperm transfer is thus indirect in the dragonflies, a unique device in the pterygote insects, but a phenomenon not unknown in other animal groups such as the spiders, squids etc. In all other Pterygota, the direct transfer of sperm is of two kinds. In one, the sperm are passed immediately to special reservoirs—the spermathecae—inside the female by means of a very highly developed male intromittent organ. In the other the sperm are enclosed within a special container, the spermatophore, which is placed inside the female's genital opening, the sperm later passing from the spermatophore to the storage reservoirs in the reproductive tract of the female. The materials for spermatophore production are secreted by the accessory glands of the male reproductive system. The liberation of the sperm packets from the spermatophore within the female reproductive system may be the result of the digestion of the materials of the spermatophore by special secretions produced in the female tract. Often, the spermatophore is so constructed that the sperm packets are ejected from the spermatophore with some force.[68, 69] The sperm packets move up the oviducts to the spermathecae usually as a result of muscular contractions of the ducts; these contractions may be initiated by a further secretion from the male.

ENDOCRINE CONTROL OF OOCYTE DEVELOPMENT

In adult pterygote insects, the thoracic glands disappear during or soon

after the last moult (p. 115). Consequently, of the organized endocrine system only the cerebral neurosecretory system and the corpora allata remain as possible sources of hormones controlling reproductive development.

In most insects, the corpora allata are essential for the full development of the eggs.[134] In the adult *Rhodnius*, a blood meal fully activates the corpus allatum and the eggs develop; development is also normal if *Rhodnius* is decapitated after the meal in front of the corpus allatum, but if it is decapitated *behind* the gland the eggs only develop to the stage at which yolk deposition (vitellogenesis) begins, when the nurse cells have completed supplying the oocytes with the materials for protein synthesis (p. 134).[269] The follicle cells then invade the oocyte and resorb it, a process known as oosorption. If the decapitated female is joined in parabiosis with a fed male, or with a female with intact corpus allatum, or if active corpora allata are implanted into the headless female, then normal oocyte development is restored (Fig. 7.9). These experiments allow three very important conclusions to be drawn:

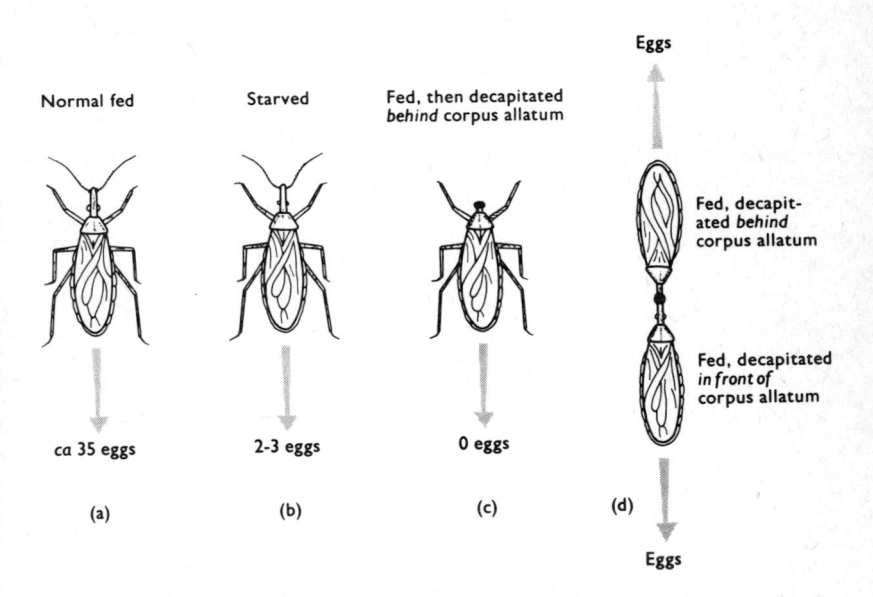

Fig. 7.9 Experiments to show the necessity of the corpus allatum for egg production in *Rhodnius prolixus*. A normal fed female (a) produces about 35 eggs. If starved (b) a very small number of eggs is produced depending upon the amount of food present in the gut. Fed, and then decapitated behind the corpus allatum (c), no eggs are produced. Fed, and joined in parabiosis with a fed female possessing its corpus allatum, both individuals produce eggs (d). (After Wigglesworth[269])

1 The corpus allatum is necessary for egg development.
2 The corpus allatum hormone is *not* sex specific.
3 In *Rhodnius*, the neurosecretory system is not necessary for oocyte development (decapitation in front of the corpus allatum removes the brain).

In *Rhodnius*, an adult corpus allatum transplanted into an allatectomized larva has morphogenetic effects identical with those of the larval gland; larval corpora allata implanted into an allatectomized adult restore normal egg development. It is thus very likely that the adult gonadotrophic hormone from the corpus allatum is identical with juvenile hormone.

Similar results have been recorded from many insects from most of the major orders. The exceptions are the stick-insects (Phasmidae) and many butterflies and moths (Lepidoptera). In these groups, egg development proceeds normally in the absence of the corpora allata. The stick-insects are unusual species, normally considered to be neotenous, i.e. the reproductive individuals are *juvenile* forms so far as their somatic structure is concerned. It may not be surprising, therefore, that endocrine control over reproduction has been lost. Many Lepidoptera are short-lived as adults, implying the need for rapid maturation. In fact, many female moths emerge with almost fully developed eggs: the loss of endocrine control of maturation in these forms is also not unexpected (and see p. 290).[134]

In insects with panoistic ovarioles, allatectomy after vitellogenesis has begun results in oosorption, as in *Rhodnius* and other insects with meroistic ovarioles.[142, 143] But allatectomy before the beginning of vitellogenesis does not cause the death and resorption of the oocytes: instead they remain quiescent, and will initiate development again if corpora allata are reimplanted. It seems that once the cellular apparatus for vitellogenesis has been established, either *de novo* in the oocytes of panoistic ovarioles, or by passage from the nurse cells in meroistic ovarioles, then development *must* proceed along its normal path. Any factor which prevents such development causes oocyte death and oosorption.

In starved *Rhodnius* adults, and in those which have been decapitated behind the corpus allatum, the oocytes develop normally up to the state at which they lose their connections with the nurse cells, and it would seem that the vitellogenic process itself is under the immediate control of the hormone. But is the endocrine control by the corpus allatum of vitellogenesis direct or indirect? In other words, can the corpus allatum hormone in the adult be considered a gonadotrophic hormone, directly affecting yolk deposition by the developing oocytes? Or is the vitellogenic

promoting action indirect, some other tissue or organ being influenced by the hormone, and the effect upon vitellogenesis itself being secondary to this other action?

From the results of parabiosis between an adult female *Rhodnius without* its corpus allatum and a fourth stage larva *with* its corpus allatum, Wigglesworth pointed out that 'in those experiments in which no egg development occurred, the characters produced by the moulting fourth stage nymph were those of the normal fifth stage nymph. Whereas when active development of eggs took place the characters of the nymphs showed a certain degree of differentiation towards the adult form. This result could be interpreted to mean that the developing eggs were absorbing juvenile hormone and so depriving the tissues of the moulting fourth stage nymph of the supplies they need.'[272] In other words, the corpus allatum hormone of the adult female is a gonadotrophic hormone in the strict sense of the term. But this conclusion is admittedly very tentative and elsewhere Wigglesworth says: 'Associated with the development of eggs in the adult female there is much more rapid digestion of the intestinal contents. Whether this is a direct effect of the hormone on digestion and metabolism . . . or whether it is an indirect effect consequent upon the demands of the developing ovaries, has not been determined.'[269] Here the possibility is admitted that the corpus allatum hormone may not strictly be gonadotrophic, but may perhaps be a metabolic hormone, and that the ovaries develop as a result of the nutrients which are thereby made available.

Blood pigments from the host can circulate in the haemolymph of *Rhodnius* as kathaemoglobin, and can be transferred to the eggs in this form.[273] In *Rhodnius*, therefore, fairly large molecules are able to pass through the membranes of the gut and those of the follicle cells and developing oocytes. A large percentage of the yolk in the oocytes consists of protein. Are these proteins synthesized by the follicle cells and oocytes from amino acid precursors in the haemolymph? Or are they synthesized elsewhere, perhaps under the influence of the corpus allatum hormone, and transferred, like the kathaemoglobin, *in toto* to the developing eggs? And may such a transfer also be controlled by the corpus allatum hormone? In effect, is the corpus allatum hormone a metabolic hormone, or a gonadotrophic hormone, or even both?

It will be seen later that a great deal of confusion exists as to the exact role of the corpus allatum hormone in adult insects generally. But in *Rhodnius*, some attempt has very recently been made to determine the gonadotrophic or metabolic function of the hormone.

Because of differences in their constitution, particularly in their possession of different proportions of polar groups, proteins in suitable solutions and on particular substrates will migrate at different rates

when subjected to a voltage difference. This movement, which can be towards either the positive or the negative electrode, is called protein electrophoresis. The substrates most used are paper, starch gel or cellulose acetate. The solvents are buffered to a pH which is kept constant for any series of determinations. The sample to be analysed is placed on the substrate and the potential difference applied for a specific time. The different proteins, or groups of proteins move at their respective rates along the substrate during this time. After fixing, the protein bands can be stained, and their positions relative to the line of origin determined.

When samples of blood from last instar *Rhodnius* larvae are analysed electrophoretically, a characteristic sequence of proteins, or protein groups, is obtained. When blood from adult females is analysed in the same way, a similar pattern of proteins is obtained, except that two additional slowly moving bands are present.[62] Blood from adult males has the same electrophoretic pattern as that from the adult females. When the egg proteins are examined electrophoretically, fewer bands are present, but those which occur in largest amounts are the two which are additionally present in adult compared with larval blood. It is therefore very likely that it is these two proteins, or protein groups, found in the blood of the adult which are incorporated into the yolk of the eggs. (Fig. 7.10).

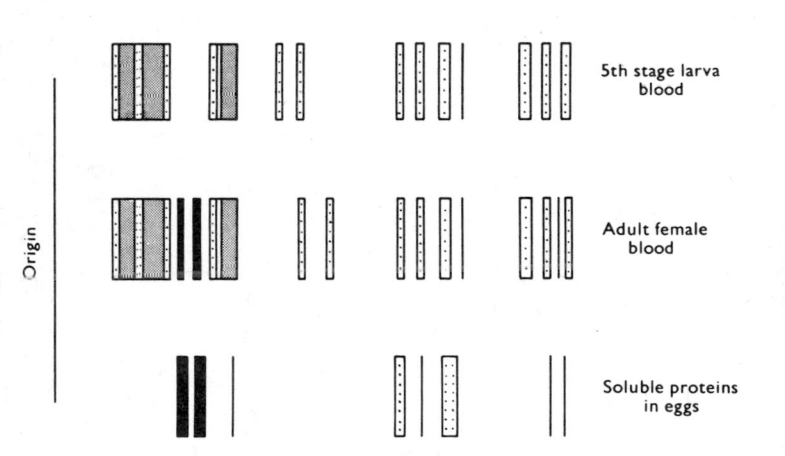

Fig. 7.10   Electrophoretic separation of the proteins of the blood and eggs of *Rhodnius prolixus*. The blood of the adult female contains two protein fractions which are not present in the blood of the fifth stage larva. These same two protein fractions constitute the larger part of the soluble proteins of the eggs. This suggests that adult specific proteins appear in the blood of the adult female and are taken up by the developing eggs. (After Coles[62])

The proteins of the adult blood can be combined with a compound, fluorescein iso-cyanate, which fluoresces strongly in ultraviolet light. When this 'labelled' blood is injected into the body of an adult female, the eggs become strongly fluorescent within 24 hours. Blood from a fifth instar larva, similarly labelled and injected into the body of an adult female, results in very little fluorescence in the eggs. This suggests strongly that the adult-specific blood proteins are passed into the eggs *in toto*.[62]

But where are the blood proteins produced? And where does the corpus allatum hormone act in promoting the process?

When radioactively labelled amino acids are injected into a female *Rhodnius*, radioactive proteins are formed in the fatbody, and these subsequently appear in the blood. The fatbody is therefore very likely the site of synthesis of the adult blood proteins.[63]

When adult female *Rhodnius* are fed and then decapitated behind the corpus allatum, no yolk deposition occurs, of course, in the oocytes. But in addition, only very small quantities of the adult-specific proteins appear in the blood. When the ovaries are removed, the adult-specific proteins reach a very high concentration in the blood. Similarly, no yolk proteins can be found in the fatbody after decapitation behind the corpus allatum. Amino acids are present in high concentration in the blood after decapitation; this means that digestion of the meal is not interrupted by the operation. Finally, when the developing oocytes are incubated with fluorescein-labelled proteins *in vitro*, there is some uptake of protein by the oocytes, but not as much as that taken up after the labelled protein is injected *in vivo*.

It is concluded from these experiments that the corpus allatum in the female *Rhodnius* directly controls the synthesis by the fatbody of specific proteins which are passed to the developing oocytes to be incorporated into the yolk. In other words, the corpus allatum hormone in the female *Rhodnius* is a metabolic hormone. It perhaps acts upon specific genes in the nuclei of the cells of the fatbody, causing these eventually to produce enzymes for the synthesis of the adult specific proteins. This action of the corpus allatum hormone in the adult is therefore thought to be similar to the way in which juvenile hormone acts during growth and development (p. 129). In addition, the difference in uptake of labelled protein between oocytes *in vitro* and *in vivo* suggests a further gonadotrophic effect of the hormone.

It is now known that when the oocytes in *Rhodnius* become vitellogenic, spaces open up between the follicle cells which allow the passage of proteins into the eggs; the spaces do not appear after allatectomy.[225a] *In vitro*, spaces between the follicle cells appear only when juvenile hormone is added to the incubation medium: the hormone produces its maximum

effect at a concentration of $10^{-4}\mu l/ml$ (about 10^{-7} M) but the follicle cells will show some response even with 10^{-17} M juvenile hormone.[75a] However, a particular previtellogenic stage must be exposed to the hormone before the vitellogenic stage responds: the hormone acts during two periods of oocyte development.[1a] The corpus allatum hormone therefore does have a gonadotrophic effect in *Rhodnius* and its site of action is the follicle cells. So both possibilities—metabolic and gonadotrophic—originally envisaged by Wigglesworth for the action of juvenile hormone in its control of oocyte development in *Rhodnius* (p. 143) have now been realised (Fig. 7.11).

Since this work in *Rhodnius*, the fatbody in a number of other insect species has been shown to synthesize proteins which are sequestered by the developing oocytes.[21f, 86a, 120d, 214m] The proteins are normally restricted to the female and are selectively removed from the haemolymph by the developing oocytes.[86c] They may therefore be properly called vitellogenic proteins or vitellogenins.

The dual action of the corpus allatum hormone in the control of vitellogenesis has been particularly well demonstrated in cockroaches. Vitellogenins are only produced by the fatbody in *Leucophaea maderae* in the presence of juvenile hormone[86a] and it is suggested that, as in *Rhodnius*, the hormone activates the fatbody genome to produce specific messenger RNAs for vitellogenin synthesis.[86d] When vitellogenins extracted from mature oocytes are injected into allatectomized *Periplaneta americana* females, they are only taken up by the host oocytes when juvenile hormone is also injected, or corpora allata are implanted.[12a] Starved cockroaches begin to resorb their oocytes after about ten days and the concentration of vitellogenins increases markedly at this time; but yolk deposition recommences almost immediately when juvenile hormone or a mimic is injected into the females.[12b] Finally, when the ovaries of *Eublaberus posticus* are transplanted into allatectomized males, yolk deposition occurs

Fig. 7.11 The endocrine control of reproduction in *Rhodnius prolixus* (a), locusts (b) and *Leptinotarsa decemlineata* (c). In *Rhodnius*, the juvenile hormone acts at three points : it controls vitellogenin synthesis by the fatbody, affects previtellogenic oocyte development and facilitates the uptake of vitellogenic protein by the oocyte by opening spaces between the follicle cells ; a neurosecretory hormone intervenes only in the final stage—oviposition. In locusts, a neurosecretory hormone affects previtellogenic oocyte growth and the same or another neurosecretory hormone affects general protein synthesis in the fatbody ; juvenile hormone switches at least part of the fatbody protein synthesis into vitellogenin production, and also facilitates protein uptake by the oocytes ; a neurosecretory hormone promotes oviposition. In *Leptinotarsa*, vitellogenin production is controlled by both neurosecretory and juvenile hormones ; juvenile hormone also facilitates protein uptake by the oocytes. Differences in the apparent endocrine control mechanisms could be explained by interactions between the cerebral neurosecretory system and the corpora allata.

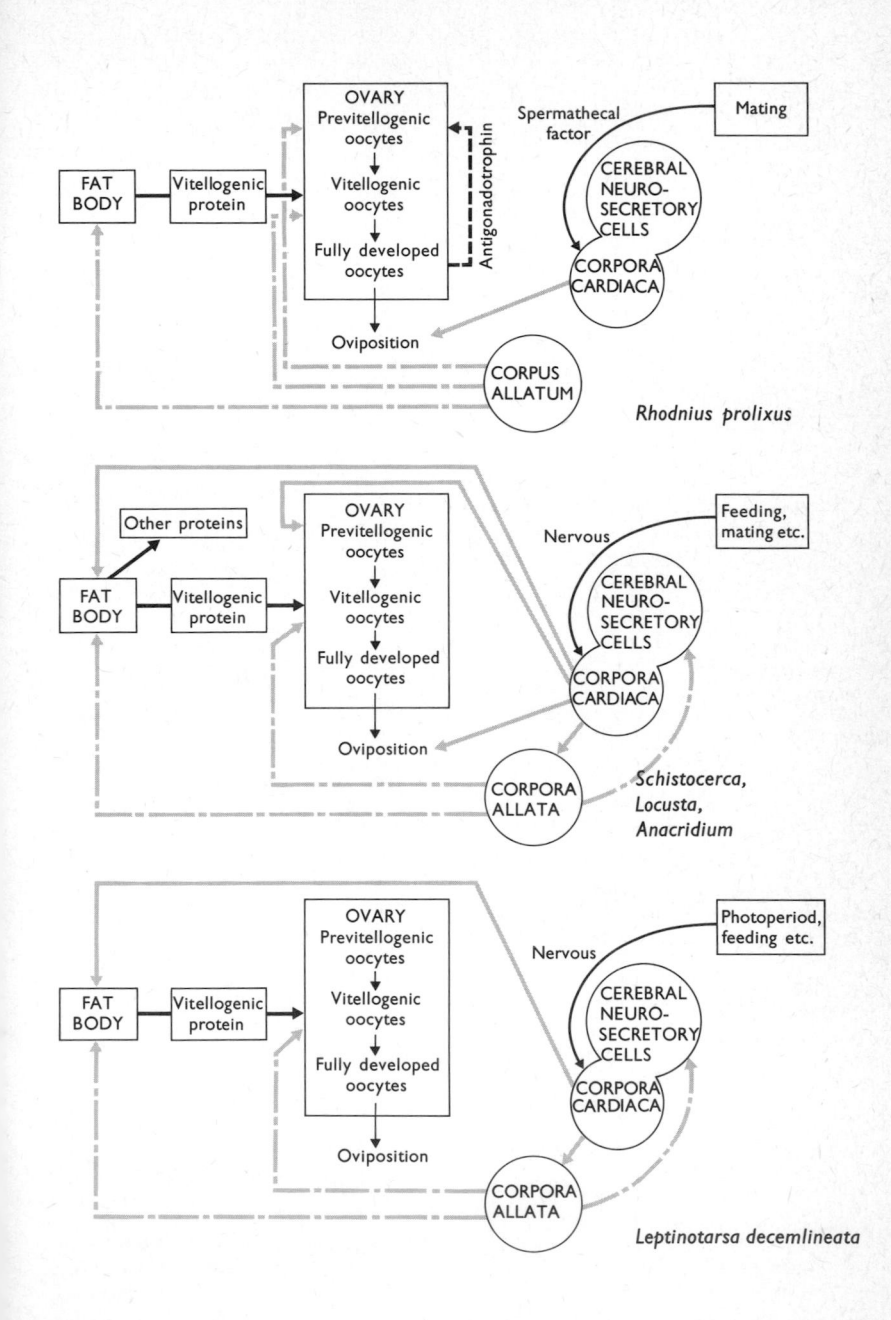

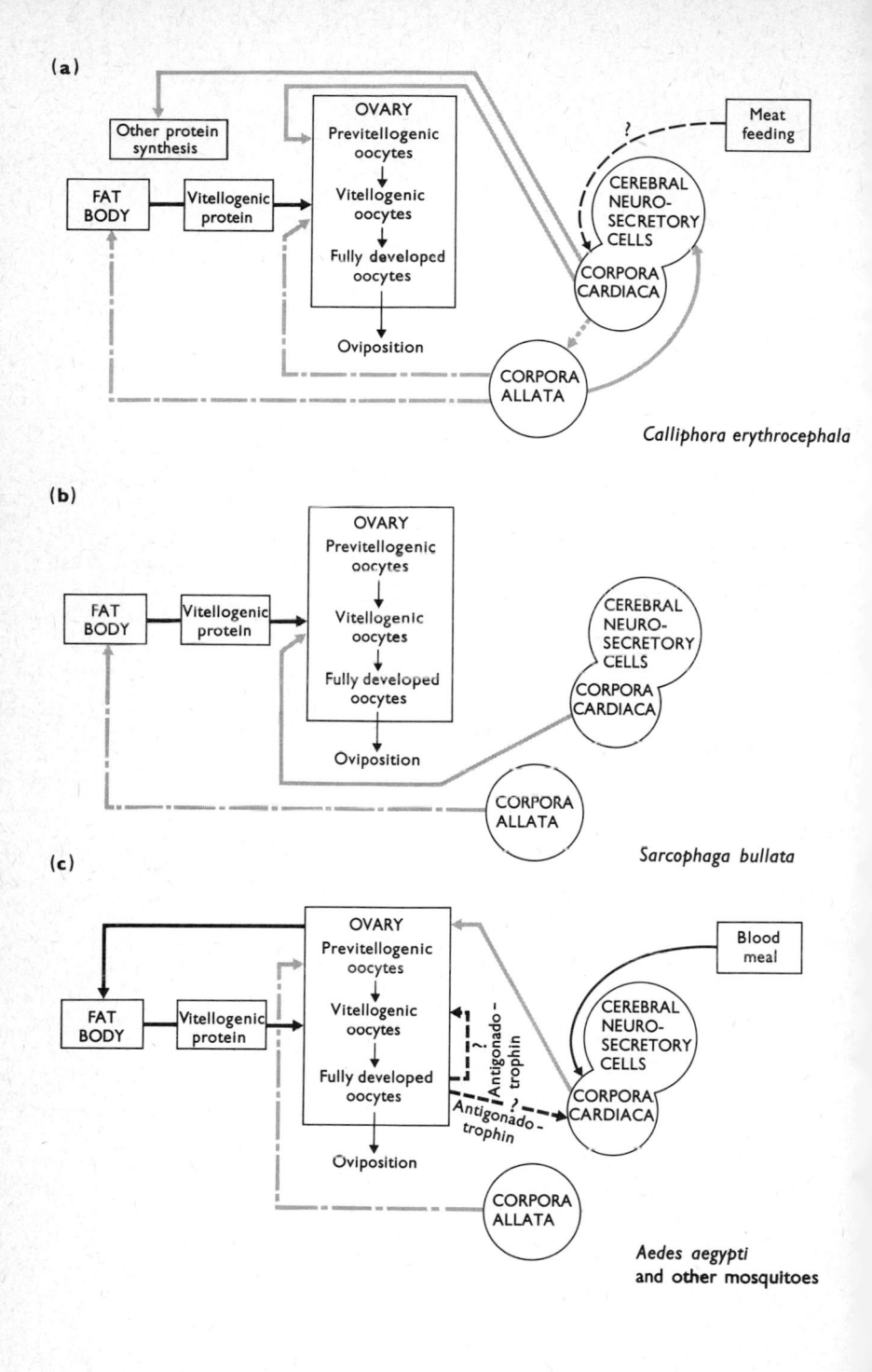

(a)

Other protein synthesis

FAT BODY → Vitellogenic protein

OVARY
Previtellogenic oocytes
↓
Vitellogenic oocytes
↓
Fully developed oocytes
↓
Oviposition

Meat feeding

?

CEREBRAL NEURO-SECRETORY CELLS

CORPORA CARDIACA

CORPORA ALLATA

Calliphora erythrocephala

(b)

FAT BODY → Vitellogenic protein

OVARY
Previtellogenic oocytes
↓
Vitellogenic oocytes
↓
Fully developed oocytes
↓
Oviposition

CEREBRAL NEURO-SECRETORY CELLS

CORPORA CARDIACA

CORPORA ALLATA

Sarcophaga bullata

(c)

FAT BODY → Vitellogenic protein

OVARY
Previtellogenic oocytes
↓
Vitellogenic oocytes
↓
Fully developed oocytes
↓
Oviposition

Antigonado-trophin ?
Antigonado-trophin ?

Blood meal

CEREBRAL NEURO-SECRETORY CELLS

CORPORA CARDIACA

CORPORA ALLATA

Aedes aegypti and other mosquitoes

only when both vitellogenins and juvenile hormone are injected, and not in the presence of either alone.[12c]

In adult female cockroaches, the cerebral neurosecretory system apparently plays no part in the control of egg development.[86c, 86e] However, the cerebral neurosecretory cells in mated *Rhodnius* females produce and release a myotropin which facilitates the laying of the fully developed eggs.[75] The neurosecretory cells are stimulated to produce the myotropin by a blood-borne factor which originates in the spermathecae of mated females[73] (Fig. 7.11). Virgin females lack the spermathecal factor and consequently do not produce the neurosecretory myotropin: as a result, virgin females begin to oviposit later, and at a lower rate, than do mated females. The retention of fully developed eggs by virgin females has a further consequence: the accumulation of an antigonadotrophin (p. 175) which reduces the *number* of eggs produced[225b] (Fig. 7.11).

In some insects (Fig. 7.12), the cerebral neurosecretory system intervenes in the reproductive process at an earlier stage than in *Rhodnius*: removal or destruction of the medial neurosecretory cells prevents oocyte growth and vitellogenesis. In the blowfly, *Calliphora*, and in locusts and grasshoppers, the size of the egg chambers or oocytes is much less after destruction of the neurosecretory cells than after allatectomy,[261, 133] (Fig. 7.13). In these insects, the corpora allata are smaller after the neurosecretory cells have been removed, so part of the effect upon the oocytes could be due to these glands being inactive. But in *Calliphora*, the fleshfly *Sarcophaga*, locusts, the Colorado beetle *Leptinotarsa decemlineata*, and several species of mosquito, the implantation of active corpora allata or the injection of juvenile hormone, does not restore full growth and vitellogenesis to the oocytes (Fig. 7.12). The neurosecretory cells must produce a factor affecting oocyte development in addition to any effect upon the corpora allata.

In *Calliphora*, which has polytrophic meroistic ovarioles, oocyte growth up to the stage of vitellogenesis involves the synthesis of structural proteins and the establishment of enzymic mechanisms derived from the

Fig. 7.12 The endocrine control of reproduction in *Calliphora erythrocephala* (a), *Sarcophaga bullata* (b) and mosquitoes (c). The system in *Calliphora* is very similar to that in locusts (Fig. 7.11). In *Sarcophaga*, the juvenile hormone controls fatbody vitellogenin synthesis, although other (unknown) factors also appear to be involved; the neurosecretory hormone apparently facilitates protein uptake by the oocytes, differing markedly from other species in this respect. In mosquitoes, juvenile hormone affects previtellogenic oocyte development, reflecting one of the actions of the hormone in *Rhodnius* (Fig. 7.11), but unlike the other species does not control vitellogenin production by the fatbody: instead, and apparently uniquely, vitellogenin synthesis is controlled by a hormone produced by the ovary, itself controlled by the neurosecretory hormone from the cerebral neurosecretory system.

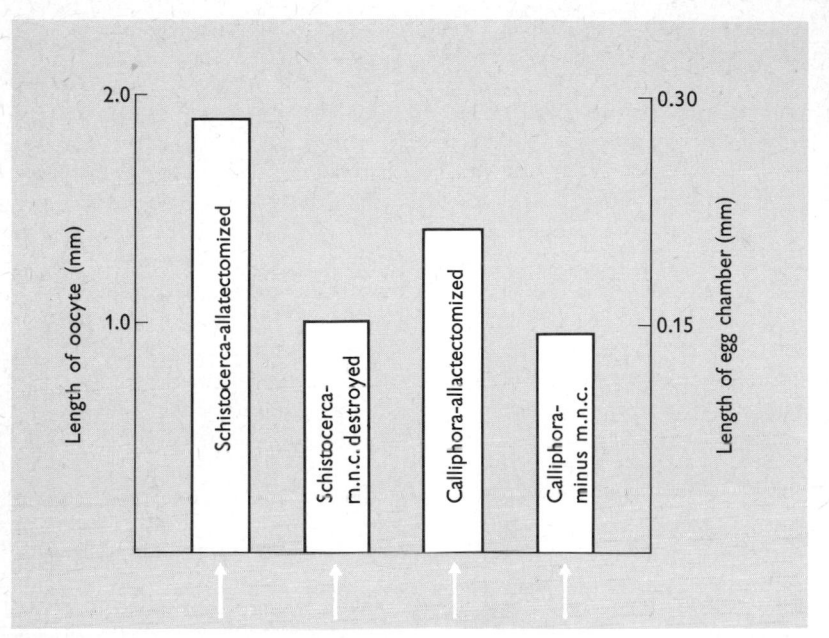

Fig. 7.13 Differential effects of allatectomy and cautery or removal of the cerebral neurosecretory cells in the desert locust, *Schistocerca gregaria*, and the blowfly, *Calliphora erythrocephala*. In both insects, the oocytes or egg chambers develop further after allatectomy than they do after destruction or removal of the cerebral neurosecretory cells. (After Thomsen[261], and Highnam[133])

nurse cells for future vitellogenesis. Since this is the stage most affected by removal of the neurosecretory cells, it is possible that the cells in some way control protein synthesis in the previtellogenic oocytes and nurse cells. How can this hypothesis be tested?

It is well known that blowflies will live for long periods when fed carbohydrate alone, but will not lay eggs. In what way is egg development affected in the absence of protein? This question can be answered very simply. When newly emerged flies are fed upon sugar and water, growth in the oocytes and accessory glands is retarded, and the corpus allatum fails to increase in size. These effects are remarkably similar to those which follow removal of the median neurosecretory cells. These observations provide strong support for the hypothesis that the neurosecretory cells are involved in some aspect of protein metabolism.

It was considered at one time that a correlation between midgut protease activity and the presence or absence of cerebral neurosecretory cells in *Calliphora* indicated a general control over protein metabolism in the body by cerebral neurosecretion, since proteases are themselves proteins.[263]

But it is now known that a secretagogue mechanism controls midgut protease synthesis in other insects: the stimulation of protease synthesis is directly related to the quantity of proteinaceous food consumed.[86b, 86f] *Calliphora* preferentially consumes protein during egg development, and its appetite for protein declines in the absence of the neurosecretory cells.[252] Thus in *Calliphora*, also, a secretagogue and not necessarily a direct neurosecretory mechanism could control the variations in midgut protease production.

In the newly emerged female *Calliphora*, the neurosecretory cells contain very little intracellular material and their nuclei are large (Fig. 7.14). When the flies are fed only sugar, the cells become stuffed with material and their nuclei are reduced in size[264] (Fig. 7.14). The most reasonable explanation for these observations is that in the newly emerged flies, the neurosecretory cells are actively synthesizing their material (as suggested by the large nuclei) which is rapidly transported to the corpus cardiacum and allatum to be released into the blood or otherwise utilized. Consequently, the cells contain little material at this time. But in the absence of meat, the release of neurosecretory material from the cells is inhibited, the cells fill up with material, and synthesis is eventually much reduced. Subsequent meat feeding reverses this process (Fig. 7.14). Consequently, it is concluded that the active synthesis and release of the brain factor which affects oocyte development is stimulated by meat feeding itself.

Similarly, it can be shown that allatectomy results in the accumulation of material within the neurosecretory cells, whose nuclei are also reduced in size after the operation. The implantation of corpora allata into allatectomized flies causes the neurosecretory cell nuclei to increase in size (Fig. 7.14), as does the implantation of corpora allata into sugar-fed flies. Thus the corpus allatum hormone itself can stimulate the synthetic activity of the neurosecretory cells.

In the light of this new evidence about the function of the median neurosecretory cells in the adult, the role of the corpus allatum during egg development in *Calliphora* may now seem rather obscure. The original experiments with the corpus allatum suggested that the gland was involved in oocyte growth. The idea that the corpus allatum hormone may be a metabolic hormone, like that in *Rhodnius*, receives some support from experiments on the oxygen uptakes of ovariectomized and allatectomized flies[260, 262] (Fig. 7.15). But it must be remembered that these experiments were carried out before the effects of the corpus allatum hormone upon neurosecretory cell activity were known. In *Calliphora*, the reduced oxygen uptake following allatectomy could consequently be a secondary effect following upon the reduced activity of the neurosecretory system. At the present time, it might be simplest to consider the corpus allatum hormone in *Calliphora* as being principally a gonado-

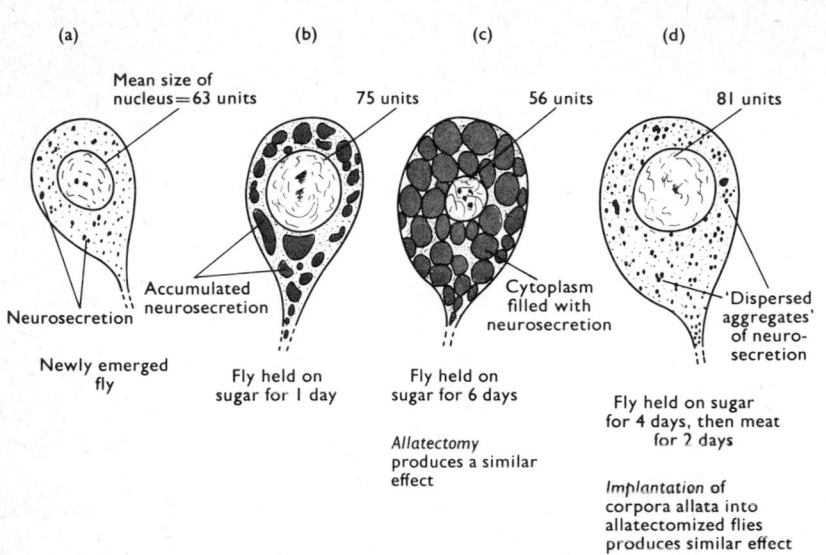

Fig. 7.14 Median neurosecretory cells of the blowfly, *Calliphora crythrocephala* in (a), the newly moulted female, (b) a female fed sugar, but no meat, for 1 day, (c), a fly fed on sugar only for 6 days, or allatectomized, and (d), a fly fed sugar and then meat, or one in which corpora allata are implanted after previous allatectomy. Nuclear sizes are given in arbitrary units. The accumulation of inclusions in (b) and (c) indicates that release of neurosecretion is inhibited, and the smaller size of the nucleus in (c) suggests that the synthetic activity of the cell is reduced. Subsequent meat feeding, or corpus allatum implantation (d), raises both the inhibition on release of material and the synthetic activity of the cell. (After Thomsen[264])

trophic hormone, directly influencing the uptake of vitellogenic materials by the oocyte.

More recently, evidence has been produced that the size of the corpus allatum in *Calliphora* is directly influenced by the concentration of yolk precursors in the blood[252, 253, 254] (Fig. 7.16). Associated with the claim that the appetite of the fly for either protein or carbohydrate is influenced by the neurosecretory and corpus allatum hormones respectively, it would seem that the activity of the corpus allatum may be *indirectly* influenced by the neurosecretory system. When *Calliphora* is forced to feed upon protein, the size of the corpus allatum increases in proportion to the amount taken, particularly in ovariectomized females (Fig. 7.16). When the yolk precursors are absorbed by the developing eggs, then the size of the corpus allatum decreases. A hypertrophied corpus allatum in ovariectomized females would consequently be the result of a con-

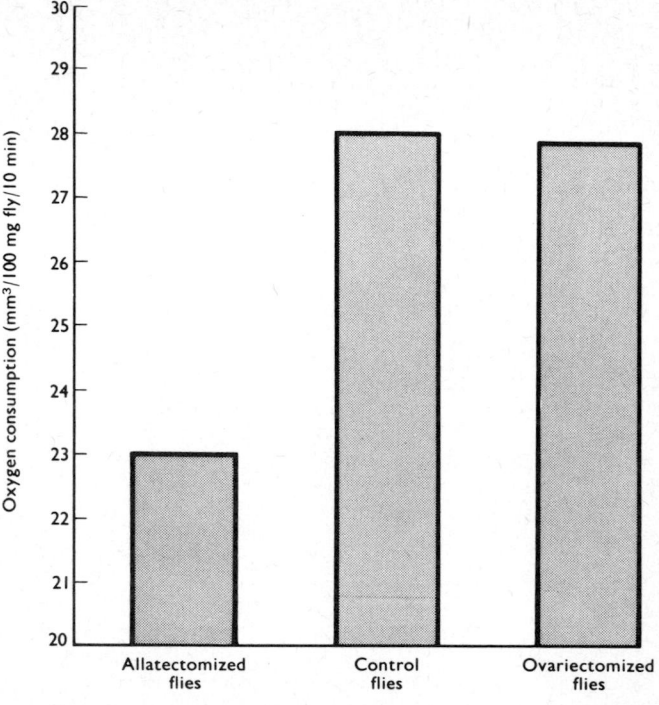

Fig. 7.15 Oxygen consumption in the blowfly, *Calliphora erythrocephala*, after allatectomy and ovariectomy. Allatectomy causes a significant decrease in oxygen consumption, whereas ovariectomy has little effect. This suggests that the corpus allatum hormone has a general metabolic effect in the adult female insect, and is not merely a gonadotrophic hormone affecting directly the development of the ovaries. (After Thomsen and Hamburger[262])

sistently high concentration of yolk precursors in the blood, because there is no tissue to remove them. This idea neatly explains the cyclic changes in volume of the corpus allatum associated with developmental cycles in the oocytes[252] (Fig. 7.17). It must be stressed, however, that the size of a corpus allatum is not necessarily related to its activity in terms of the production and release of hormone.[156a, 156b, 185g, 208b]

In *Calliphora*, the cerebral neurosecretory cells must be removed very shortly after adult emergence to inhibit previtellogenic growth. But in the fleshfly *Sarcophaga*, the operation can be delayed for as long as four days after adult emergence when the flies are held on sugar, or for one day after meat feeding, and oocyte development is prevented.[285b] It is therefore suggested that, in *Sarcophaga*, a neurosecretory hormone controls vitello-

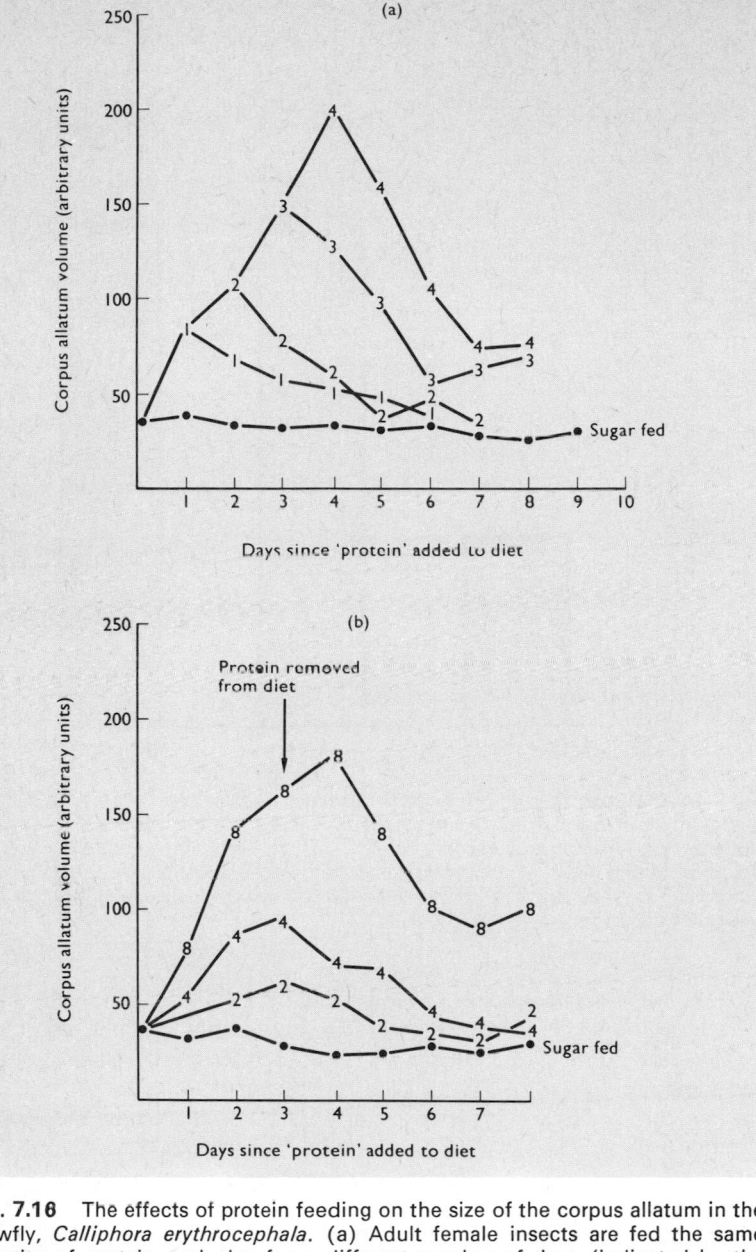

Fig. 7.16 The effects of protein feeding on the size of the corpus allatum in the blowfly, *Calliphora erythrocephala*. (a) Adult female insects are fed the same quantity of protein each day for a different number of days (indicated by the figures which constitute the points on the graph). The longer the time protein remains in the diet, the larger the corpus allatum becomes. In sugar fed flies, the gland remains uniformly small. (b) Flies fed for the same length of time upon different quantities of protein (indicated by the figures constituting the points on the graph). The greater the quantity of protein in the diet, the larger are the corpora allata. (After Strangways-Dixon[254])

genesis in the oocytes rather than previtellogenic growth (Fig. 7.12). The corpus allatum hormone is considered to control fatbody vitellogenin production in *Sarcophaga*,[86e] but the results of allatectomy are somewhat equivocal in this species: allatectomy within 6 hours of adult emergence still allows 70% of the operated animals to deposit yolk, and allatectomy of pupae allows *all* the subsequent adults to lay down yolk.[285a] The vitellogenic protein is present in all the allatectomized females which deposit yolk in the oocytes, and is also present in about 40% of the allatectomized females which do not deposit yolk.[285b] But when juvenile hormone is applied to allatectomized *Sarcophaga* females, the vitellogenic protein is produced in all.[86e] It is possible, therefore, that in *Sarcophaga* females the corpus allatum hormone is not absolutely essential for the production of the vitellogenic protein, or that other unknown factors contribute to its formation[86e] (Fig. 7.12).

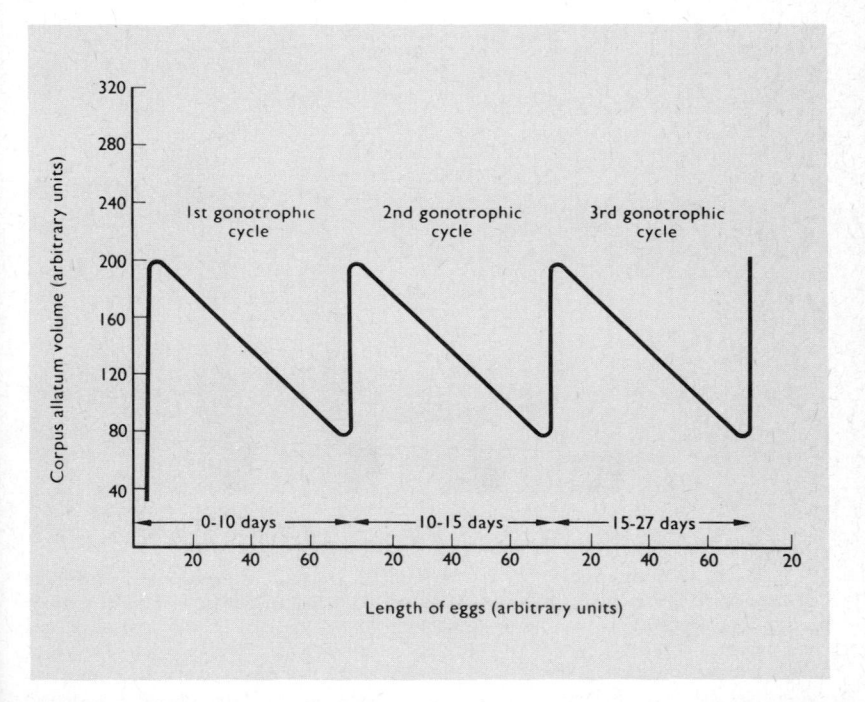

Fig. 7.17 Cyclic variation in the volume of the corpus allatum of the blowfly, *Calliphora erythrocephala*, during the first three gonotrophic cycles. The corpus allatum volume increases during the initial phase of yolk deposition and decreases as the oocytes grow, reaching a minimum value when the eggs are fully developed. (After Strangways-Dixon[252])

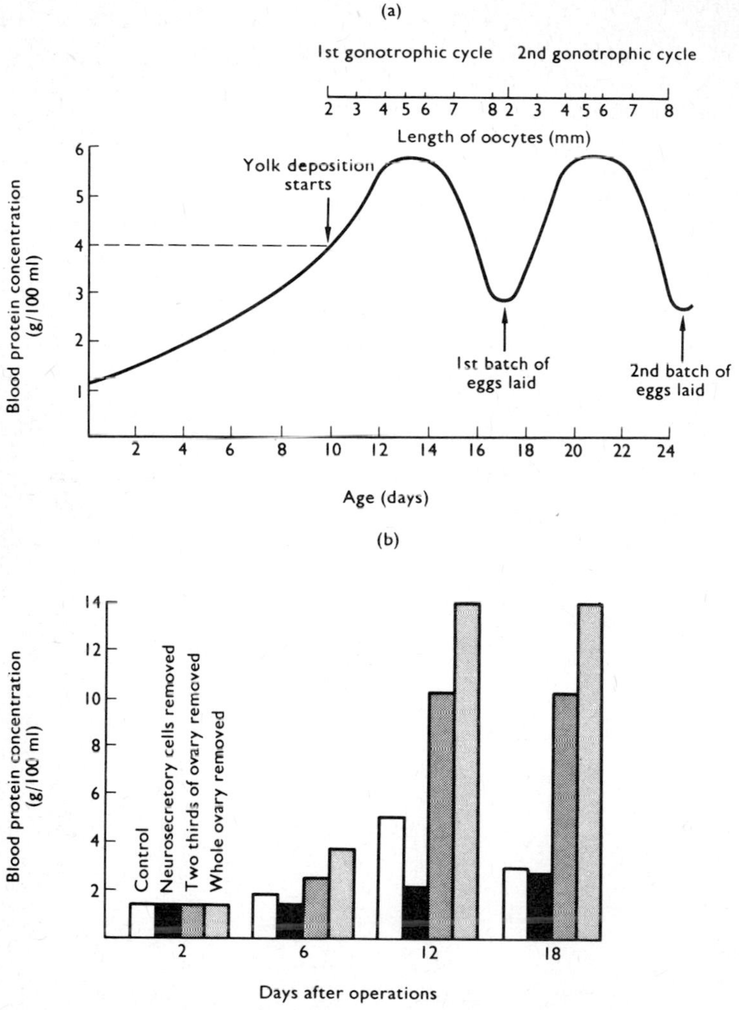

(a)

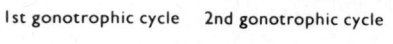

Fig. 7.18 Blood protein concentrations in the adult female desert locust, *Schistocerca gregaria*. (a) Variations during normal egg development. The blood protein concentration increases slowly during the first 8–10 days of adult life, and when it reaches 4 g/100 ml yolk deposition begins in the oocytes. The blood protein concentration continues to increase until the oocytes reach a length of 6 mm, and then decreases to a minimum as the fully developed eggs are ovulated. The pattern is repeated during the subsequent gonotrophic cycles. (b) Variations after cautery of the cerebral neurosecretory cells and after ovariectomy. After the neurosecretory cells are destroyed, the blood protein concentration remains low; after removal of at least two-thirds of the ovaries, the blood proteins increase to a concentration far above normal. (After Hill[144])

As in *Rhodnius* and the cockroaches, the dual role of the corpus allatum hormone in locusts has been clearly defined. In the desert locust, *Schistocerca gregaria*, the total blood protein concentration undergoes cyclic changes correlated with egg development (Fig. 7.18).[144] A female specific protein has been isolated from *Schistocerca gregaria*,[771] and immunoelectrophoresis of the blood of *Schistocerca vaga* shows that a female specific protein is present during egg development, but is absent before vitellogenesis begins. The female specific protein is identical with the major yolk protein in the eggs, and is therefore properly a vitellogenin; it is not produced in allatectomized females but appears after the topical application of juvenile hormone (Fig. 7.19).[86e] When radioactively labelled amino acids are injected into maturing females, labelled protein appears first in the fatbody, then in the blood and finally in the developing oocytes (Fig. 7.20),[145] indicating that the fatbody synthesizes the vitellogenic protein. This is supported by the massive increase in blood protein which occurs after ovariectomy (Fig. 7.18).

In the desert locust, removal of the corpora allata once vitellogenesis has begun results in rapid oosorption[142, 143] (Fig. 7.21). The effect of allatectomy upon the oocytes is obvious within two days; six days after the operation, all the oocytes have been resorbed (Fig. 7.21). This comparatively rapid effect of removal of the corpora allata suggests two things: first, the hormone has a short biological life in the blood, and secondly, because the oocytes respond quickly to its absence, or to its diminished concentration, it is likely that the hormone acts directly upon the oocytes, i.e. it is a gonadotrophic hormone. In some other insects, corpora allata implanted into the ovaries of allatectomized females will induce vitellogenesis only in those oocytes in close proximity to the implanted glands[157]—suggesting strongly a gonadotrophic function of the corpus allatum hormone.

The corpus allatum hormone in locusts clearly controls both the synthesis of the vitellogenic protein and its uptake by the developing oocytes (Fig. 7.11). But what is the role of the cerebral neurosecretory system in the reproductive process? Removal of the cerebral neurosecretory cells in the desert locust and in the Egyptian locust, *Anacridium aegyptium*, inhibits previtellogenic oocyte growth (Fig. 7.13), and the reimplantation of whole brains restores such growth.[133] But in addition, after the removal of the cerebral neurosecretory cells, the total blood protein concentration remains low, vitellogenic protein is not produced, and the protein synthetic activity of the fatbody is reduced (Figs. 7.18, 7.20). These effects must be partly due to the absence of juvenile hormone, since removal of the neurosecretory cells prevents the normal increase in activity of the corpora allata. But since the protein synthetic activity of the fatbody is lower after removal of the neurosecretory cells than after allatectomy, the neurosecre-

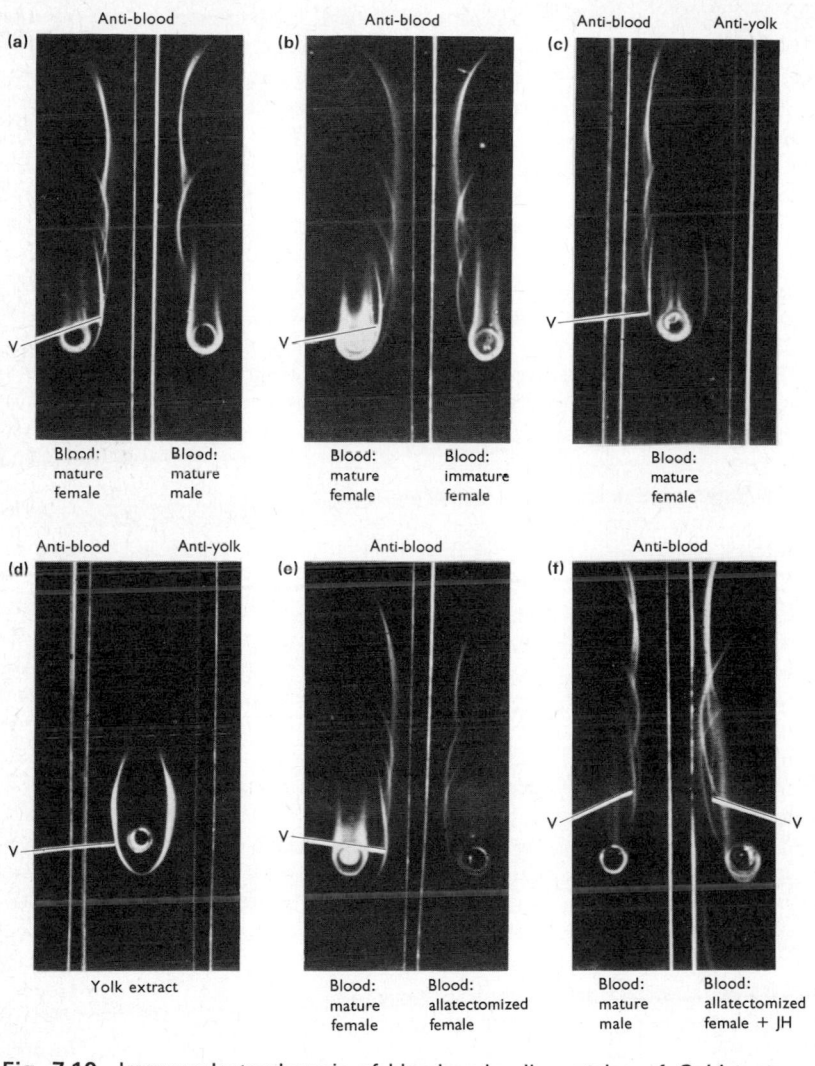

Fig. 7.19 Immunoelectrophoresis of blood and yolk proteins of *Schistocerca vaga*. The technique of immunoelectrophoresis consists of coating a microscope slide with agar gel, placing blood or yolk extracts in a circular hole in the gel, and separating the proteins by electrophoresis. Rabbit serum containing antibodies made by injecting blood from insects with developing eggs (anti-blood) or with yolk extracts (anti-yolk) is then placed in troughs in the gel. The separated proteins and the antibodies diffuse towards each other, and when they meet a precipitation arc is formed as antigen meets antibody. In this way, immunologically

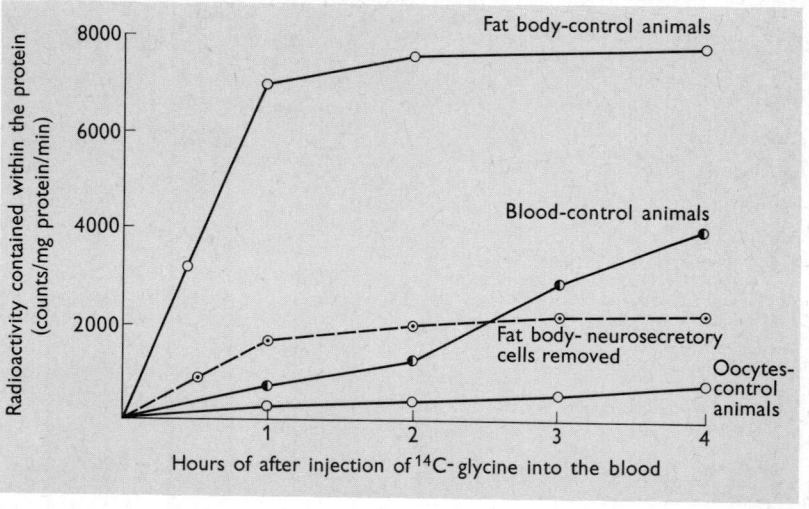

Fig. 7.20   Rates of incorporation of ^{14}C-labelled glycine into proteins of the fat body, blood and ovaries of the female desert locust. The rate of incorporation of amino acids into the proteins of the fatbody is used as a measure of the protein synthetic activity of the tissue. The high incorporation rate of control animals suggests a tissue actively synthesizing proteins, while the low rate in animals *minus* their cerebral neurosecretory cells suggests that little synthesis is taking place. Radioactivity does not appear in the blood proteins until after it has appeared in the fatbody proteins, and it finally appears in the ovarian proteins. This sequence suggests that proteins are synthesized in the fatbody, released into the blood, and then taken up from the blood by the ovaries. (After Hill[145])

tory hormone must be exerting a direct effect. A similar situation exists in *Locusta migratoria*,[112a] and perhaps the most likely explanation is that the neurosecretory system controls the overall fatbody synthesis of protein, and that the corpus allatum hormone 'switches' at least part of this synthetic activity into the production of vitellogenic protein[208] (Fig. 7.11). Allatectomy of the desert locust not only inhibits the synthesis of vitellogenic protein, but also reduces the concentrations of several other blood proteins (Fig. 7.22),[146a] which could be the consequence of the reduced activity of the neurosecretory system after allatectomy.[203a, 203b, 203c]

identical proteins can be demonstrated. (a) and (b) show the precipitation arc of vitellogenic protein, v, found only in female insects with developing eggs; it is absent in males and in immature females. (c) shows that the vitellogenin in the blood is identical with a component of the yolk protein; anti-yolk serum precipitates vitellogenin. (d) shows that yolk protein consists predominantly of a single protein identical with the vitellogenin in the blood. (e) shows that the vitellogenin is absent from the blood of allatectomized females. (f) shows that the injection of juvenile hormone into allatectomized females induces the appearance of vitellogenin in the blood.

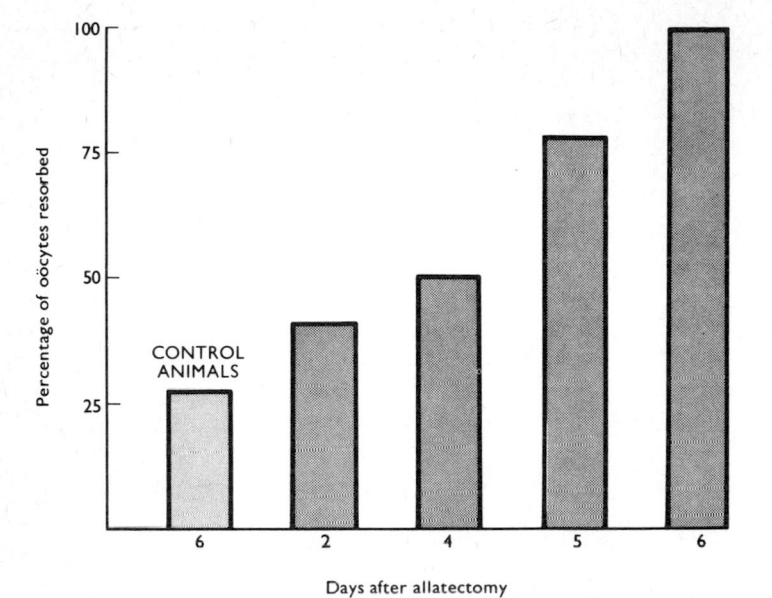

Fig. 7.21 Effects of removing corpora allata in mature female desert locusts, after the initiation of vitellogenesis. Yolk deposition in the oocytes ceases, and that already present is broken down and the oocyte resorbed. Six days after the operation all the developing oocytes in the ovaries have been resorbed. (After Highnam, Lusis and Hill[142])

In some insects, even under the most favourable conditions, some oosorption always occurs. Thus the desert locust, with about 50 ovarioles per ovary, could produce a maximum of 100 eggs in every gonadotrophic cycle. But in practice, about 22% of the oocytes are resorbed some time during their development, the greatest oosorption occurring during the later stages of vitellogenesis (Fig. 7.23). The haemolymph protein concentration begins to fall at this time, and it is possible that oosorption results from competition between the developing oocytes for available protein. When female desert locusts are *partially* ovariectomized, the remaining oocytes develop in conditions of increased protein 'availability'—in fact the haemolymph protein concentration can be very considerably increased (Fig. 7.18). Under these conditions, the percentage oosorption is reduced by about half, compared with normal animals (Fig. 7.23). But even with three-quarters of the ovaries removed, about 10% of the remaining oocytes are resorbed. Therefore, competition between oocytes for protein can account for only some of the oosorption which occurs in normal animals. But when such partially ovariectomized females are treated with corpus allatum hormone mimics (see p. 277),

then oosorption is completely eliminated (Fig. 7.23), and in fact the penultimate and even the antepenultimate oocytes in the ovarioles lay down yolk (Fig. 7.24). In unoperated females, the percentage oosorption is reduced—but not eliminated—by treatment with such corpus allatum hormone mimics. It is therefore concluded that in the desert locust, the

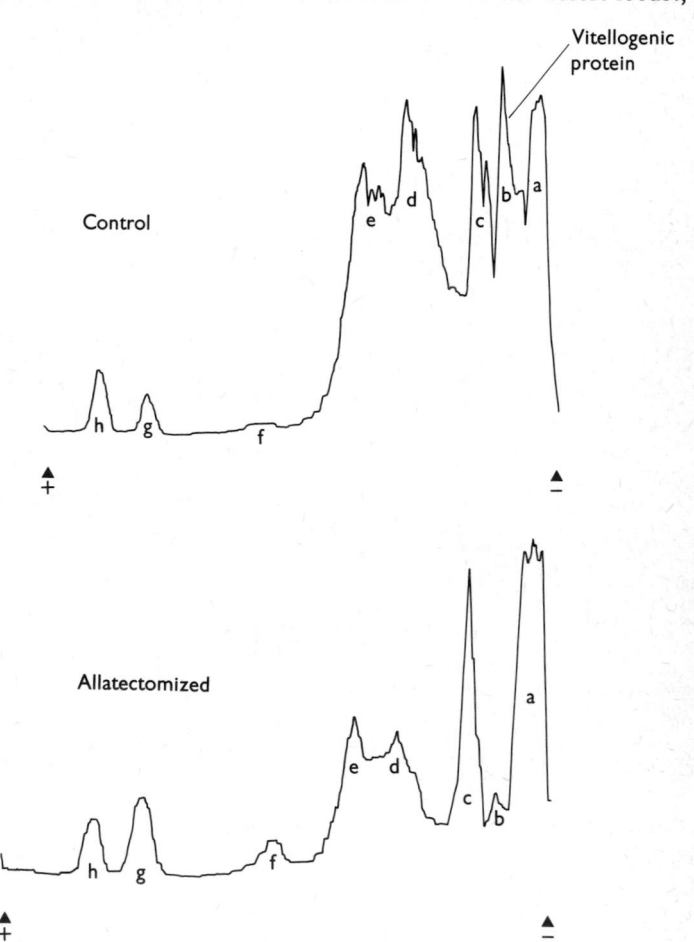

Fig. 7.22 The effects of allatectomy on the blood proteins of the adult female desert locust. The proteins are separated by gel electrophoresis, and after staining the densities of the different protein bands are measured. Each peak represents a protein band and the quantity of each protein is related to the height of its peak. In normal female locusts the vitellogenic protein is present in large quantity and is contained within band (b). In allatectomized female locusts, the vitellogenic protein is absent and band b is very much reduced. Bands (e) and (d) are also reduced in allatectomized females.

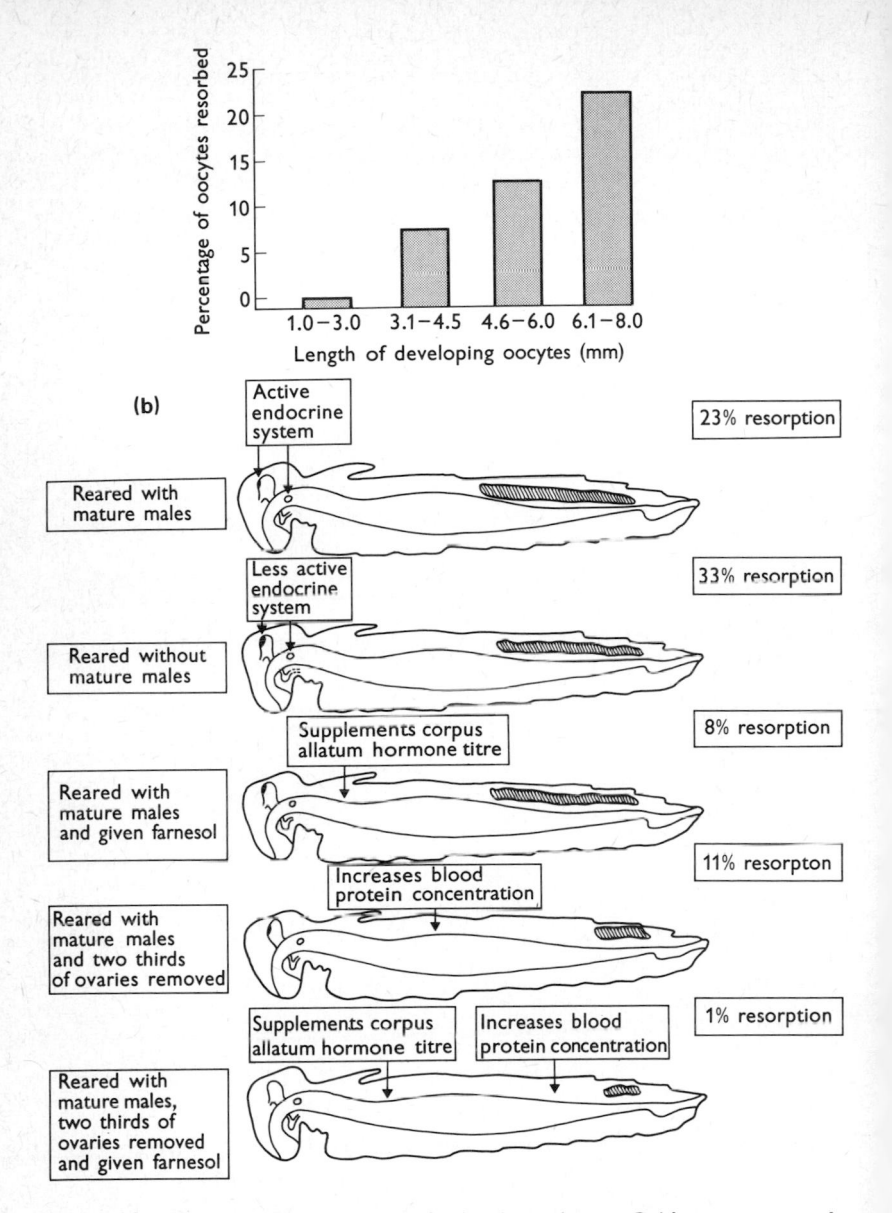

Fig. 7.23 The resorption of oocytes in the desert locust, *Schistocerca gregaria*. (a) A proportion of the oocytes are resorbed even during normal oocyte development: 8% during the early stages and a further 15% when the oocytes are between 4·5 and 8·0 mm in length. (b) If *either* vitellogenic proteins *or* corpus allatum hormone are made available in larger quantity, oocyte resorption is decreased; if *both* are increased oocyte resorption is virtually eliminated. It is concluded that oocyte resorption in the desert locust is due to competition by the developing oocytes for vitellogenic proteins and corpus allatum hormone.

Normal locust

Partially ovariectomized
locust plus farnesol

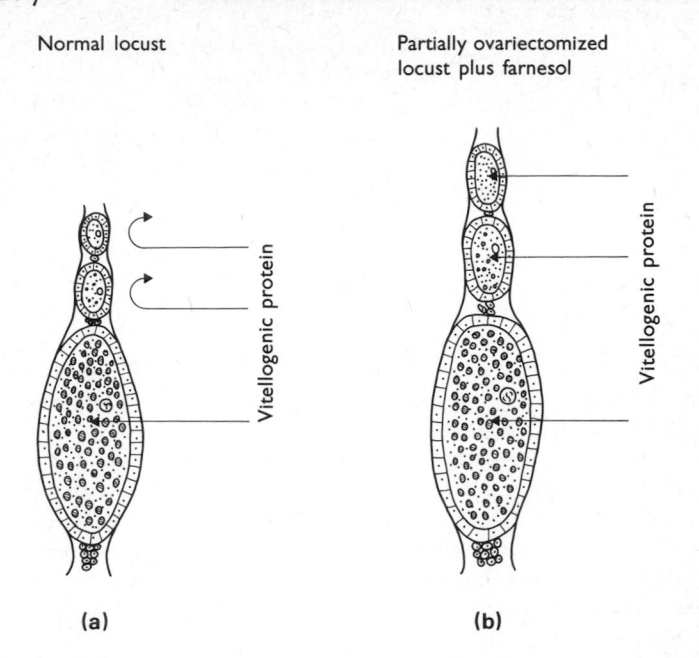

(a)   (b)

Fig. 7.24 Yolk deposition in the oocyte of the desert locust, *Schistocerca gregaria*. In the normal locust (a) yolk is deposited only in the terminal oocyte (that nearest the oviduct) in each ovariole. But in a female with two-thirds of her ovaries removed, with the juvenile hormone mimic farnesol added (b), not only the terminal oocytes but the penultimate and ante-penultimate oocytes also deposit yolk. In such females, the haemolymph protein concentration and juvenile hormone (or its mimic) concentration are both high and presumed competition for these materials by the developing oocytes is consequently greatly reduced. Compare Fig. 7.23.

oosorption which occurs normally is the result of oocyte competition for *both* protein and corpus allatum hormone.[143] Since the materials from a resorbed oocyte are presumably made available to the developing oocytes, this system has the obvious advantage of enabling a proportion of the oocytes to develop completely when conditions are not entirely favourable.

The Colorado beetle, *Leptinotarsa decemlineata*, enters a reproductive diapause (p. 180) when reared under short day conditions. The immediate cause of the reproductive arrest is the inactivity of the corpora allata— the females may be considered to be 'physiologically allatectomized'. Females reared under long day conditions, on the contrary, develop active corpora allata and produce eggs normally. The haemolymph of long day females contains a vitellogenic protein which is absent from that of short day females; the vitellogenin accumulates in the blood after ovariec-

tomy.[193a] It might therefore be supposed that the activa corpora allata of the long day females control the production of the vitellogenic protein. However, allatectomy of long day female beetles does not prevent the synthesis of the vitellogenic protein, although it does induce the production of other (non-vitellogenic) proteins which are characteristic of the short day females and males. Vitellogenin production in long day females is prevented when both the corpora cardiaca and the corpora allata are removed. But when corpora cardiaca from long day females are implanted into short day beetles, vitellogenin production is not initiated; nor is it induced in any large amount when juvenile hormone is injected into the short day females. But when corpora cardiaca from long day females are implanted into short day individuals, and the latter are also injected with juvenile hormone, the production of vitellogenic protein is markedly increased.[193b] Implanted brains have effects similar to those of implanted corpora cardiaca. Thus in the Colorado beetle, both neurosecretory and corpus allatum hormones are necessary for the production of the vitellogenic protein (Fig. 7.11). Juvenile hormone can also exert a localized effect upon the ovary, stimulating vitellogenesis on the injected side of the body to a greater extent than on the opposite side; the hormone is therefore likely to have a gonadotrophic effect in addition to its participation in the control of vitellogenin synthesis[193b] (Fig. 7.11).

In mosquitoes, both the corpus allatum and the cerebral neurosecretory system are again absolutely necessary for oocyte development,[185c, 185d] and it might be thought that the control mechanisms would follow one of the patterns already described. But the control of oocyte development in mosquitoes appears to be unique. Most species require a blood meal before vitellogenesis begins: the oocytes reach a resting stage after adult emergence, and development will continue only after the meal is taken. For allatectomy to be effective in preventing egg development, the operation has to be performed very shortly after adult emergence.[185e] Thus unlike the situation in *Rhodnius* (p. 141), it is not the blood meal itself which stimulates hormone production by the corpus allatum. In such allatectomized females, the early development of the oocytes stops short of the normal resting stage and the oocytes subsequently degenerate when the females feed on blood.[118m] In *Aedes aegypti*, if allatectomy is delayed until 3 days after adult emergence, or if early allatectomy is followed by the reimplantation of corpora allata or the topical application of juvenile hormone or its analogues, the oocytes develop normally to their resting stage and will subsequently complete their development when the female takes a blood meal.[118m] Thus in *Aedes aegypti*, the corpus allatum hormone controls previtellogenic oocyte development (Fig. 7.12), in contrast to the situation in *Calliphora* and the locusts, where a neurosecretory hormone is involved.

What, then, is the role of neurosecretion in controlling oocyte development in mosquitoes? In several species, removal of the median neurosecretory cells shortly after adult emergence prevents egg development, but if the operation is delayed, the eggs develop normally once the females take a blood meal.[185d] So the *production* of the neurosecretory hormone, like that of the corpus allatum hormone, is independent of blood feeding. However, if the corpora cardiaca are removed from females before blood feeding, the further development of the oocytes is arrested. This indicates that the neurosecretory hormone controlling egg development is synthesized by the cerebral neurosecretory cells for a period after adult emergence, and is transported to the corpora cardiaca where it is stored. When the female takes a blood meal, release of the hormone from the corpora cardiaca is stimulated. Release of the hormone can occur from denervated corpora cardiaca and from isolated transplanted glands, which suggests that the stimulus for release is carried in the haemolymph and may be either a specific chemical or perhaps a change in the haemolymph composition as the blood meal is digested and absorbed.[185f] Moreover, transplantation experiments show that the corpora cardiaca store sufficient of the neurosecretory hormone to mature several batches of eggs: release is metered in such a way that only sufficient hormone is allowed to circulate to mature a single batch of eggs.[185f]

Since the blood meal releases the neurosecretory hormone, and is followed by intensive vitellogenesis in the oocytes, it can be assumed that the hormone is in some way involved with the production of vitellogenic proteins. In *Aedes aegypti*, vitellogenins are produced by the fatbody,[120a, 120c] as they are in other insects (p. 145), but the neurosecretory hormone does not act directly upon the fatbody to control vitellogenin synthesis (Fig. 7.12). When *Aedes aegypti* females are ovariectomized shortly after a blood meal, no vitellogenins are produced; ovariectomy at progressively later times after blood feeding results in the decline of vitellogenin synthesis.[120b] Vitellogenin synthesis is restored when ovaries are transplanted into ovariectomized females, and will also occur when ovaries are incubated with fatbody *in vitro*; moreover, since vitellogenins are produced by fatbodies from *unfed* females incubated *in vitro* with ovaries from fed females, the initiation of vitellogenin synthesis as well as its maintenance depends upon the ovaries.[120b] Thus unlike other insects (p. 157), the synthesis of vitellogenic protein by the fatbody in mosquitoes does not take place in the absence of the ovaries (Fig. 7.12).

All these results point to the control of vitellogenin synthesis in the fatbody by a hormone—now called the vitellogenin stimulating hormone—produced by the ovary. The vitellogenin stimulating hormone is itself produced by the action of a hormone—the egg development neurosecretory hormone—released from the corpora cardiaca after the female has

	Longevity (days)	Egg yield (total number of eggs produced)
Control animals	88·3 ± 14·1	1263·4 ± 252·4
Animals without cerebral neuro-secretory cells	78·3 ± 12·0	573·5 ± 121·2

Fig. 7.25 Neurosecretion and reproduction in the milkweed bug, *Oncopeltus fasciatus*. Removal of the cerebral neurosecretory cells does not prevent egg development in this species, but the fecundity of the female is considerably reduced (After Johansson[155a])

taken a blood meal (Fig. 7.12). Vitellogenin synthesis in *unfed Aedes aegypti* females can be induced by the injection of ecdysone, and also by incubating fatbodies with ecdysone *in vitro*; it is suggested, therefore, that the vitellogenin stimulating hormone may in fact be ecdysone itself.[86g]

HORMONAL INTERACTIONS DURING OOCYTE DEVELOPMENT

In *Rhodnius prolixus*, cerebral neurosecretion plays no part in the control of oocyte development, and the corpus allatum hormone affects both the synthesis of vitellogenic protein and its uptake by the oocytes. But cerebral neurosecretion can control total egg production indirectly through the production of an ovarian antigonadotrophin. In some stick insects (Phasmidae), and the milkweed bug, *Oncopeltus fasciatus*, removal of the cerebral neurosecretory cells again has no immediate effect upon oocyte development, but the total fecundity of the operated females is considerably reduced (Fig. 2.75).[155a] In cockroaches, also, it is uncertain whether neurosecretory hormones play a direct role in oocyte development, but the corpus allatum hormone, as in *Rhodnius*, has both vitellogenin producing and gonadotrophic effects.

In other insects in which the control of oocyte development has been examined in detail, both neurosecretory and corpus allatum hormones are involved, but in a variety of ways. Thus in the blowfly *Calliphora erythrocephala*, and the locusts *Schistocerca gregaria* and *Anacridium aegyptium*,

cerebral neurosecretion affects previtellogenic growth of the oocytes, whereas in mosquitoes, this function is taken over by the corpus allatum hormone. In locusts, cerebral neurosecretion also controls overall protein metabolism by the fatbody, the corpus allatum hormone affecting vitellogenin production by 'switching' protein synthesis into its particular pathway. This resembles the situation in the Colorado beetle, *Leptinotarsa decemlineata*, where both neurosecretory and corpus allatum hormones are involved in vitellogenin production. In locusts, *Calliphora* and the Colorado beetle, the corpus allatum hormone retains its gonadotrophic function. In mosquitoes, cerebral neurosecretion also controls vitellogenin production by the fatbody, but uniquely by an indirect route involving the production of a special hormone by the ovary. Gonadotrophins have not been described in mosquitoes, although since the control of previtellogenic growth by the corpus allatum hormone terminates with the oocytes in a state ready to take up proteins, the hormone may also be considered to be gonadotrophic. In *Sarcophaga*, the cerebral neurosecretory hormone appears to be gonadotrophic while the corpus allatum hormone controls vitellogenin production. Finally, an antigonadotrophin (or oostatic hormone) occurs not only in *Rhodnius*, but in mosquitoes and other Diptera (pp. 174–175).

In those insects in which both cerebral neurosecretion and corpus allatum hormone combine to control the progress of oocyte development, the hormonal relationship is complicated by an interaction between the two endocrine centres. In the blowfly, the activity of the cerebral neurosecretory cells, as measured by their nuclear volumes, is enhanced by the presence of the corpus allatum hormone (p. 151, Fig. 7.14). In the desert locust, similarly, synthesis and release of neurosecretion is retarded after allatectomy, and accelerated when the glands are reimplanted or juvenile hormone or its mimics are injected.[203b, 203c] Moreover, in *Locusta migratoria* the activity of the corpora allata is said to be inhibited by the secretion from the A- and B-cells in the brain, and stimulated by the secretion from the C-cells[113] (Fig. 7.27). Such interactions between the cerebral neurosecretory system and the corpora allata obviously make it difficult to interpret the effects of removal and reimplantation of one or the other endocrine centre. Indeed, it is possible that some of the metabolic effects ascribed to the corpus allatum hormone may be the results of alterations in neurosecretory activity consequent upon allatectomy and reimplantation of the corpora allata. The effects of neurosecretory and corpus allatum hormones upon other metabolic and physiological processes may also affect reproduction in different ways in different species. The hyperglycaemic, adipokinetic, and diuretic hormones (pp. 192ff) are obvious candidates, but the role of juvenile hormone as a metabolic hormone may also extend beyond the control of vitellogenin production. Thus after

allatectomy of both male and female locusts (and other insects), abnormally large quantities of lipids accumulate in the fatbody. In intact animals, fatbody lipid synthesis decreases when the corpora allata become active and so no accumulation of lipids occurs. It is thus possible that the corpus allatum hormone acts both to stimulate fatbody vitellogenin synthesis and to inhibit fatbody lipid synthesis (Fig. 7.26). This can be viewed as a switch

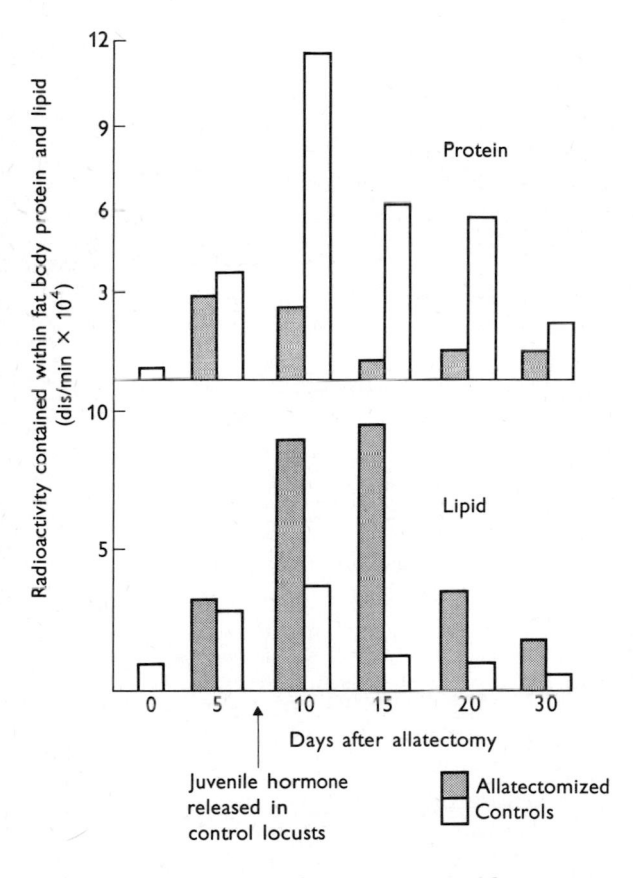

Fig. 7.26   Effects of allatectomy on the incorporation of ^{3}H-leucine into fatbody proteins and ^{14}C-acetate into fatbody lipids in the adult female desert locust. After allatectomy, the protein synthetic activity of the fatbody remains low, but in control animals it increases when the corpora allata become active. The lipid synthetic activity of the fatbody increases after allatectomy, and in control animals the release of juvenile hormone from the corpora allata prevents this increase. (After Hill and Izatt[146b])

mechanism to divert the metabolic resources of the insect away from the accumulation of energy reserves towards reproductive activity.[146b] The significance of this switch in the male insect is not clear, but little is yet known of the metabolic demands of reproduction in the male.

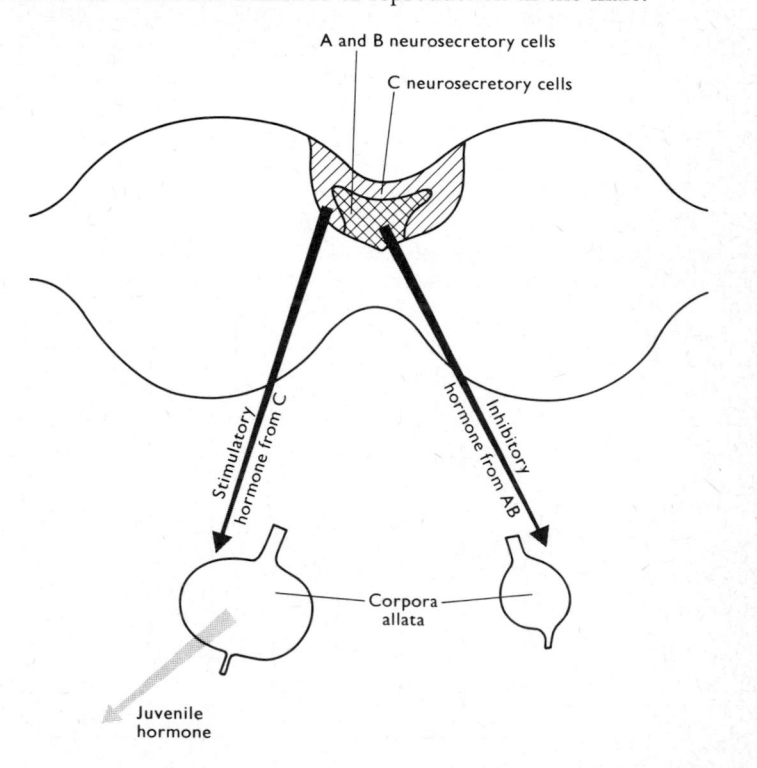

A and B neurosecretory cells

C neurosecretory cells

Stimulatory hormone from C

Inhibitory hormone from AB

Corpora allata

Juvenile hormone

Fig. 7.27 Possible control of the corpora allata by neurosecretory factors in the migratory locust, *Locusta migratoria*. The C neurosecretory cells in the brain stimulate the corpora allata to produce juvenile hormone, the A and B cells inhibit the glands. *Isolated* corpora allata can be stimulated or inhibited, so the neurosecretory factors do not necessarily pass along the nerves to the glands, but can be released into the haemolymph. These conclusions have been reached by selective destruction of the A, B and C cells: their arrangement in *Locusta* makes this possible. (After Girardie[112])

ENDOCRINE CONTROL OF ACCESSORY GLAND DEVELOPMENT

In many female insects, the accessory glands are small and thin in the newly emerged adult, increasing in diameter and length as the oocytes

develop and producing secretions which are poured onto the eggs during ovulation and oviposition. After allatectomy, the accessory glands remain small, only growing if corpora allata are subsequently reimplanted (Fig. 7.28).

But it must not immediately be assumed that the corpus allatum hormone directly affects accessory gland development. Growth of the accessory glands is associated with the development of the oocytes, and these do not develop after allatectomy. Consequently, the retarded accessory gland development after allatectomy *could* be due to the absence of a necessary factor from the undeveloped oocytes.

The obvious way to test whether the corpora allata control accessory gland development directly, or indirectly through the ovaries, is to ovariectomize the newly emerged females. In such females, the accessory glands develop more or less normally. Consequently, it is concluded that

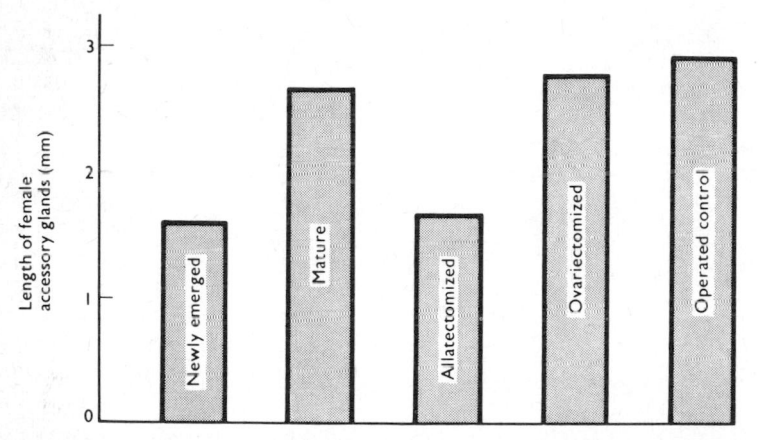

Fig. 7.28   Comparison of the lengths of the accessory glands of the blowfly, *Calliphora erythrocephala*, after allatectomy and ovariectomy. Allatectomy prevents the normal increase in length of the glands which accompanies maturation (see Fig. 7.7), but ovariectomy does not. Consequently, the ovaries themselves do not influence accessory gland size. The glands and ovarian development are controlled independently by the corpus allatum.

in insects the development of the accessory glands and the oocytes is correlated by the corpus allatum hormone, and not by the accessory glands being dependent upon ovarian secretions for their growth. This contrasts markedly with the situation in the vertebrates, where glandular and other changes in the reproductive tract are controlled by ovarian hormones. In fact, there is no indication that the insect ovary is a source of any hormone.

In some insects, the female accessory glands respond to a lower concentration of corpus allatum hormone than do the oocytes. Consequently as the corpora allata become active after adult emergence, the accessory glands are developmentally slightly in advance of the oocytes, ensuring that their secretions are present in quantity when the oocytes are ovulated.

CYCLIC DEVELOPMENT OF OOCYTES

In many insects, the terminal oocytes in all the ovarioles develop synchronously and all the eggs are laid at one time; after ovulation from the ovarioles, their places are taken by the previously penultimate oocytes, which continue development and are laid. The originally antepenultimate oocytes then become terminal and begin their development, and so on. Consequently, batches of eggs are laid at more or less regular intervals, and each laying represents the end of a particular gonotrophic cycle. After ovulation, the follicle cells which surrounded each oocyte contract to form a ring of tissue, often pigmented, around the base of each ovariole.[198] This ring of pigmented tissue is sometimes called the *corpus luteum*, although it must be clearly understood that it has no functional analogy whatsoever to the corpora lutea of vertebrates. The number of corpora lutea at the base of each ovariole can often be used to determine how many gonotrophic cycles have occurred (Fig. 7.29).

The volumes of the corpora allata, the size of their nuclei, and the volume of cytoplasm per nucleus fluctuate in relation to each gonotrophic cycle[252] (Fig. 7.18). These fluctuations are said to reflect varying secretory activity in the glands, the largest glands being most active. However, in some insects under certain conditions, it has recently been shown that the relationship between size and activity does not always hold true. A large corpus allatum, for example, could be *storing* hormone and not releasing it into the blood. Now that the corpus allatum hormone has been identified, it should be possible in the future to measure its concentration in the blood directly.

Variations in the activity of the cerebral neurosecretory system are also associated with each gonotrophic cycle. The criteria used to determine activity are again largely histological, depending upon the amounts of stainable material present in the neurosecretory system at different times. Neurosecretions are synthesized by the cells in the brain, transported along the neurosecretory cell axons and stored for a longer or shorter period of time in the corpora cardiaca. Clearly, the content of neurosecretion within the system at any time will depend upon its relative rates of synthesis, transport and release, which can all vary independently.

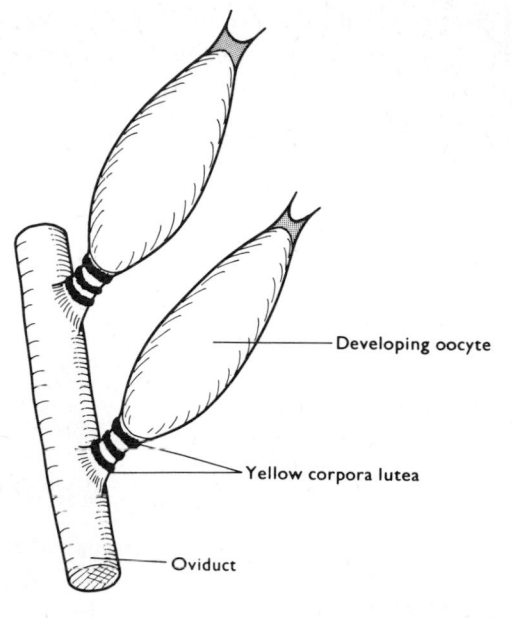

Fig. 7.29   Diagram of the corpora lutea at the bases of two ovarioles in the desert locust, *Schistocerca gregaria*. Three gonotrophic cycles have been completed, and the developing oocytes therefore represent the fourth cycle.

Histological pictures from material fixed at an instant in time are consequently of little use in estimating variations in the essentially dynamic process of turnover of neurosecretion [133, 136] (Fig. 7.30).

Neurosecretory proteins are rich in the sulphur containing amino acids cystine and cysteine. Consequently, radioactively-labelled S-amino acids injected into the blood, are incorporated preferentially into neurosecretory proteins. The rate of incorporation of the amino acids by the cerebral neurosecretory cells, and their rate of release from the corpora cardiaca, can be measured by autoradiographic and other means, and the *relative* activity of the neurosecretory system can then be estimated at different times during the gonotrophic cycle. [133, 137, 143a, 143b, 208a]

What causes these cyclic changes in the cerebral neurosecretory system and the corpora allata? A large number of factors influence the activity of the cerebral neurosecretory system (see Chapter 14) amongst which feeding and mating have marked stimulatory effects. The first gonotrophic cycle in the desert locust could therefore be initiated and maintained by the voracious feeding in which the females indulge after emergence, together with copulation and other factors (Fig. 14.4) which

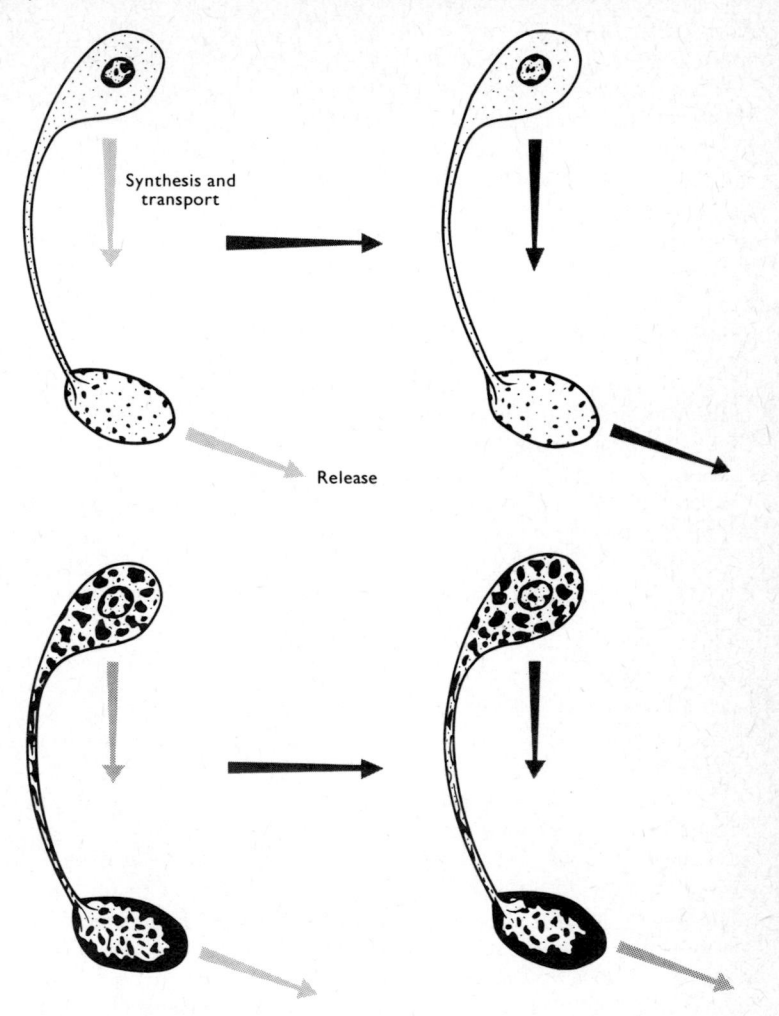

Fig. 7.30 Histological differences between insect neurosecretory systems (diagrammatic). The density of shading of the broad arrows represents degrees of synthesis and transport, and release of neurosecretion.
Upper row : the neurosecretory systems contain relatively little histologically demonstrable material, but synthesis and release of neurosecretion can increase considerably without altering the appearance of the system so long as they remain in balance.
Lower row : the systems are full of material, but again synthesis and release of neurosecretion can vary widely without altering the histological appearance so long as synthesis exceeds release. (After Highnam[136])

combine to stimulate the production and release of neurosecretory hormones. Because of their relationship with the cerebral neurosecretory system, activation of the corpora allata would follow. In the blowfly, meat feeding is necessary for the release of material from the cerebral neurosecretory cells (p. 151; Fig. 7.14). In *Rhodnius*, a blood meal activates the corpus allatum, although whether this is direct or mediated through the neurosecretory system is not known.

But what causes the decline in activity of the parts of the endocrine system towards the end of a gonotrophic cycle? At present, the answer to this question is unknown. In some insects, ovariectomy causes hypertrophy of the corpora allata, either because the glands now store their hormone or because they are hyperactive—striving to achieve a result which is no longer possible. Ovariectomy can also inhibit the release of neurosecretion from the corpora cardiaca. It is possible that the increasing concentrations of haemolymph metabolites which follow ovariectomy in some way act back on the source of the metabolic hormone which is responsible for their production. A similar situation arises in normal animals when the oocytes have completed vitellogenesis and no longer abstract metabolites from the haemolymph at their previous rate. It is also possible that in some insects the *removal* of metabolites from the blood will itself affect endocrine activity. But it is not inconceivable that nervous impulses from an abdomen stretched by its content of large oocytes could affect the endocrine system. In the viviparous cockroach, *Leucophaea maderae*, the presence of developing young in the 'uterus' stimulates sensory receptors whose signals eventually inhibit the activity of the corpora allata.[85, 86]

By analogy with the situation in the vertebrates, the production of a hormone by the developing oocytes which reduces the activity of the cerebral endocrine system might seem the obvious way to achieve the known result.[212] In the housefly, *Musca domestica*, a factor can be extracted from mature females which retards ovarian development when injected into young individuals;[1c] it is suggested that this 'oostatic hormone' prevents the release of juvenile hormone from the corpus allatum.[1b] In *Lucilia cuprina*, a mature terminal oocyte in an ovariole prevents further development of the penultimate oocyte either directly or *via* the cerebral endocrine system.[61a] Mosquitoes retain their mature eggs until they find a suitable site for oviposition, and may take a blood meal if oviposition is delayed. If this meal matured another batch of eggs, the female could become too heavy to fly. Consequently, the processes which normally follow a blood meal (p. 165) are interrupted, and penultimate oocytes are prevented from maturing until the terminal oocytes have been laid[84b, 185f, 206b] (Fig. 7.12). The inhibiting factor from the mature oocytes could prevent release of the egg development neurosecretory hormone from the

corpora cardiaca, or could suppress vitellogenin synthesis by the fatbody, or even inhibit the production of vitellogenin stimulating hormone by the ovaries.[120a] The last possibility is unlikely, since in *Aedes aegypti*, a second injection of ecdysone fails to restimulate or prolong the high rates of vitellogenin synthesis induced by an initial injection of the hormone.[86g] The ovarian inhibiting factor is thus most likely to act upon the fatbody to arrest vitellogenin synthesis. This is certainly not so in *Rhodnius prolixus*, for *in vitro* experiments with antigonadotrophin extracts have shown conclusively that the factor acts more or less locally to antagonize the action of juvenile hormone in opening up spaces between the follicle cells[75a, 151d] (Fig. 7.11). Such an effect is unlikely in the locust, *Schistocerca gregaria*, since penultimate and even antepenultimate oocytes can lay down yolk in partially ovariectomized females (Fig. 7.24). The antigonadotrophic factor, where it exists, must operate in different ways in different insect species.

In the desert locust, the act of egg-laying itself stimulates the cerebral neurosecretory system once more into activity, the maintenance of this activity subsequently depending upon continued feeding and perhaps other factors. Thus the terminal process of one gonotrophic cycle triggers off the beginning of the next. It is likely that neurosecretory hormones released during oviposition facilitate the oviducal and other movements necessary for the movement of the eggs out of the body. In *Rhodnius*, neurosecretory factors are also necessary for egg-laying[75]; the process has not been examined in detail in any other insects.

ENDOCRINE CONTROL OF REPRODUCTION IN THE MALE INSECT

The endocrine control of oocyte development applies strictly to previtellogenic growth and vitellogenesis—oogenesis within the germarium is not affected by hormones. It is not surprising, therefore, that in the adult male insect spermatogenesis proceeds independently of any hormonal influence, although differentiation of the testes, including the development of spermatids from spermatogonia, proceeds rapidly during the last moult. However, in male Saturniid moths, a macromolecular factor, probably produced by particular blood cells, is necessary for meiosis and spermatogenesis within the germinal cysts of the testes[158j] (Fig. 7.31). But although this factor is present in the haemolymph of diapausing pupae, spermatogenesis is not initiated until ecdysone is released by the prothoracic glands, or is introduced experimentally. In other words, *intact* testes require the prior action of ecdysone before the macromolecular factor can operate.[158k] Radioactively labelled ecdysone is selectively bound to the

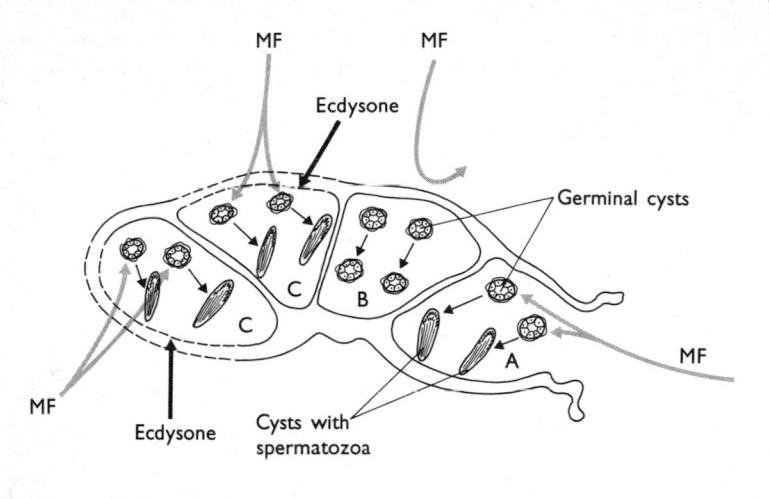

Fig. 7.31 The action of ecdysone and the macromolecular factor (MF) produced by blood cells during spermatogenesis in *Samia cynthia*. The diagram represents a cross section of the testis containing germinal cysts (not to scale). A indicates the situation where the testis wall has been ruptured and MF has free access to the germinal cysts : the primary spermatocytes undergo meiosis and metamorphose into spermatozoa. When the testis wall is intact (B), MF has no effect upon the germinal cysts. But if ecdysone is present just before, or at the same time as, MF (C), meiosis and the production of spermatozoa proceeds normally : ecdysone makes the walls of the testis permeable to MF.

testis walls, presumably to specific receptors, and even brief exposure to the hormone has long-lasting effects when testes are subsequently incubated in a medium containing the macromolecular factor.[1581] The evidence points strongly to ecdysone making the testis walls permeable to the macromolecular factor, allowing the factor to penetrate inside the testis and exert its effect upon the germinal cysts. This view of the action of ecdysone accords well with what is known of its effects upon other tissues during metamorphosis (p. 129). This process is part of the general metamorphic transformation of the body, and it is likely, therefore, that it is controlled by the same combination of hormones that effects metamorphosis.

Accessory glands in the male insect secrete protein and other materials which form the spermatophore, in those species where it occurs, and also the bulk of the seminal plasma which carries the sperm. In many insects growth and development of the male accessory glands are controlled by the corpus allatum hormone in an exactly similar manner to that of the female accessory glands (p. 169*ff*).

8

Endocrine Mechanisms in the Insecta—III

Other Endocrine-controlled Processes

HORMONES AND DIAPAUSE

Many insects avoid unfavourable periods by entering diapause, a state of arrested development. The most familiar examples are the over-wintering chrysalids, or pupae, of many British butterflies and moths, but diapause may also occur in the egg, in the larval stages and also in the adult. Usually only one diapause intervenes during the life of an individual, but species with more than one diapause in their life-cycles are not unknown.

Diapause is characterized by a considerably diminished metabolism, by the presence of large reserves of fat, sometimes associated with protein and carbohydrate, by a marked resistance to desiccation (important particularly for aestivating insects in arid tropical regions) and by extreme cold-resistance, reflected in an ability to withstand low temperatures (from 0° to −10°) that are usually lethal to an insect, and even temperatures as low as −40°. Diapause must therefore be anticipated by the building-up of special reserves, and the formation of protective devices such as wax layers to minimize evaporation. Consequently, it is easily distinguished from the torpor induced by low or high temperatures or desiccation, which cause pathological changes resulting in death.

In insects with a long period of post-embryonic growth, lasting a year or more, diapause may intervene at the correct time regardless of environmental conditions. Such diapause is termed **obligatory** and is normally associated with univoltine species, i.e. those having one generation a year. In polyvoltine species, which have two or more generations a year, diapause occurs only in that generation which spans the unfavourable period. Thus the cabbage white butterfly, *Pieris brassicae*, lays eggs in early summer, larvae emerge from these and develop directly into adults through several instars and through a pupal stage. But the eggs from these adults hatch into larvae which develop later in the summer and produce a diapausing pupa which overwinters. Such a diapause is called *facultative*, since it depends for its appearance upon the appropriate environmental conditions.

Many hypotheses have been proposed to account for the intervention of diapause in insect life-histories, the most popular at one time being self-poisoning or auto-intoxication of the insect by a gradual accumulation of metabolic waste materials. But it is now known that post embryonic diapause results from the temporary cessation of activity of those parts of the endocrine system which control growth and development (Chapter 6).

In the pupa of the giant silkmoth, *Hyalophora cecropia*, diapause can be terminated by a prolonged period of chilling, for example 6 weeks at 3°C, followed by 2 weeks at a higher temperature such as 25°C.[286] The period of adverse conditions is the time required for specific physiological changes to occur, which make possible the recommencement of normal development. It is therefore called the period of diapause development and is characteristic of most insect diapauses.[2]

If the brain of *Hyalophora* is removed at any time during pupal diapause development, then the decerebrate individual will not terminate diapause. But if the brain from an individual chilled for 6 weeks is implanted into an unchilled diapausing pupa, then development is eventually resumed in the normal way. Thus the effect of chilling is solely upon the brain, which develops a competence to terminate diapause.

If a chilled brain is implanted into the isolated pupal abdomen of *Hyalophora*, diapause is not terminated (Fig. 8.1). But if the chilled brain is implanted into a pupa containing the thoracic glands, or if the glands are implanted together with the chilled brain then normal development will ensue[287] (Fig. 8.1). Clearly, diapause is due to failure of the cerebral neurosecretory cells to produce thoracotrophic hormone after pupation in *Hyalophora*, the absence of ecdysone preventing further development. Convincing evidence for this theory is that implanted activated thoracic glands will terminate diapause in pupae of *Hyalophora*, as will injections of pure ecdysone. During diapause development in

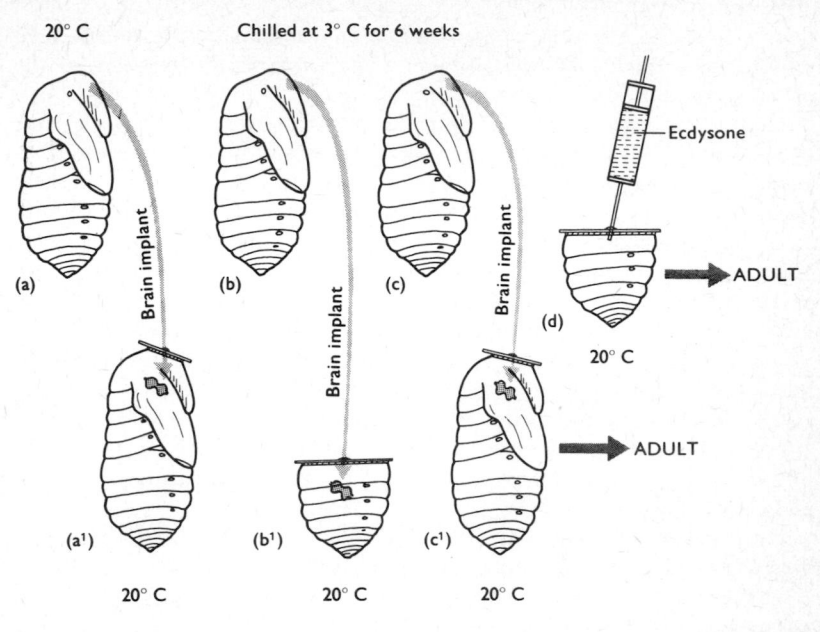

Fig. 8.1 Termination of pupal diapause in *Hyalophora cecropia*. The brain from an unchilled pupa (a) implanted into an unchilled pupa (a¹) does not end diapause and initiate development. A brain from a chilled pupa (c) will terminate diapause when implanted into an unchilled individual (c¹). But a chilled brain (b) will not terminate diapause when implanted into an unchilled isolated abdomen (b¹), but ecdysone injected into such an abdomen (d) will terminate diapause and produce an adult abdomen. These results suggest that the effect of chilling is upon the brain, which subsequently releases thoracotrophic hormone to stimulate the thoracic glands to produce ecdysone. (After Williams[287])

Hyalophora, therefore, the cerebral neurosecretory cells initiate the production of thoracotrophic hormone which exerts its effect when higher temperatures prevail.

It is likely that the majority of both pupal and larval diapauses result from the failure of the cerebral neurosecretory cells to produce thoracotrophic hormone. But adult diapause, characterized by lack of development of the gonads, and particularly the ovaries, often occurs without any associated quiescence of the parents and can be due to other factors. Thus the water beetles *Dytiscus marginalis* and *Dytiscus semisulcatus* enter reproductive diapause in March and October respectively. During diapause, the oocytes with their accompanying nurse cells surrounded by follicle cells, continually pass down the ovarioles. But when the

oocyte is about 1–2 mm in diameter, it is resorbed by the follicle cells which then form a corpus luteum. The same thing happens to the oocyte next in line in each ovariole.[157]

Oosorption during diapause in these water beetles occurs at the time the oocytes lose connection with their nurse cells. This, of course, mimics exactly the effects of allatectomy upon other insects with meroistic ovarioles (p. 141), and it might be suspected that reproductive diapause in *Dytiscus* was the result of failure of the corpora allata to secrete its gonadotrophic hormone. In fact, the implantation of 10 pairs of active corpora allata into the diapausing beetles induces normal egg development and oviposition, and allatectomy of non-diapausing *Dytiscus* results in oosorption rather similar to that which occurs during diapause, except that eventually the ovarioles themselves regress. Reproductive diapause in *Dytiscus* is thus due to the relative inactivity of the corpora allata,[157] although because the changes in the ovaries during diapause are not as pronounced as those following allatectomy, the glands cannot be *completely* inactive.

During reproductive diapause in the Colorado beetle, *Leptinotarsa decemlineata*, the ovaries show changes which mimic exactly the effects of allatectomy, and again the inactivity of the corpora allata must be of importance in controlling diapause.[285] But in this species, the implantation of corpora allata into the diapausing females will only induce full oocyte development for a short time after the induction of diapause. If the environmental conditions which bring about diapause are continued, brain implants from active individuals are necessary to break diapause. In the Colorado beetle, therefore, the onset of diapause is brought about by a rapid inhibition of the corpora allata, but the diapause is strengthened and maintained by the subsequent inactivity of the cerebral neurosecretory cells. In the bug *Pyrrhocoris apterus*, reproductive diapause is similarly maintained by the absence of circulating neurosecretory and corpus allatum hormones.[247]

During the reproductive diapause in the Egyptian grasshopper, *Anacridium aegyptium*, which has panoistic ovarioles, there is no turnover of oocytes—development to a certain stage followed by oosorption—like that found in species with meroistic ovarioles (see above). Instead, the oocytes remain small and undeveloped and never show signs of vitellogenesis. This condition again recalls the different effects of allatectomy in insects with panoistic or meroistic ovarioles, and in *Anacridium* inactivation of the corpora allata is certainly one aspect of reproductive diapause. But the very small oocytes of *Anacridium* during diapause resemble more closely the oocytes of non-diapausing females which have had their cerebral neurosecretory cells removed, than those of allatectomized non-diapausing females in which the oocytes are much larger

although still lacking yolk.[104] In *Anacridium*, therefore, failure of the cerebral neurosecretory cells is likely to be the primary cause of reproductive diapause, inactivity of the corpora allata being a secondary consequence.

Postembryonic diapause in insects is thus due to the *absence* of developmental hormones. But embryonic diapause, on the contrary, is induced by the *presence* of an inhibitory hormone, secreted by the female and affecting her developing oocytes.

Embryonic diapause is facultative in many races of the silkmoth, *Bombyx mori*. According to how they are reared (see below) adults will lay either diapause or non-diapause eggs, i.e., in which embryonic development is interrupted at an early stage or is direct. In *Bombyx*, oocyte development begins in the pupal stage, and this is when the diapause hormone acts.[97] One of the first functions of the hormone is to cause the accumulation of 3-hydroxykynurenine, the precursor of the ommochrome pigment which makes eggs in which embryonic diapause is to occur easily distinguishable from those in which development is direct.

When the brain is removed from diapause-producing pupae 60 hours after pupation, only about 17% of the eggs eventually produced enter diapause. This proportion increases progressively the longer brain removal is delayed after pupation. Removal of the brain shortly after pupation in non-diapause producing pupae causes about 57% of the eggs eventually laid to *enter* diapause. This proportion diminishes progressively the longer brain removal is delayed.

Removal of the sub-oesophageal ganglion from diapause producing pupae shortly after pupation completely eliminates diapause from the eggs eventually produced; when the operation is delayed, a progressively larger proportion of eggs enter diapause. The same operation on non-diapause producing pupae has no apparent effect; all the eggs are without diapause just as they would have been if the ganglion had not been removed.

How can these results be explained? The effects of brain removal from a diapause producing pupa might suggest that the brain is the source of the diapause hormone, but this cannot be so because the same operation on non-diapause producing pupae causes some of the eggs to enter diapause. When brains from either diapause producing or non-diapause producing pupae are transplanted into either kind of pupa, they are absolutely without effect. But the lack of diapause eggs after removing the sub-oesophageal ganglion suggests that this is the source of the diapause hormone in diapause producing pupae, and that it is absent from non-diapause producing pupae. This is confirmed by transplanting sub-oesophageal ganglia from diapause producing to non-diapause producing pupae: a large proportion of the eggs laid enter diapause (Fig.

8.2). In fact, the diapause hormone is secreted by a single pair of large neurosecretory cells in the sub-oesophageal ganglion.[97]

But the brain is clearly involved also in diapause induction. A clue to its function is given by the experiment described above: when a sub-oesophageal ganglion from a diapause producing pupa is transplanted to a non-diapause producing individual, not *all* the eggs produced enter diapause. Moreover, when a sub-oesophageal ganglion from a non-diapause producing pupa is transplanted to a similar non-diapause producer, then a proportion of eggs *enter* diapause as though the ganglion

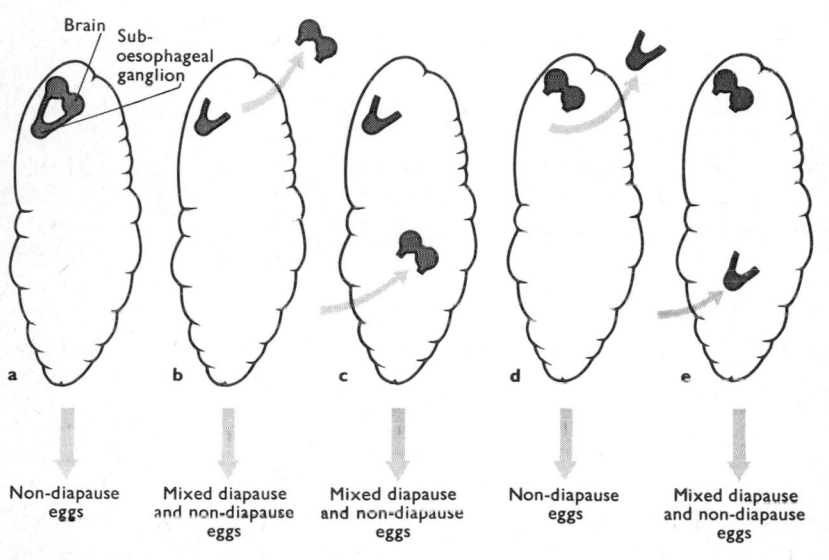

Fig. 8.2(a) The effects of brain and sub-oesophageal ganglion removal and implantation in pupae of a non-diapause producing generation of *Bombyx mori*. (a) Control pupae, brain and ganglion *in situ*: no eggs eventually produced have diapause. (b) Brain removed: although no diapause eggs are normally produced, when the brain is removed early in pupal life, mixed diapause and non-diapause eggs are produced. (c) When the brain is reimplanted to another site, mixed diapause and non-diapause eggs are still formed. The connections between the brain and the ganglion must be intact, as in a, for all non-diapause eggs to be formed. (d) Sub-oesophageal ganglion removed: all non-diapause eggs produced. (e) Sub-oesophageal ganglion reimplanted to a different site: mixed diapause and non-diapause eggs produced as in c.

Thus a proportion of diapause eggs are formed in the non-diapause pupae only when the sub-oesophageal ganglion is present either without the brain, or when the connectives with the brain are severed. This suggests that the source of the diapause factor is the sub-oesophageal ganglion, but that normally its release is prevented by the brain *so long as the nervous connections between the two are intact.*

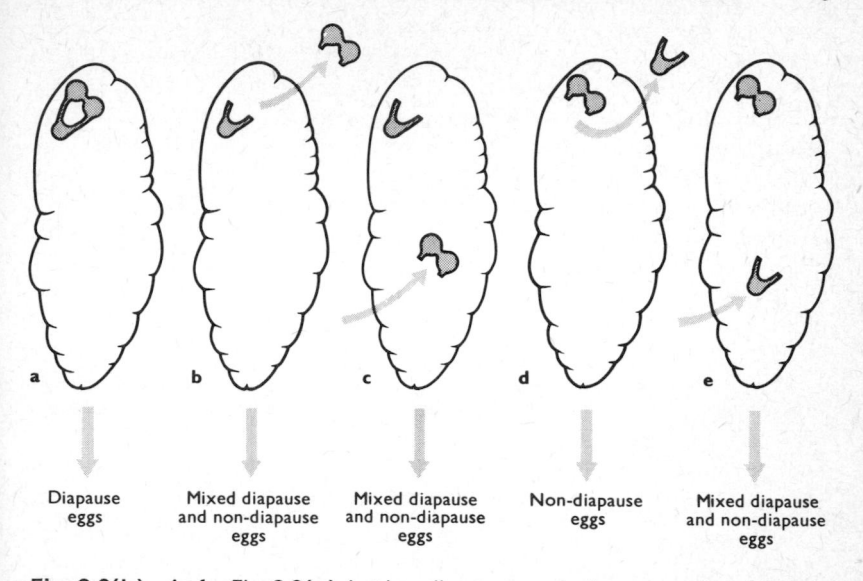

| Diapause eggs | Mixed diapause and non-diapause eggs | Mixed diapause and non-diapause eggs | Non-diapause eggs | Mixed diapause and non-diapause eggs |

Fig. 8.2(b)   As for Fig. 8.2(a), but in a diapause producing generation. (a) Brain and sub-oesophageal ganglion intact (control): all diapause eggs produced. (b) Brain removed: a proportion of non-diapause eggs produced (compare Fig. 8.2). (c) Brain reimplanted to a different site: a proportion of non-diapause eggs produced. (d) Sub-oesophageal ganglion removed: non-diapause eggs produced. (e) Ganglion reimplanted to a different site: a proportion of non-diapause eggs produced.

Thus again it may be concluded that the sub-oesophageal ganglion is the source of a diapause factor, but that it produces its maximal effect *only when the connectives between ganglion and brain are intact.* In effect, the brain stimulates the production of a diapause factor by the sub-oesophageal ganglion. (After Fukuda[97])

came from a diapause producer (Fig. 8.2). The function of the brain is now clear: in diapause producers the brain stimulates the production and/or release of hormone from the sub-oesophageal ganglion; in non-diapause producers the brain inhibits hormone production and/or release from the sub-oesophageal ganglion[97] (Fig. 8.3). Merely cutting the circumoesophageal connectives between the brain and sub-oesophageal ganglion will produce mixed diapause and non-diapause eggs in both diapause producing and non-diapause producing pupae, because stimulation of hormone production is interrupted in the former, and inhibition removed in the latter. The brain will still exert its stimulatory or inhibitory effect upon the sub-oesophageal ganglion when transplanted into other individuals, so long as the connectives remain intact. A brain and a

sub-oesophageal ganglion implanted separately act as if the ganglion alone were implanted, producing mixed diapause and non-diapause eggs.

The nature of the control exercised by the brain upon the neuro-secretory cells of the sub-oesophageal ganglion is unknown. It is pre-sumably nervous, from some endogenous centre, because the brain can act in the absence of any afferent input—as for example, when it is transplanted—since all its nerves are necessarily severed during such an operation. But an endocrine control, perhaps by neurosecretory axons travelling directly to the sub-oesophageal ganglion neurosecretory cells from some centre in the brain, cannot be entirely ruled out, although no trace of such axons has yet been found.

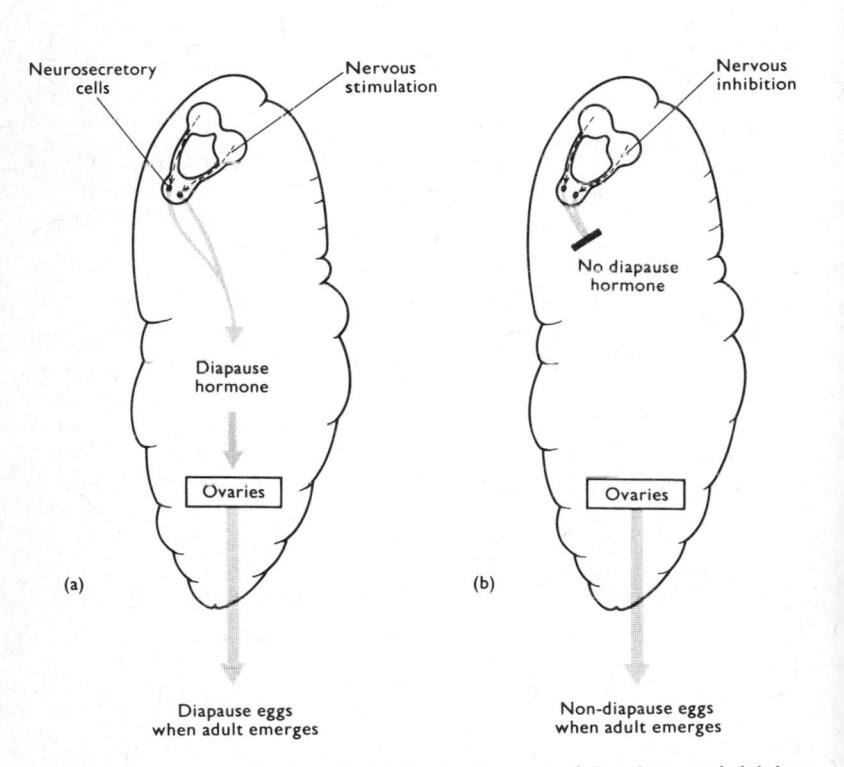

Fig. 8.3   Summary of diapause induction in the eggs of *Bombyx mori*. (a) In a diapause producing pupa, the brain stimulates the production of diapause hormone by neurosecretory cells in the sub-oesophageal ganglion. The hormone is absorbed by the developing eggs, eventually to cause diapause when the eggs are laid. (b) In a non-diapause producing pupa, the brain inhibits the sub-oesophageal ganglion neurosecretory cells from producing diapause hormone, and in its absence the eggs develop without diapause. (After Fukuda [97])

The diapause hormone is probably a lipoprotein (p. 264), and one of its major functions is to elevate trehalase activity in the ovaries, possibly by *de novo* synthesis of the enzyme.[293a] Trehalase increases the synthesis of glycogen in the ovaries at the expense of blood trehalose and fatbody glycogen, which are consequently reduced to low levels in diapause producing pupae. Although such glycogen accumulation in diapause eggs may be considered an essential prerequisite for a long period of dormancy, it is not necessarily the *cause* of diapause. A particular isozyme of esterase (esterase A) has been implicated in the growth processes of the eggs of many animals: in diapausing eggs of *Bombyx*, esterase A activity is masked or inhibited in some way, and this inhibition is another consequence of the presence of the diapause hormone.[158f, 158g, 158h] When yolk cells from diapause eggs are cultured in an artificial medium with yolk material from eggs after the termination of diapause, the cell membranes are disrupted and yolk lysis occurs; the same happens when the yolk cells are incubated with esterase A.[158i] The unmasking or disinhibition of esterase A during the termination of diapause would thus make nutrients available to the embryo through the effect of the enzyme upon the yolk, and thus development is re-initiated. How the diapause hormone acts initially to inhibit esterase A is unknown.

What determines whether *Bombyx* pupae will be diapause producers or non-diapause producers? The most important factor turns out to be the temperature at which the eggs are incubated. When the embryos develop at a temperature of about 15°C, then the emergent larvae and their pupae are non-diapause producers—the adult laying eggs without diapause. But incubating the eggs at 25°C gives diapause producing larvae and pupae—the adults will lay all diapause eggs. Incubation at 20°C produces three kinds of larvae and pupae—exclusively diapause producers, exclusively non-diapause producers, and also individuals which will eventually produce mixed diapause and non-diapause eggs. The proportions of these three types vary with the conditions of larval rearing. But most important, long periods of light during egg incubation at 20°C increase the proportion of diapause producers; their absence increases the proportion of non-diapause producers.

The significance of this mechanism for a bivoltine species in nature is obvious (Fig. 8.4). Embryonic development in the low temperatures and short photoperiods of early spring produces eventually adults which lay non-diapause eggs. These develop directly into the second generation, which takes advantage of the favourable conditions of summer. But the embryos in the non-diapause eggs develop in the high temperatures and long photoperiods of summer—and therefore the adults eventually produced will lay diapause eggs to avoid the unfavourable winter. What is less obvious, of course, is how the incubation temperature of the embryo

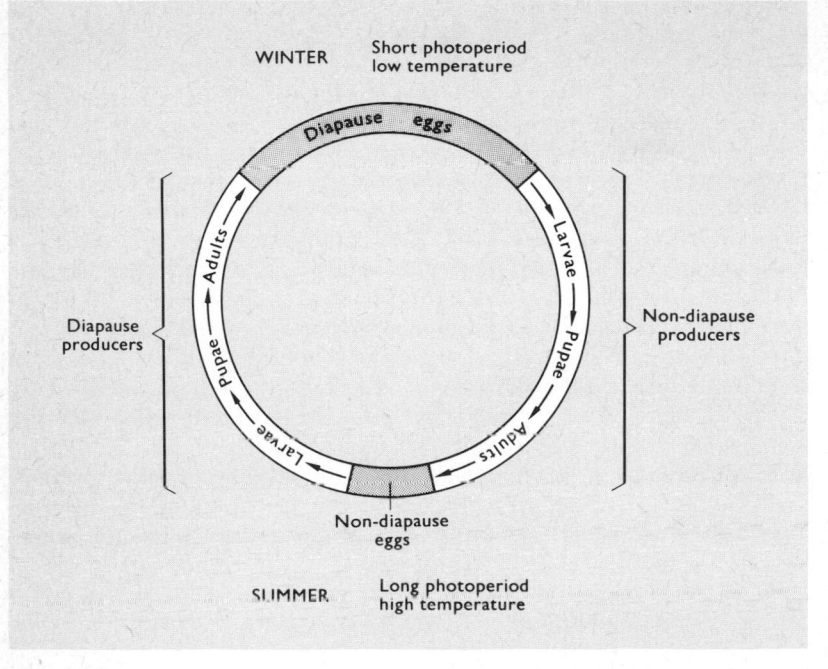

Fig. 8.4 Life cycle of a bivoltine race of *Bombyx mori* with egg diapause. Environmental conditions for the embryos developing in the summer induce diapause in the winter eggs ; those for winter embryos prevent diapause in the summer eggs.

affects an endogenous nervous centre in the brain which then either stimulates or inhibits the production of a neurosecretory hormone in the pupa.

Although *Bombyx* is rather unusual in that the conditions which determine diapause act upon a stage almost an entire generation before development is arrested, yet it still illustrates an important principle: that when an unfavourable period is avoided by diapause, prior preparation has to be made, and consequently diapause is determined by conditions which invariably precede the unfavourable period and not by the period itself.

Diapause can be triggered by a number of environmental factors, including temperature and the quality or quantity of food, which vary seasonally.[187] But these factors are intrinsically variable, and are not very good indicators of season. But one environmental factor varies predictably during the course of the year: this factor is photoperiod, the

number of hours of light in 24 hours. It is not surprising that many insects react to particular photoperiods which determine the onset of diapause (Fig. 8.5). Even so, the action of photoperiod is often closely linked with temperature, a particular day length inducing diapause only when the temperature lies between rather close limits[67] (Fig. 8.6).

In many insects, diapause results from exposure of the sensitive stages to *short* photoperiods (Fig. 8.5). *Bombyx* is exceptional in that long photoperiods induce diapause. High temperatures usually avert diapause, while low temperatures induce it (Fig. 8.6), although *Bombyx* (see above) is again exceptional in this respect.

The stage in the life-history which is sensitive to these factors which induce diapause is usually fixed and characteristic for each species (Fig. 8.7). In *Bombyx*, the egg is most sensitive to photoperiod, although the early larval instars also show some sensitivity. In the giant silkmoths

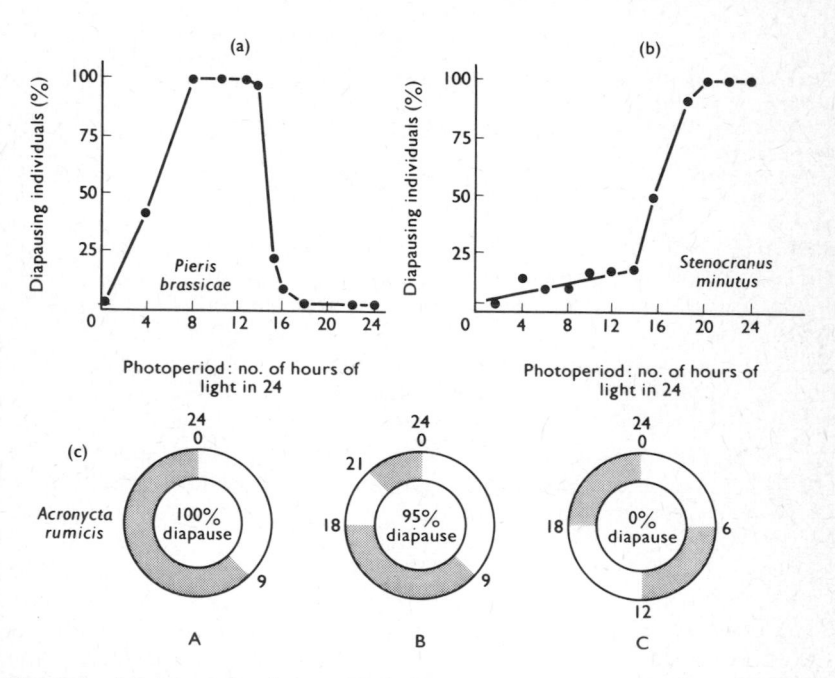

Fig. 8.5 (a) A long day insect, *Pieris brassicae*, in which short day conditions induce diapause. (b) A short day insect, *Stenocranis minutus*, in which long days induce diapause. (c) In the noctuid, *Acronycta rumicis*, pupal diapause is induced by short photoperiods (A), but the dark phase must last for at least 9 *continuous* hours, or not all individuals enter diapause (B). When a 12 hour dark period is interrupted by 6 hours light, diapause is prevented (C). (a and b after Danilevskii[67]; c after Lees[187])

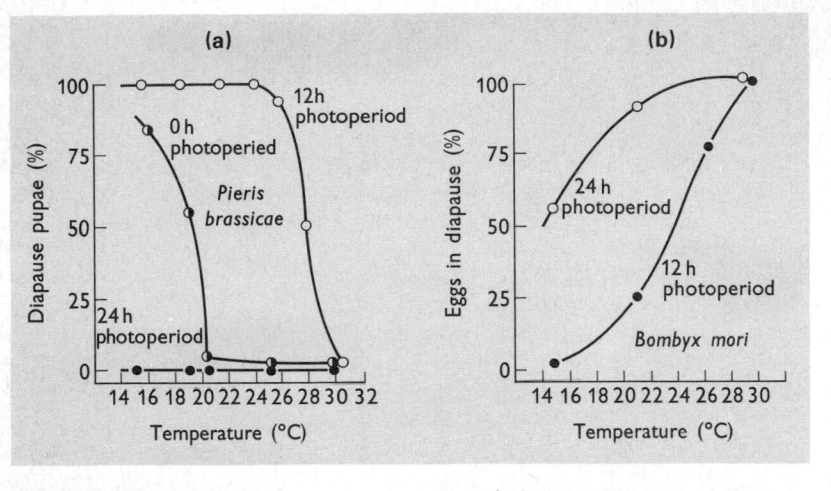

Fig. 8.6 Effects of temperature and photoperiod upon diapause. (a) In the pupa of *Pieris brassicae* when reared at different temperatures and three light regimes. Increasing temperature opposes short photoperiods in inducing diapause. (b) Egg diapause in *Bombyx mori* reared at different temperatures under two light regimes. Increasing temperature *supplements* the longer photoperiod in inducing diapause.

Telea polyphemus and *Philosamia cynthia*, only the 4th and 5th instars react to the short photoperiods which induce pupal diapause. In *Anthereae pernyi*, all the larval stages are sensitive to some extent, but the later ones are more so and the effects are cumulative. In the tomato moth *Diataraxia oleracea*, the most sensitive period is no more than 2 days at the end of the penultimate larval instar. Unexpectedly, the compound eyes and ocelli are usually not the receptor sites for photoperiod: the epidermis, the brain or even the cerebral neurosecretory cells themselves may react to the stimuli.

When *Pieris brassicae* larvae are reared in a 6 hours light: 18 hours dark regime, all the subsequent pupae enter diapause. But when the dark period is interrupted by 2 hours of light at different times, the incidence of diapause is very much reduced if the light interruption comes 16 hours after the beginning of the primary photoperiod (Fig. 8.8)—in effect, the insects react as if they were in long days. This suggests that there is a light-sensitive period, the photo-inducible phase, at a particular time during the daily cycle. Many such experiments on different species have revealed the presence of one, and sometimes two, photo-inducible phases which occur at different times in different insects. The way in which this cycle works in inducing or preventing diapause is now clear: if the photoperiod overlaps

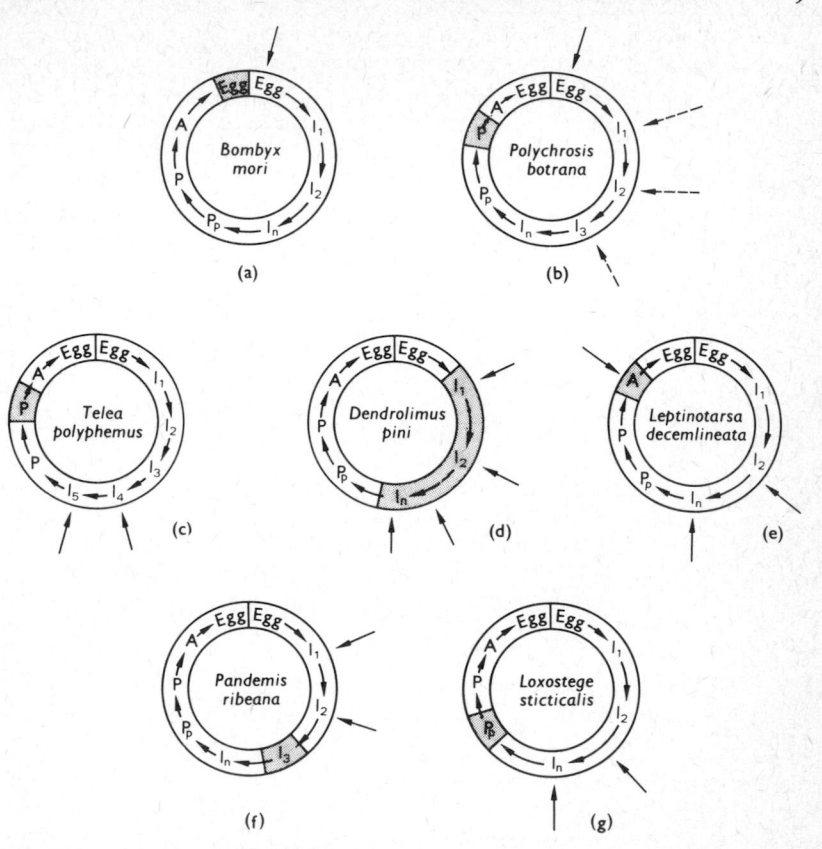

Fig. 8.7 Sensitive stages (shown by arrows) in the life histories of various insects during which environmental stimuli (particularly photoperiod) induce diapause (stippled). In *Bombyx mori* (a) almost a whole generation intervenes between the sensitive stage and the diapause stage. Usually the diapause stage follows the sensitive stage(s) (b, c, e, f, g) although they may be the same (d).

the peak of the photo-inducible phase, diapause is prevented; but if the photoperiod is so short that the photo-inducible phase comes in darkness, then diapause is induced (Fig. 8.9).

The photosensitive diapause mechanism is only one aspect of the functioning of an endogenous rhythm, or internal clock, common not only to insects, but to almost all animals and plants, with a periodicity of about 24 hours. Called a circadian (*circa*: about, *diem*: day) rhythm, it manifests itself in many overt behavioural or physiological phenomena which recur about every 24 hours. The rhythm persists for some time in continuous

light or darkness before it becomes irregular, suggesting that the 'clock' is reset normally by dawn and/or dusk or, in experimental situations, by 'lights on' and 'lights off'. The nature of the internal clock is uncertain. It may be associated with a sequence of biochemical events which requires a period of light for its initiation, and probably a period of darkness for its completion. The clock may be cyclic, or it may be an interval measurer like an hour-glass. Since both devices have become adapted to the normal

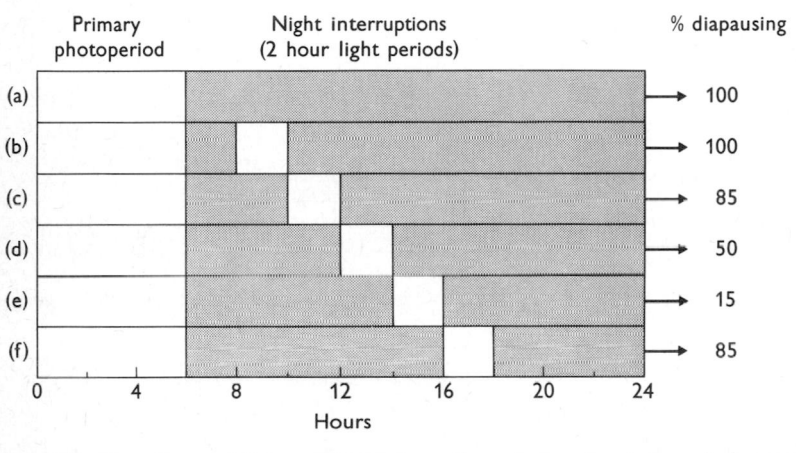

Fig. 8.8 The effects of 2 hour light interruptions during the dark period upon diapause in *Pieris brassicae*, reared under short day conditions (6 hour primary photoperiod). When larvae are reared under a 6 hours light : 18 hours dark regime, all the subsequent pupae enter diapause (a) ; when a 2 hour light interruption begins 12 hours after the commencement of the primary photoperiod, the incidence of pupal diapause is reduced by 50% (d) ; diapause is almost eliminated when the light interruption is 14–16 hours after the commencement of the primary photoperiod (e) ; light interruptions before or after these periods have no or little effect upon the incidence of diapause (b, c, f).

twenty four hour cycle of light and dark on the earth over evolutionary time, it is very difficult to distinguish between them in any species.

Photoperiod is the major factor inducing diapause in many insects; temperature is of the greatest importance in its termination. The effect of low temperature in diapause development in *Hyalophora cecropia* has already been described (p. 178). In *Bombyx*, development recommences at 22·5°C if the diapausing eggs experience 80 days at a temperature of 2·5°C, about 65 days at 12·5°C or 160 days at 15°C. Diapause is not broken even after long periods at 0°C or 17·5°C. The temperature of 5°C is therefore optimal for diapause development.

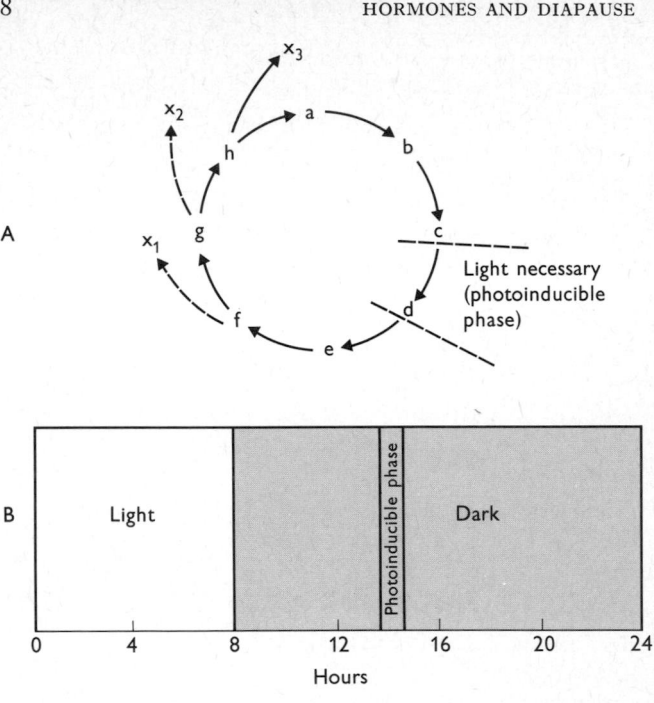

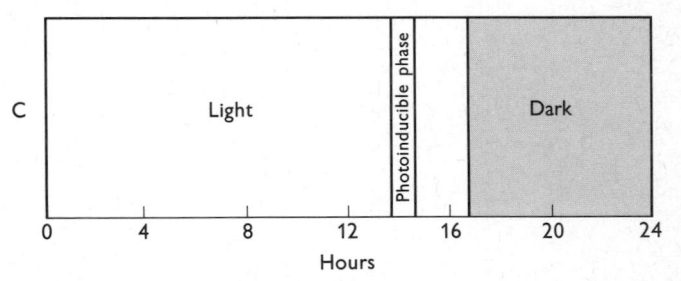

Fig. 8.9 A simple theoretical model for the photoperiodic control of diapause in insects. In (A), a series of (biochemical ?) events leads to the production of factor x (or series of factors x_1, x_2, x_3) necessary for direct development or the prevention of diapause. The sequence takes about 24 hours to complete and it is assumed that at least one of the steps is light-dependent—the photoinducible phase. During short days (B), the photoinducible phase falls during the night, the sequence is therefore not 'driven', factor x is not produced, and diapause results. During long days (C), the photoperiod overlaps the photoinducible phase, the sequence proceeds and diapause is prevented. It is also possible that a dark period of a particular length, following the photoinducible phase, is necessary for some of the subsequent steps in the sequence.

The optimum temperature for diapause development is closely related to the environment in which the insect lives.[67] Species in cold climates usually require low temperatures: in the emperor moth, *Saturnia pavonina*, it lies between -15 and $+7°C$. Most insects in temperate regions have optima between o and $7°C$. But in insects which aestivate to avoid hot, dry summers, the optimum temperatures for diapause development are correspondingly high—as much as $35°C$ for the diapausing eggs of the South African brown locust. In general, those conditions which the diapause is intended to avoid are largely instrumental in bringing about its eventual termination.

Environmental factors which switch off the endocrine system are often quite different from those which switch it on again. In postembryonic diapauses, the cerebral neurosecretory cells play a central part in the insect's reaction to these exogenous stimuli. Little is known of the internal pathways which transmit these stimuli to the cells (see Chapter 14). It is quite likely that the neurosecretory cells, or neurones in close proximity, react to photoperiod and temperature,[1g, 290, 291] although how they can go through a number of moults producing thoracotrophic hormone normally before they shut down is unknown. The possibility that the cerebral neurosecretory cells react to the presence of another hormone produced by the hindgut[8e] or by the sub-oesophageal ganglion[2a] rests upon rather equivocal experimental evidence. Still more mysterious is the way in which environmental factors can terminate egg diapause, which often occurs before the embryonic endocrine centres are established. Possibly the egg tissues have a general sensitivity to such stimuli, a sensitivity which becomes confined and limited to special cells in the postembryonic stages.

HORMONES AND DIURESIS

Terrestrial insects usually have to conserve water, and it is now axiomatic that the cuticular wax layer, spiracular closing mechanisms and uricotelic nitrogen excretion are all adaptations to this end. But circumstances often arise when the insect is presented with an excessive water load—taken in with the food, for example—which must be quickly disposed of if the blood is not to be so diluted that vital functions are impaired.

In blood-sucking insects such as *Rhodnius prolixus*, large quantities of unnecessary fluid are imbibed with each meal, which with other waste materials are excreted through the malpighian tubules. An *in vitro* preparation of malpighian tubules, together with a piece of attached rectum, will function normally for a long time in a physiological saline covered

with liquid paraffin. The rectal cuticle has an affinity for the paraffin, and consequently the rectum opens out onto the liquid paraffin layer[199] (Fig. 8.10). Droplets of fluid from the malpighian tubules pass out onto the paraffin through the rectum, and therefore the rate of excretion of the malpighian tubules can be measured, together with the total volume of fluid filtered through the tubules (Fig. 8.11).

Malpighian tubules bathed in haemolymph from a fed *Rhodnius* produce a much more copious urine in a shorter time than when they are exposed to haemolymph from an unfed bug (Fig. 8.12). This immediately suggests that diuresis is accelerated by a hormone present in the fed

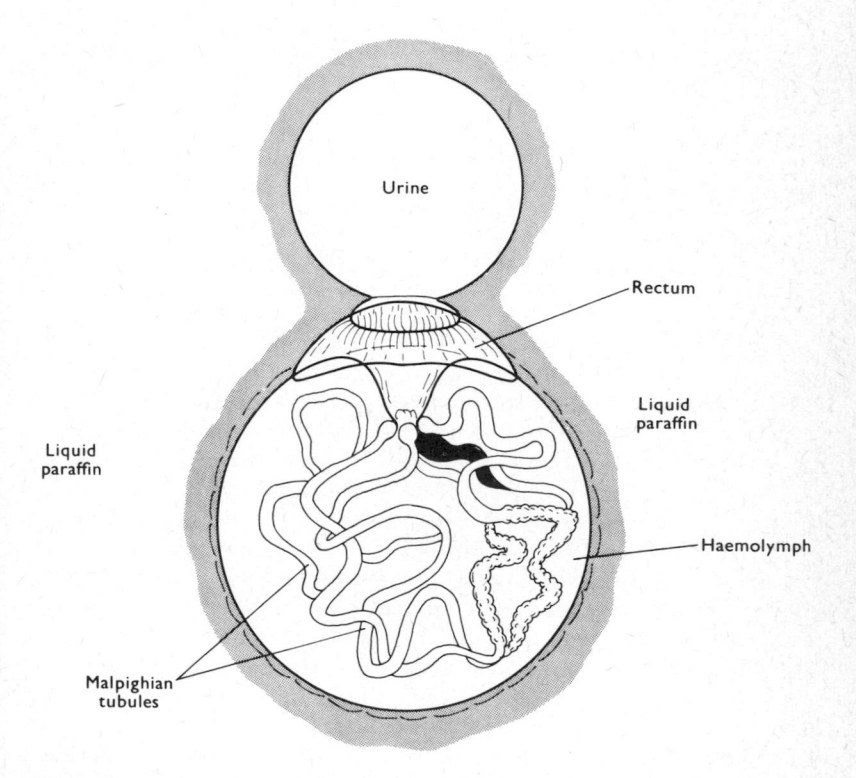

Fig. 8.10 A preparation of malpighian tubules of *Rhodnius prolixus* in liquid paraffin. The tubules, with part of the rectum attached, lie within a bubble of haemolymph in the liquid paraffin. The cuticular lining of the rectum, with an affinity for the paraffin, causes the rectum to spread out along the interface between haemolymph and paraffin. Urine produced by the malpighian tubules forms a droplet where it leaves the rectal lumen. The diameter and hence volume of this droplet can be measured periodically. (After Maddrell[199])

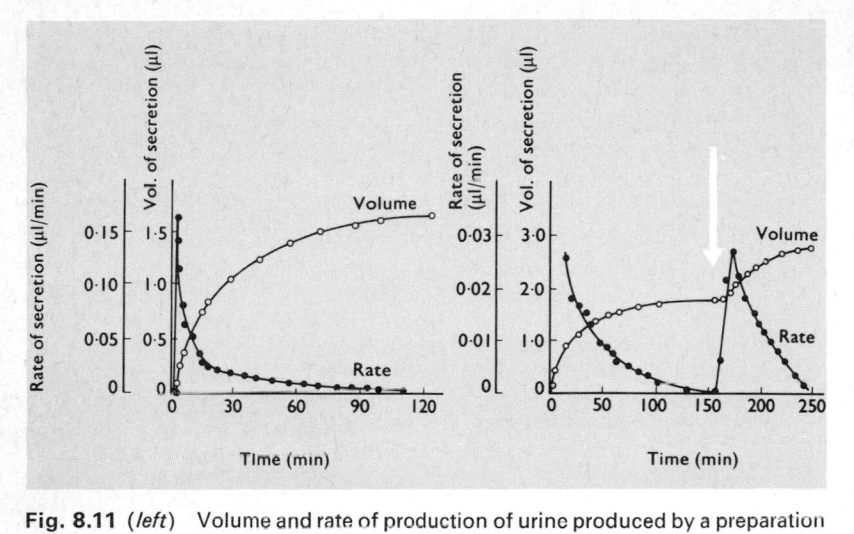

Fig. 8.11 (*left*) Volume and rate of production of urine produced by a preparation shown in Fig. 8.10.

Fig. 8.12 (*right*) Volume and rate of production of urine produced when the malpighian tubules are bathed in haemolymph of unfed and fed *Rhodnius*. An arrow marks the time when 'unfed' haemolymph was replaced by 'fed' haemolymph.

insects, which because it has a positive effect upon urine production, is a diuretic hormone. But extracts of the known hormone-producing tissues in *Rhodnius* have little effect upon malpighian tubule activity (Fig. 8.13); in fact only extracts of the fused thoracic and 1st abdominal ganglia have diuretic activity[199] (Fig. 8.13). All the activity is contained in the posterior part of the ganglionic mass, where large neurosecretory cells are located (Fig. 8.14). In *Rhodnius*, therefore, the diuretic hormone is neurosecretory in origin, and has no relationship with the major cerebral endocrine system. Diuresis begins about 30 seconds after *Rhodnius* has fed, so the hormone must be released very rapidly after feeding. Hormone release is inhibited when the ventral nerve cord is cut: it is likely, therefore, that the abdominal expansion caused by feeding, which stimulates stretch receptors whose signals eventually lead to activation of the cerebral endocrine system, also releases diuretic hormone from the ventral ganglionic neurosecretory cells.[199] However, the threshold of response of the neurosecretory cells in the ventral ganglia must be lower than that of neurosecretory cells in the brain, since small blood meals will not elicit the release of brain hormone, but will release the diuretic hormone.[199a]

Rhodnius, and perhaps all blood-sucking insects, are unusual in that there are periodic crises in water-loading: the diuretic mechanism may be a special adaptation to this manner of feeding. In another bug, the cotton stainer *Dysdercus fasciatus*, a diuretic hormone is produced by the cerebral neurosecretory cells[17] (Fig. 8.15). In the locusts, *Schistocerca*

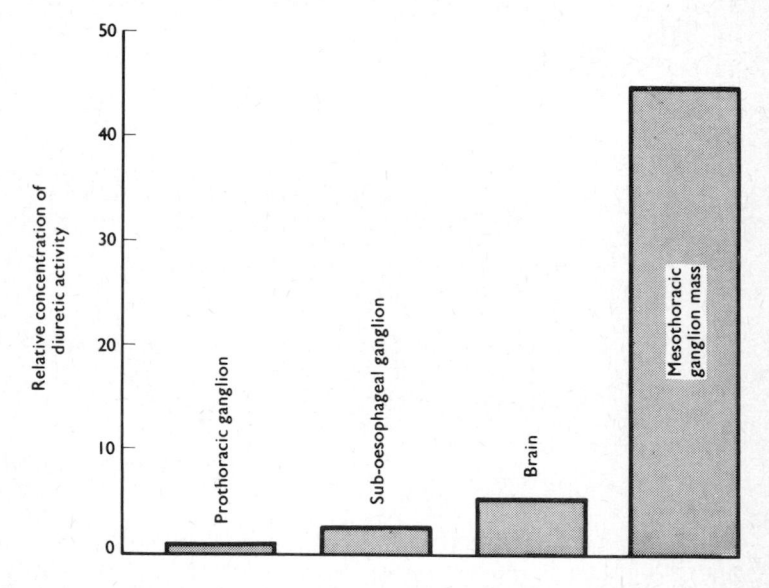

Fig. 8.13 Relative diuretic activity of different parts of the central nervous system of *Rhodnius*.

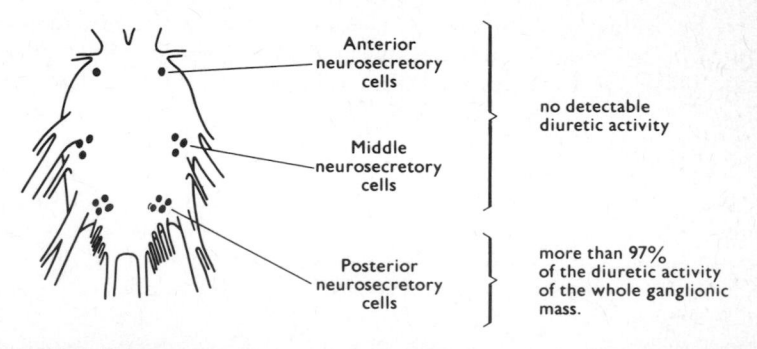

Fig. 8.14 Relative diuretic activity of different parts of the mesothoracic ganglionic mass (containing groups of neurosecretory cells) of *Rhodnius*. (Figs. 8.12–8.14 after Maddrell[199])

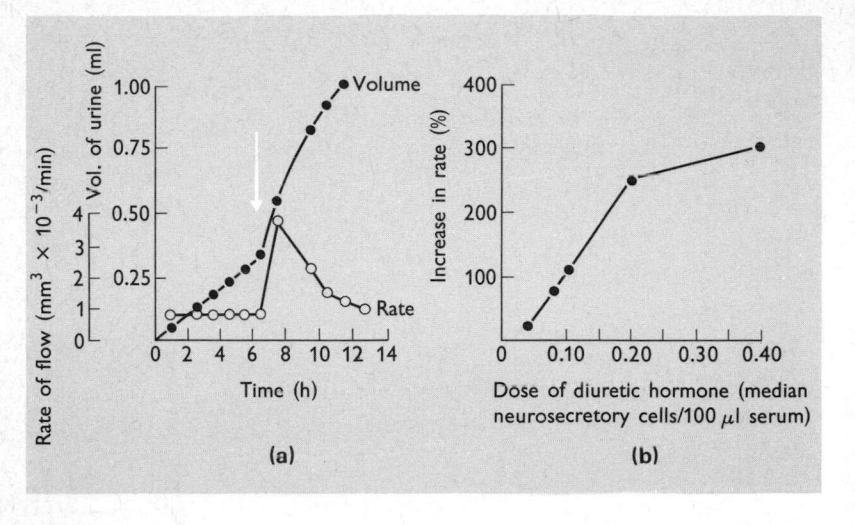

Fig. 8.15 Diuresis in *Dysdercus fasciatus*. (a) The volume and rate of flow of urine from an isolated malpighian tubule preparation (similar to that in Fig. 8.10). The arrow marks the addition of 'diuretic hormone' (0·1 of a group of median neurosecretory cells/100 μl serum). (b) A dose-response curve for 'diuretic hormone' activity. (After Berridge[17])

gregaria and *Locusta migratoria*, the cerebral neurosecretory cells also produce a diuretic hormone. Its existence in these locusts was first suspected when the endocrine control of reproduction was being investigated. Destruction of the cerebral neurosecretory cells resulted in a large increase in the haemolymph volume—sometimes to as much as three times its normal value. The abdomens of the operated individuals were so swollen that at first they appeared to be carrying fully developed eggs; only dissection revealed the true reason for the enlargement. Clearly, destruction of the cerebral neurosecretory cells prevented the normal removal of water taken in with the food.[139]

In *Schistocerca*, movement of fluid through the malpighian tubules can be estimated by injecting a neutral dye such as amaranth into the haemolymph. Small samples of blood are taken at regular intervals after injections, and the concentration of amaranth determined by comparison with a known concentration of the dye, for example by means of an absorptiometer. The rate of removal of amaranth from the haemolymph gives a measure of malpighian tubule activity[209] (Fig. 8.16). Extrapolating the graph back to the time of injection of the known quantity of amaranth gives the concentration of dye in the blood at this

time—from which the blood volume of the insect can be calculated (Fig. 8.16.)

After cautery of the cerebral neurosecretory cells, the activity of the malpighian tubules is low and the blood volume increased (Fig. 8.16). After allatectomy, malpighian tubule activity is reduced, though not so much as in individuals without cerebral neurosecretory cells, and the blood volume is slightly greater than normal (Fig. 8.16). This again suggests an effect of the corpus allatum hormone upon neurosecretory activity (see p. 167): a direct effect of the hormone upon diuresis is unlikely, since extracts of the corpora allata do not affect the rate of removal of amaranth. But extracts of corpora cardiaca injected into locusts with their cerebral neurosecretory cells destroyed have a marked stimulatory effect upon malpighian tubule activity (Fig. 8.18). Increasing concentrations of corpus cardiacum extracts stimulate malpighian tubule activity in a linear manner (Fig. 8.17). Brain extracts also stimulate the malpighian tubules, suggesting that it is the contained neurosecretion of the corpora cardiaca which is the diuretic hormone. In the locust, the intrinsic glandular cells of the corpora cardiaca can be relatively easily separated from the 'neurosecretory storage' parts of the glands (Fig. 2.14). Extracts of the glandular regions have little effect upon diuresis when injected into locusts; extracts of the neurosecretory storage regions have the same effect as extracts of whole glands (Fig. 8.18), confirming that the diuretic hormone originates in the cerebral neurosecretory cells.

When amaranth is injected into the locust some time after the injection of corpus cardiacum extract, it is removed more slowly than when it is

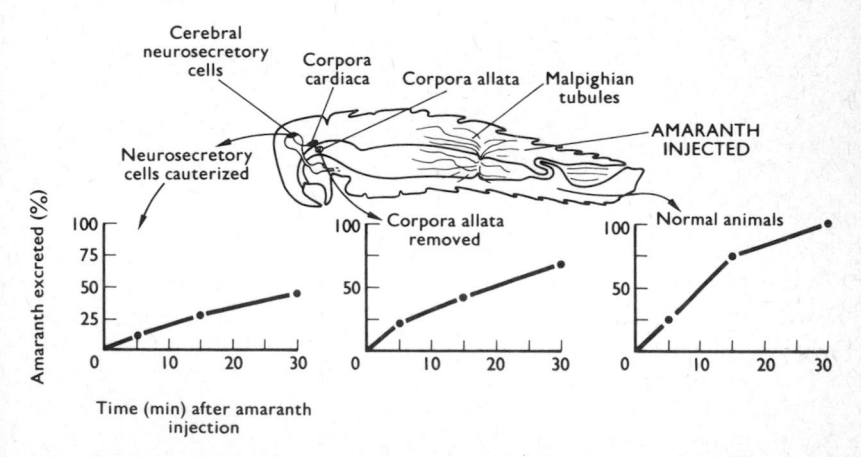

Fig. 8.16 Amaranth clearance from the haemolymph of neurosecretory cell cauterized, allatectomized and normal desert locusts (*Schistocerca gregaria*).

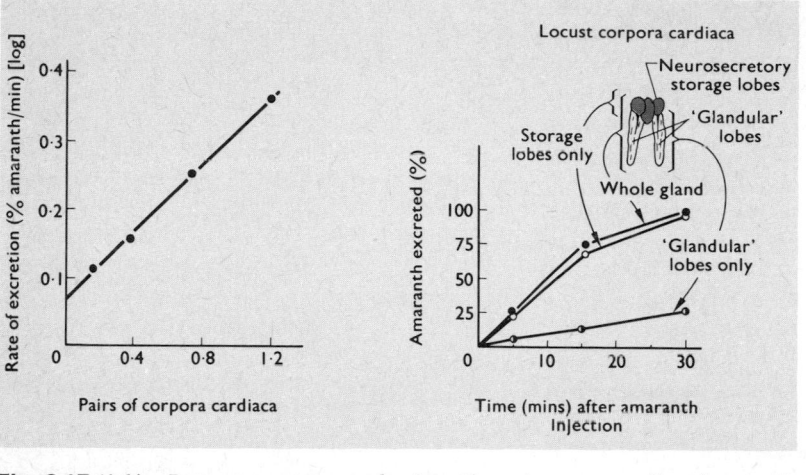

Fig. 8.17 (*left*) Dose response curve for desert locust corpora cardiaca/amaranth clearance rate in the desert locust.

Fig. 8.18 (*right*) The effects of storage and glandular regions, and whole corpora cardiaca upon amaranth clearance in the desert locust. The storage lobes possess almost all the diuretic activity of the whole glands.

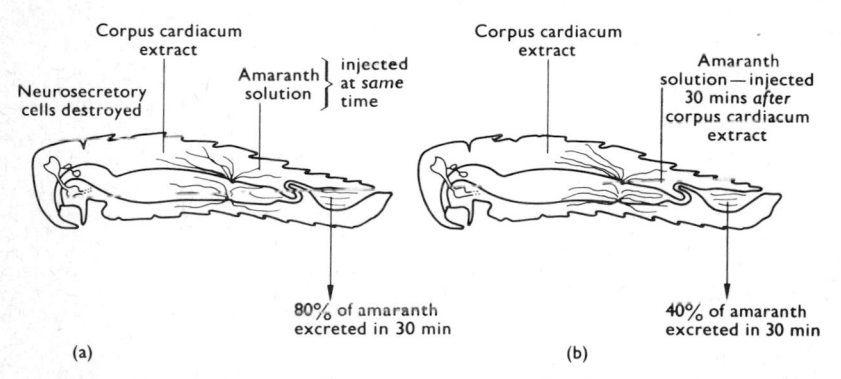

Fig. 8.19 An experiment to demonstrate the biological life of diuretic hormone in the desert locust. The extract loses about half its activity in 30 minutes: its effect upon amaranth clearance when the dye is injected simultaneously with the extract (a) is twice that when the dye is injected 30 minutes after the extract (b).

injected simultaneously with the extract (Fig. 8.19). This is because some of the hormone has been excreted or destroyed or even used up in the period between the two injections. This technique in fact measures the biological life of the hormone in the blood. The diuretic hormone loses about 50% of its activity every hour.[209] Measuring malpighian

tubule activity in a previously starved locust at different times after the recommencement of feeding indicates an increasing concentration of diuretic hormone in the blood (Fig. 8.20). Knowing its biological life, the release of diuretic hormone from the cerebral neurosecretory system can be calculated. This proves to be quite close to the estimates made from histological changes in the neurosecretory system and illustrates again the remarkable effects of feeding upon neurosecretory activity.

The desert locust, like most terrestrial insects, has a very efficient water reabsorbing mechanism in the hind gut, localized particularly in the rectal glands.[223] The diuretic hormone increases the movement of fluid through the malpighian tubules, but this cannot be reabsorbed by the rectum, otherwise the total water content of the insect would remain unchanged. Mordue has developed a technique to measure the effect of the locust diuretic hormone upon rectal reabsorption.[209] The rectum is removed, turned inside out and both its ends ligatured, so that it forms a sac with its cuticle outside. When suspended in saline, water moves into the sac (i.e. from rectal lumen side to haemolymph side of the rectal wall, as it would when reabsorbed from a rectum *in situ*) and the rate of movement can easily be measured by weighing the sac at intervals, the cuticle on the surface being easily dried. When corpus cardiacum extracts are included in the sac, i.e. are placed on the haemolymph side of the

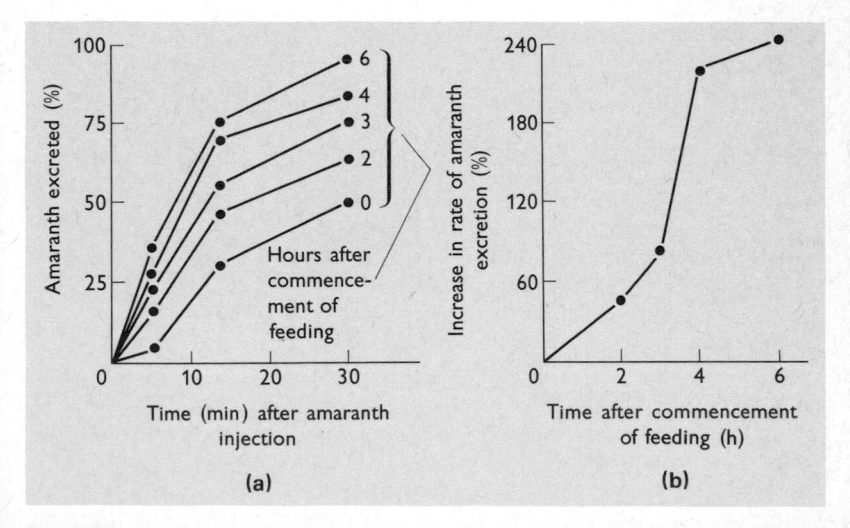

Fig. 8.20   (a) Amaranth clearance rates at different times after feeding previously starved locusts. (b) The percentage increase in the rate of amaranth clearance at these times compared with that in a starved locust. The increasing rates indicate increasing concentrations of diuretic hormone in the blood.

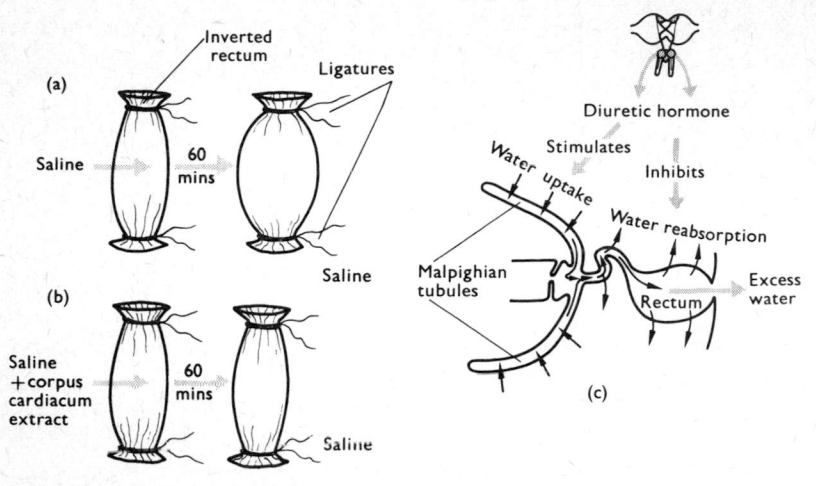

Fig. 8.21 Diuretic hormone and rectal reabsorption in the desert locust. The rectum is turned inside out and its absorption of water measured periodically by the increase in weight of the preparation. Weight increases rapidly when the rectum is filled with saline (a), but is prevented when 'hormone' extract is included (b). (c) shows diagrammatically the dual effects of the locust diuretic hormone, stimulating malpighian tubule activity and inhibiting rectal reabsorption. (Figs. 8.16–8.21 after Mordue[209])

rectal wall, water movement into the sac is *inhibited* (Fig. 8.21). The lowest concentrations of corpus cardiacum extracts which stimulate the malpighian tubules have no effect upon rectal reabsorption: the threshhold of the response of the rectum is higher than that of the malpighian tubules. So in the desert locust, the diuretic hormone has a double effect: it *stimulates* fluid movement through the malpighian tubules, and inhibits its reabsorption by the rectum. Its net result, therefore, is the loss of water by the insect.

Diuretic hormones seem to be of general occurrence in insects. But in the phantom midge larva, *Chaoborus*, the thoracic ganglia release an antidiuretic hormone, in addition to a diuretic hormone from the abdominal ganglia.[104b] In the cockroach *Periplaneta americana*, antidiuretic factors have been demonstrated in the haemolymph.[113a] In locusts, an antidiuretic principle can be extracted from the corpora cardiaca[42a, 209, 209a, 209b] and from the perivisceral neurohaemal organs,[17c] but its release under normal conditions has not yet been proved. The antidiuretic hormone could stimulate the rectal reabsorption of water,[113a] and in locusts could exercise a fine control over excretory function to enable the insects to cope with a variety of environmental situations.[209c]

HORMONES AND PERISTALSIS

In the cockroach *Periplaneta americana*, extracts of the corpora cardiaca will accelerate the rate and increase the amplitude of the beating of the isolated heart, and increase both gut peristalsis and the twisting movements of the malpighian tubules. The hormone(s) responsible do not act directly, but stimulate pericardial cells in the vicinity of the heart, and analogous cells in the gut, to secrete a tryptamine compound (related to a compound found in the vertebrates: serotonin or 5-hydroxytryptamine) which is the *immediate* cause of the effects.[70, 71] The corpus cardiacum hormone is a peptide, and is released when the cockroach feeds. Because of the close juxtaposition of glandular and neurosecretory storage regions in the cockroach, it is impossible to decide in which part the factor originates. In the desert locust, glandular extracts are more potent than storage extracts of the corpora cardiaca in their effects upon the isolated heart:[136] it is possible that two hormones are involved, one from each region. In the locust, there also seems to be some interaction between the storage and glandular regions in the production of the factor: the effects of feeding upon its release tend to support such a suggestion. It must be emphasized that the effects of corpus cardiacum factors upon heart beat are apparent only with isolated heart preparations. *In situ*, the insect heart has a complex innervation and has myogenic, cholinergic and adrenergic properties. Corpus cardiacum extracts have little effect upon heart beat in the intact insect.[210, 213c, 229b] Thus the factors from the corpora cardiaca which affect the beating of the *isolated* heart could be pharmacological effects of peptide hormones whose main functions are elsewhere.[115j] It is an attractive hypothesis that feeding releases a hormone which increases gut movement, blood circulation and malpighian tubule contractions, thus affecting the digestion of food, transport of metabolites and the secretion of waste products,[71, 72] but the concept is probably an oversimplification.

A neurosecretory hormone is involved in the complex movements of muscles involved in oviposition. In *Rhodnius*, the presence of the cerebral neurosecretory cells are certainly necessary at this time,[75] and in the locust changes in the histology of the cerebral neurosecretory system during egg-laying suggest that neurosecretion is involved[133] (p. 175). But an interaction between neurosecretion and the production of an intrinsic hormone by the cells of the corpora cardiaca cannot be ruled out.

HORMONES AND HYPERGLYCAEMIA

In *Periplaneta*, an extract of the corpora cardiaca elevates the blood sugar concentration of the haemolymph. Two hormones are involved,

one very much more potent than the other, and both have a similar effect: to convert the enzyme phosphorylase in the fatbody from its inactive to its active form[251] (Fig. 8.22). The active phosphorylase catalyses an important step in the breakdown of glycogen and the net result is an increased production of the disaccharide trehalose (Fig. 8.22).

In the locust, similarly, two hyperglycaemic factors are released from the corpora cardiaca, with the same effect upon the phosphorylase enzyme. In the locust, the ease of separation of neurosecretory storage from glandular regions of the corpora cardiaca allows the particular origins of the two hormones to be determined: the less potent of the two comes from the neurosecretory region, and the more potent from the gland cells. It can be assumed that their origins are similar in cockroaches. But even so, the effects of starvation, feeding, hyperactivity and stress upon the release of *both* hormones in the locust again suggests some interaction

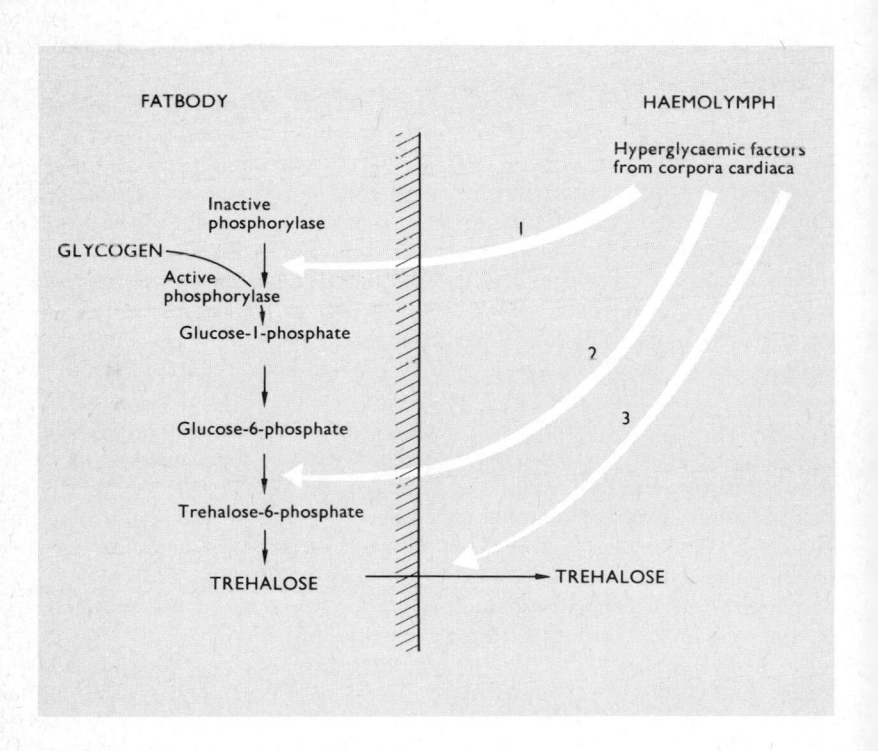

Fig. 8.22 Sites of action of the cockroach hyperglycaemic factors. Two enzyme reactions are affected (1,2) as well as the transfer of trehalose from fat body cells to haemolymph (3). (After Steele[251])

between the two parts of the corpora cardiaca. A trophic effect of neuro-secretion upon the intrinsic glandular cells of the corpora cardiaca is not at all unlikely.

Hyperglycaemic hormones have been identified also in many other insects. Their physiological significance is obscure. It has been suggested that the hormones play a role in the rapid supply of energy during periods of high metabolic activity or vigorous exercise.[251] However, in the cock-roach periods of sustained activity sufficient to require the release of hyperglycaemic hormones are unlikely, particularly in view of the normally large amounts of trehalose present in the haemolymph.[115j] In the locust, although the fatbody phosphorylase is activated by corpus cardiacum extracts,[210, 115f] the glycogen reserves are insufficient to produce significant elevations of the trehalose concentration in the haemolymph.[115e, 115f] It is likely, therefore, that the hyperglycaemic hormones may be more important in their capacity to influence other metabolic pathways indirectly by their effects upon carbohydrate metabolism.[115j]

HORMONES AND LIPID TRANSPORT

In locusts, a hormone is produced by the glandular lobes of the corpora cardiaca which stimulates the release of diglyceride from the fatbody.[9a, 202a] Because of its function, it is called the adipokinetic hormone. It is released during flight and is essential for the maintenance of prolonged flight activity[115g, 115h] (Fig. 8.23). The diglyceride is mobilized at the expense of triglyceride in the fatbody and is transported in the haemolymph as a lipoprotein. Locust flight muscles possess the enzymes necessary for the utilization of fat as a fuel, and the respiratory quotient of the insects during flight indicates that fat is indeed the fuel used. It is also likely that the locust adipokinetic hormone may additionally have a direct effect upon flight muscle metabolism, stimulating the muscles to utilize lipid in preference to carbohydrate[227h] (Fig. 8.23).

Extracts of cockroach corpora cardiaca, when injected into locusts, stimu-late diglyceride release from the fatbody.[115i] But when extracts of corpora cardiaca from locusts or cockroaches are injected into cockroaches, they *reduce* haemolymph lipid concentrations, stimulating the transport of triglyceride and diglyceride *into* the fatbody.[77e, 77f] The significance of this hypolipaemic response in the cockroach is obscure, and the release of the factor *in vivo* has not been demonstrated. The hypolipaemic factor appears to be different from the hyperglycaemic hormone[77g] and is not necessarily the same as the locust adipokinetic hormone.[115j] Clearly, much more information is needed from a far wider range of species before

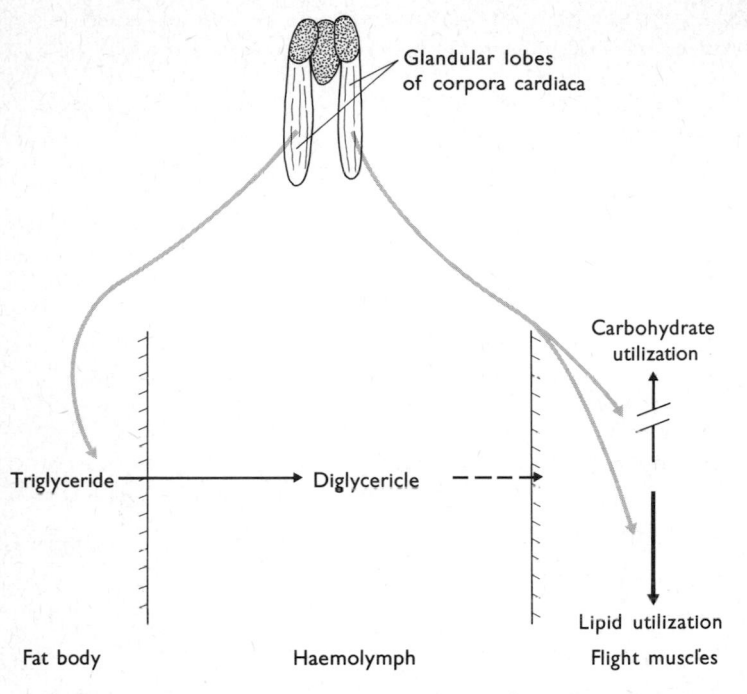

Fig. 8.23 Functions of the adipokinetic hormone in locusts. The hormone originates in the glandular lobes of the corpora cardiaca, and its release raises the level of diglyceride in the haemolymph at the expense of fatbody triglyceride ; at the same time the hormone affects the flight muscles, switching a carbohydrate-based to a lipid-based metabolsim.

general conclusions can be made about the hormonal control of lipid transport in insects.

HORMONES AND ECLOSION

Many holometabolous adult insects emerge from the pupal cuticle by a stereotyped sequence of movements which enable them first to free themselves and then to push and pull themselves out of the pupal case. Thus the saturniid moth, *Hyalophora cecropia*, escapes from the pupal cuticle by first rotating the abdomen rapidly for about 30 minutes, and then after a further 30 minutes period of relative quiescence, there are 15 minutes of rapid anteriorly-directed peristaltic movements of the abdomen which assist the final emergence of the adult. If the moth is removed from the

pupal case before the usual time (see below) for this eclosion behaviour, it remains quiescent until that time arrives, and then goes through the full sequence of abdominal movements as though it were still confined within the pupal cuticle. This shows clearly that the movements do not depend upon reactions between the moth and its restricting pupal cuticle, but originate entirely within the pharate adult.

If the brain is removed from the pharate adult, the eclosion behaviour is aberrant or non-existent, but is restored to normal by the implantation into the abdomen of a brain, or of homogenates of brains or corpora cardiaca.[265d] Eclosion behaviour is thus elicited by the release of a hormone from the cerebral neurosecretory system (Fig. 8.24). The hormone is not present in the brains and corpora cardiaca of larvae or pupae, and appears only in the last two-thirds of adult development.[265g] The hormone is distinct from bursicon (p. 207) which is also associated with eclosion.

Eclosion movements can be elicited in the isolated abdomens of pharate adults, so their neural basis must be associated with the abdominal and not with more anterior ganglia. When the second and third abdominal ganglia are isolated from peripheral innervation, and their dorsal nerves

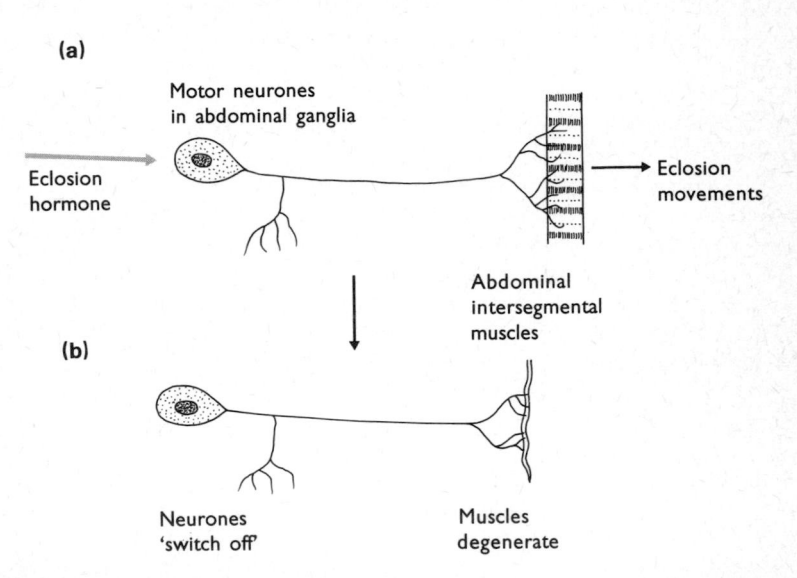

Fig. 8.24 The effect of the neurosecretory eclosion hormone upon motor neurone activity in Saturniid moths. In (a) the hormone initiates nervous activity in particular neurones in the abdominal ganglia which cause contractions of the abdominal intersegmental muscles : the activity of the neurones is coordinated so that the muscles produce peristalsis and twisting of the abdominal segments. The neurones become electrically silent after their activity, and the muscles degenerate (b).

206 ENDOCRINE MECHANISMS IN THE INSECTA—III CHAP. 8

(which innervate the intersegmental muscles responsible for the eclosion movements) monitored for electrical activity, volleys of impulses are recorded only in the presence of the eclosion hormone. The nervous activity follows exactly the time sequence of the eclosion behaviour. Moreover, in the first period, the volleys of impulses show a right-left alternation in the paired dorsal nerves—which would be responsible for rotary movements of the intact abdomen—and in the final stage, synchronous volleys occur in each pair of nerves, beginning posteriorly and moving anteriorly, and therefore clearly responsible for the normal abdominal peristaltic movements. These elegant experiments allow the firm conclusion that eclosion results from a programme of motor behaviour which is built into the abdominal ganglia and is triggered into overt activity by the eclosion hormone.

The intersegmental muscles of the abdomen in saturniid moths degenerate after emergence of the adult, apparently because the motor neurones which innervate them become electrically silent.[191e, 191f, 191g] But when the abdomen of a pharate *Antheraea polyphemus* is isolated before eclosion, the muscles remain.[191d] When eclosion hormone is injected into the isolated abdomen, the muscles degenerate as usual.[265c] Thus, in addition to triggering eclosion behaviour, the hormone provides the signal for the 'turning off' of the motor neurones supplying the intersegmental muscles (Fig. 8.24).

Adults of different species of saturniid moths emerge at different times of the day. Thus *Hyalophora cecropia* adults emerge during the morning and *Antheraea pernyi* during the late evening. When brains are interchanged between individuals of these species, the adult emergence times are transposed: *Hyalophora* with an *Antheraea* brain emerges in the evening, and *Antheraea* with a *Hyalophora* brain emerges in the morning.[265h] The timing of release of the eclosion hormone is different in these species, and must be associated with an eclosion 'clock' which determines the specific emergence times. Under the appropriate conditions, the clock shows an innate circadian periodicity (p. 189); light: dark cycles entrain the clock and determine the time of eclosion. In *Anthereaea pernyi* subjected to a 17 hours light: 7 hours dark regime, the clock begins each new cycle at 'lights off', and when transferred to continuous darkness, eclosion hormone is released 22 hours later and at 22 hour intervals thereafter. Light interruptions lasting 30 minutes which 'scan' the dark period (p. 190) indicate that the cycle has an initial synchronization period lasting about 4 hours which has an absolute requirement for darkness. Light interruptions during this period do not merely stop the clock—the cycle restarts from the beginning after the light interruption. So the reactions which occur during the synchronization period are not merely dark-requiring but are photo-reversible.[265e, 265f] The synchronization period

is followed by a dark decay period of about 20 hours before the cycle is re-initiated. The internal clock associated with the release of the eclosion hormone has therefore the nature of an interval timer (p. 190).

The way in which the eclosion hormone 'switches on' behaviour patterns which are built into the central nervous system and remain latent therein until the release of the hormone is of considerable interest, since many radical changes in the behaviour of insects are initiated by hormones.[8, 135] One example is the appearance of courtship behaviour in female grasshoppers controlled by the corpus allatum hormone.[192a, 193] Since the hormone also controls oocyte development, courtship and mating thus occur at the appropriate developmental time. The evocation of different behaviour patterns by hormones, instead of by complex neural mechanisms, means that a wide range of behaviours can be built into a small and relatively simple nervous system. It is not possible to describe the wide variety of hormonally controlled behaviours here: recent reviews are available.[8a, 265j]

HORMONES AND CUTICULAR TANNING

Despite the importance of ecdysone in inducing the *de novo* synthesis of dihydroxyphenylalanine (DOPA) decarboxylase and perhaps other enzymes associated with cuticular tanning (p. 119), it is not the only hormone involved. In many insects, it has been shown that another blood-borne factor plays a role in the regulation of tanning and darkening in both larval and adult cuticles. The hormone is called bursicon, and is produced by the cerebral neurosecretory cells although it is primarily released from the ganglia of the ventral nerve cord.[64c, 64d, 94a, 94b, 206e, 266g]

When the release of bursicon is prevented experimentally, endocuticle formation and normal melanization are prevented. In *Sarcophaga bullata*, the incorporation of leucine into the epidermis is increased by bursicon, but there is no concomitant increase in the uptake of uridine.[93a] Actinomycin D and puromycin inhibit the formation of endocuticle, but do not prevent melanization; thus the action of bursicon does not affect DOPA decarboxylase activity. Thus it is likely that the effect of bursicon upon the protein synthesis related to cuticle formation is not upon the epidermis but elsewhere—probably the fatbody. Bursicon also increases the permeability of blood cells to tyrosine,[207a, 224e] and is involved in the hydroxylation of tyrosine to DOPA.[245a]

Bursicon may thus act in concert with ecdysone, first to produce the substrate for the action of DOPA decarboxylase, the latter enzyme being induced *de novo* by ecdysone (p. 120), and secondly by influencing the

formation of protein which will react with the tanning quinone to produce sclerotin in the cuticle (Fig. 6.14).

Such a complicated control of the tanning process in insects may be necessitated by the functions of ecdysone: the hormone is released relatively early in the moult cycle and it influences many processes in addition to DOPA decarboxylase synthesis. If there is a need for some delay in the tanning process relative to ecdysis, the intervention of a control system separate from ecdysone becomes imperative. This is clearly shown by the adults of certain blowflies, which have to reach the surface of the substrate in which pupation occurred, and also perhaps by the adults of aquatic insects, which have to climb out of the water before the wings and body can be fully expanded. The necessity for such a control in larval insects is less clear, but it could be related to the vulnerability of all insects at ecdysis which causes them often to become cryptic in their behaviour at this time. That bursicon release is under nervous control is obviously connected with the receipt of the appropriate stimuli during the delay period and at its end.

9

Endocrine Mechanisms in Crustacea—I

Many developmental and physiological processes in Crustacea are controlled by hormones, but often the exact mechanisms are confused. This is due to a combination of factors, such as the great difficulty in maintaining large numbers of animals in controlled conditions in the laboratory, the long life cycles of some of the large species which are popular for experimentation, and a frequent lack of detailed information about the normal habitat and behaviour of the animals. Furthermore, almost all the observations on crustacean endocrine mechanisms refer to the Malacostraca, and mainly to the Decapoda at that. All this must be borne in mind when general conclusions are drawn about the endocrine control of moulting, hydromineral regulation, sexual differentiation and reproduction, colour change, migration of retinal pigments and heart beat in Crustacea.[41]

HORMONES AND MOULTING

The nature of the moult cycle

Information about the nature of the moult cycle and its control is limited to the Decapoda, and in fact often refers specifically only to the crabs (Brachyura). Knowledge of the process in other groups of Crustacea is fragmentary.

In the Crustacea, as in insects, moulting is a part of the mechanism of growth. Change in form and increase in size can only occur when the hard calcareous exoskeleton is shed and before the new cuticle is hardened. Periodic ecdysis, by splitting the old cuticle through expansion of the new instar, brought about by taking up water through the digestive tract, is as characteristic of the crustacean as it is of the aquatic insect. The increase in size and weight of the crustacean during ecdysis does not constitute growth. This must be defined as the increase in dry weight of the body which occurs in the periods between moults, when the absorbed water is gradually replaced by protein. Thus although ecdysis, increase in size and increase in total weight are all markedly discontinuous in Crustacea, growth itself is a continuous process (Fig. 9.1), as it is in insects (Fig. 6.18) and almost all other animals.

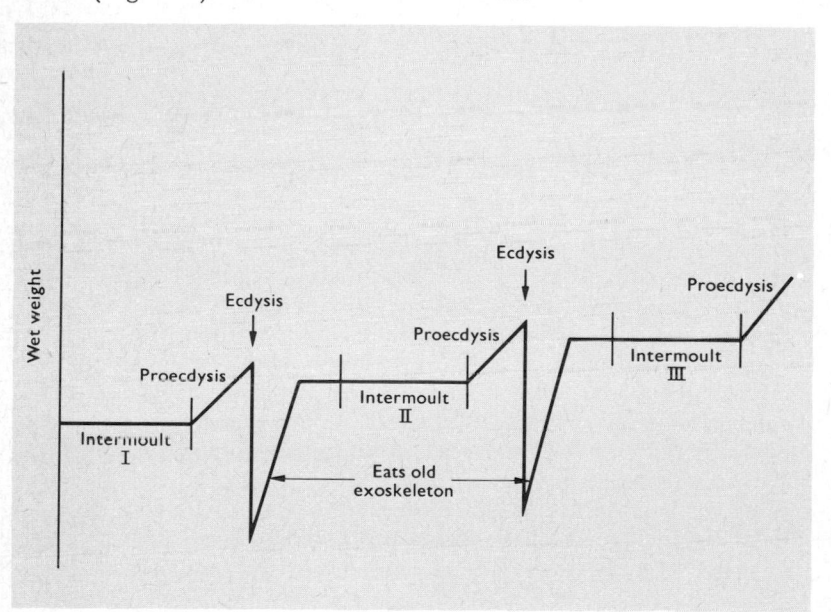

Fig. 9.1 Changes in wet weight during the growth of a crab. Growth is apparently discontinuous: water is taken up during proecdysis resulting in a weight increase. The shedding of the old cuticle causes a sudden decrease in weight, but this is regained in many crabs which eat their cast skins. Although the wet weight remains constant after the moult, growth actually occurs during this period, the water taken up at proecdysis being replaced by protein.

Moulting dominates the life of the crustacean. Ecdysis cannot be considered as a brief interruption of the normal life of the crustacean, but rather as a process which has far reaching effects upon the whole

physiology. In this respect, many decapods differ profoundly from the pterygote insects in that moulting proceeds throughout adult life. In fact, very little is known about the control of larval growth and moulting in the decapods: when the endocrine control of growth and moulting in Crustacea and insects is compared, it must always be remembered that in the former the process refers almost exclusively to postmetamorphic stages, whereas in the latter metamorphosis is usually an intrinsic part of the mechanism.

Post-metamorphic development in Crustacea follows one of several patterns. The shore crab, *Carcinus maenas*, moults several times after metamorphosis before the puberty moult, at which the sexual appendages appear and the animal becomes sexually mature and is able to reproduce; moulting then continues for up to 3 years after the puberty moult. In the spider crab, *Maia squinado*, on the contrary, the puberty moult is the final one, and the animal therefore never moults after becoming

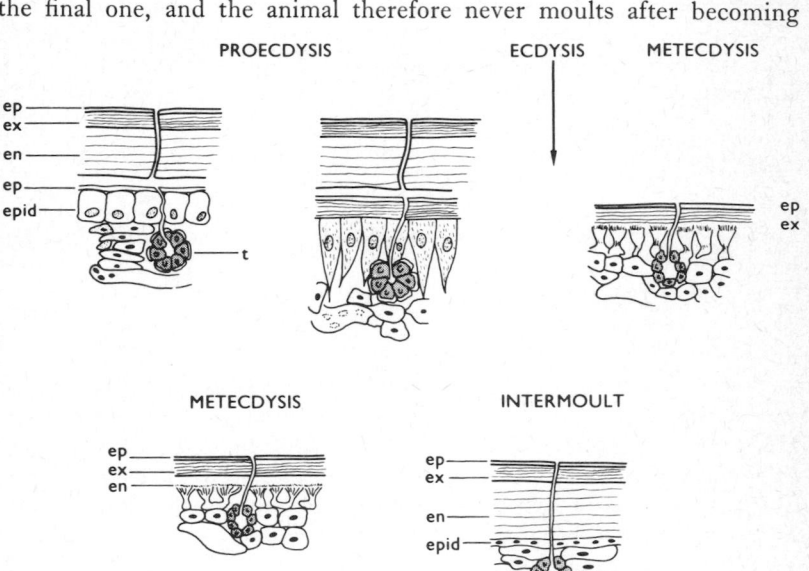

Fig. 9.2 Structure of the integument of the land crab, *Gecarcinus lateralis*, at each stage of the moult cycle. During early proecdysis the epidermal cells enlarge, separate from the old cuticle (apolysis), and secrete a new epicuticle. By late proecdysis the epidermal cells have enlarged still further and secretion of the new exocuticle has begun. After ecdysis, exocuticle secretion is complete and endocuticle production begins, to be completed during metecdysis. The epidermal cells then decrease in size and remain small during the intermoult period.
Abbreviations: *en*, endocuticle; *ep*, epicuticle; *ex*, exocuticle; *epid*, epidermal cells; *t*, tegumental gland. (After Carlisle and Knowles[41])

sexually mature. In other decapods, such as the edible crab, *Cancer pagurus*, and the lobster, *Homarus vulgaris*, growth and moulting appear to continue until death.[39, 41]

The moulting cycle can be divided into four stages (Fig. 9.2) which facilitates analysis of its control.[40] No sharp divisions exist between these stages.

Stage 1. Proecdysis

This is the preparation for moulting. The first signs of proecdysis are activation of the epidermal cells and hepatopancreas. The epidermal cells separate from the cuticle, a process known as **apolysis**, and then divide. Almost immediately the epidermal cells begin to secrete the new exoskeleton. At the same time calcium is removed from the old cuticle, resulting in an increased blood calcium concentration. As these processes continue, the animal stops feeding and becomes inactive: during this time the reserves of the hepatopancreas are utilized. Splitting of the old cuticle marks the end of proecdysis.

Stage 2. Ecdysis

This is the short period during which the animal sheds the remains of the old cuticle. There is a rapid uptake of water. The animal does not feed.

Stage 3. Metecdysis

This begins with the newly moulted animal, its exoskeleton still soft and extensible as water uptake continues. Mineral deposition begins in the exocuticle, and later endocuticle secretion starts. Initially, the animal still does not feed, continuing to utilize reserves in the hepatopancreas. But during the latter half of metecdysis feeding recommences, the production of the exoskeleton is completed, and tissue growth occurs, replacing the absorbed water. Both protein and DNA have high turnover rates during this time, and the tissues double their dry mass, losing water in direct proportion.

Stage 4. Intermoult

Both skeletal formation and tissue growth have been completed, but feeding continues and metabolites in excess of current requirement are stored in the hepatopancreas. Lipid is the major reserve, but some glycogen and protein are also stored. This intermoult period is often referred

to as the period of normality, but the specific accumulation of reserves in preparation for the next moult is no more normal than any other part of the moult cycle.

If the animal moults seasonally, the intermoult period is long, and is known as *anecdysis*; if moults are much more frequent, occurring throughout the year, the period is called *diecdysis*, hardly distinguishable as metecdysis merges gradually into proecdysis.

Endocrine control of moulting[218]

The initiation of proecdysis

Over 60 years ago, it was discovered that removal of the eyestalks during the intermoult period accelerated ecdysis and initiated precocious growth. But the significance of these effects was not realized until later work on colour change in Crustacea (p. 240ff) revealed the presence of the eyestalk neurosecretory system (Fig. 2.19).

Subsequent experiments showed that removal of the eyestalks during the intermoult period results in the premature onset of proecdysis, but their removal during proecdysis has no effect. These results suggested that the eyestalks are the source of a hormone which normally *inhibits* the onset of proecdysis. But the converse of this experiment, the implantation of the source of the hormone into eyestalkless animals, initially yielded variable results: sometimes proecdysis was delayed, at other times there was no effect. The criteria for demonstrating a hormonal effect (p. 7) were not always fulfilled.[218] In these experiments, sinus glands were implanted as the probable source of the hormone; it was not realized at this time that the glands are neurohaemal organs, storing perhaps only small quantities of hormone produced elsewhere. But when the true nature of the sinus glands was recognized, whole eyestalk neurosecretory systems were implanted into eyestalkless animals, and the accelerated onset of proecdysis was delayed.[217] Later, the selective destruction of either the sinus gland, or the X-organ, confirmed that the eyestalk neurosecretory system produces a moult-inhibiting hormone.

But it was soon discovered that the neurosecretory moult-inhibiting hormone did not act directly upon the tissues to delay proecdysis. When the glandular paired Y-organs (p. 26) were removed from *Carcinus* during the intermoult period, or very early proecdysis, ecdysis was prevented. Removal of the Y-organs later in proecdysis had no effect upon the subsequent moult, but the *next* ecdysis was blocked. Reimplantation of Y-organs into such a blocked *Carcinus* initiated the moult cycle again.

And when supernumerary Y-organs were implanted into normal animals, the intermoult period was shortened.

The progress and control of growth during proecdysis can be followed by analysing the regeneration of autotomized legs in crabs. After a leg is shed, a period of growth—basal leg growth—follows almost immediately until a growth plateau is reached, when the leg bud remains more or less constant in size until the next proecdysis. Basal leg growth is thus localized and is not due to any mechanisms controlling the overall growth of the animals[19a, 19c] (Fig. 9.3).

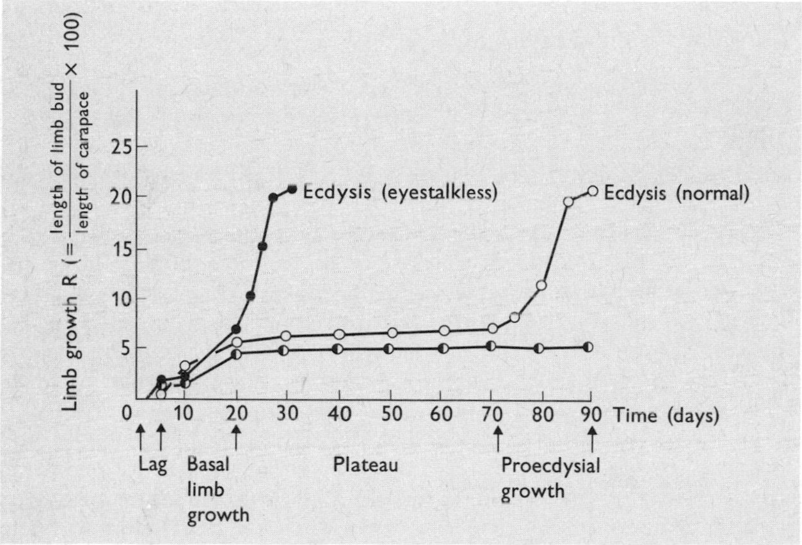

Fig. 9.3   Limb growth after autotomy in crustaceans : ○ normal, ◑ after Y-organ removal, ● after eyestalk removal. During normal growth, autotomy is followed by a short lag period and then a burst of 'basal limb growth'. A plateau is reached, terminated by the initiation of proecdysial growth. Proecdysial growth does not occur after Y-organ extirpation, but basal limb growth is unaffected. After eyestalk removal, basal limb growth merges into proecdysial growth without an intervening plateau. (After Highnam[136a])

If the eyestalks are removed when the leg is shed, regeneration is rapid with no growth plateau, and basal growth merges imperceptibly into proecdysial growth[218a] (Fig. 9.3). But if the Y-organs are removed, either alone or together with the eyestalks, the growth plateau of the leg bud persists, and is not followed by any proecdysial growth; the implantation of Y-organs into these animals causes renewed leg growth.[218a]

These results all point to the production by the Y-organs of a hormone, with a positive effect upon growth and moulting.[99, 82, 83] The crustacean Y-organs are thus analogous to the insect thoracic glands: both produce a moulting hormone. But whereas the neurosecretory thoracotrophic hormone of insects *stimulates* the thoracic glands into activity in preparation for the moult, the neurosecretory moult-inhibiting hormone of Crustacea *inhibits* the activity of the Y-organs during the intermoult period, its absence before proecdysis allowing the Y-organs to secrete (Fig. 9.4). But

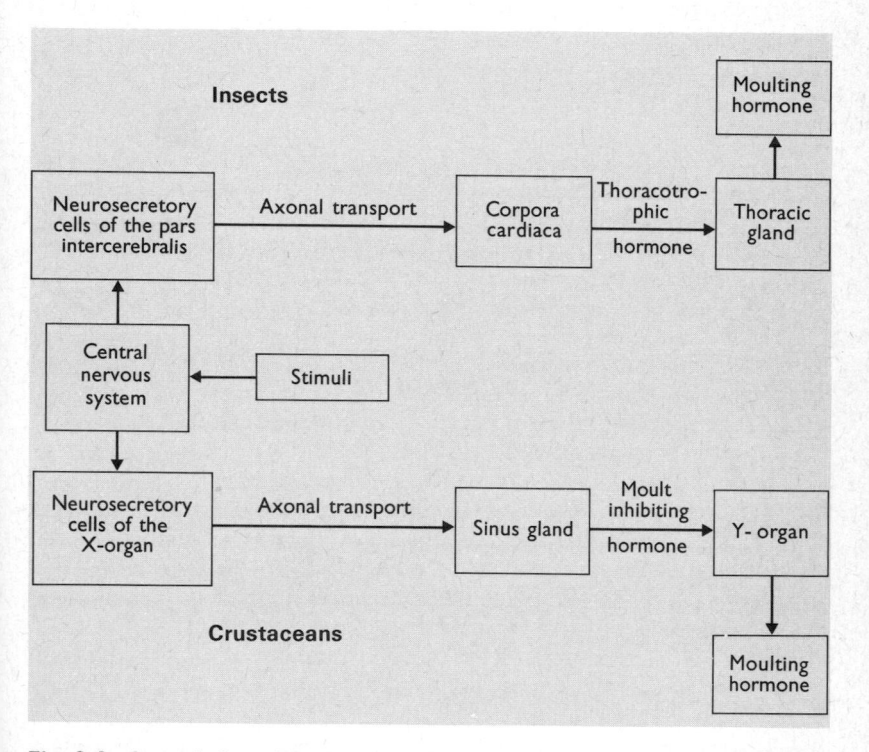

Fig. 9.4 Control of moulting in crustaceans and insects. In both groups moulting is controlled by a two-step hormonal sequence: the moulting hormone is secreted by an epithelial endocrine gland whose activity is controlled by a neurosecretory hormone. In crustaceans, the neurosecretory hormone is produced in the X-organs and released from the sinus glands and *inhibits* the activity of the Y-organs: moulting hormone is produced when secretion of the moult inhibiting hormone ceases. In insects, the neurosecretory hormone from the pars intercerebralis of the brain is released from the corpora cardiaca and *stimulates* the thoracic glands to produce the moulting hormone.

in spite of this fundamental difference, in both Crustacea and insects moulting is controlled by a two-step endocrine sequence: control over an epithelial endocrine gland exercised by neurosecretory hormones, and control over the body tissues exercised by hormones from the epithelial glands.

There is little information available on the fate of the Y-organs in those crustaceans which cease to moult, but in *Maia* the Y-organs atrophy when the animal reaches maturity. *Maia* is thus similar to the insects where the thoracic glands degenerate after the final moult. In *Carcinus*, however, the Y-organs remain functional after the final moult although reduced in size, and moulting may be induced by eyestalk removal; the Y-organs are presumably permanently restrained by the continuous production of moult-inhibiting hormone.

Other hormones in the moult cycle

The endocrine control of moulting in Crustacea, outlined above, rests upon a firm experimental foundation. But other hormones have been claimed to be involved in the moult cycle. In the prawn, *Leander serratus*, eyestalk removal from individuals in some populations results in the delayed onset of proecdysis; in other populations proecdysis is brought forward, the intermoult period being shortened, when the eyestalks are removed. The latter situation closely parallels that in *Carcinus*, the operation removing the source of the moult inhibitory hormone. But *delayed* proecdysis following eyestalk removal can only mean that the source of a moult *accelerating* hormone has been removed.[36]

The moult-accelerating hormone is said to be produced by neurosecretory cells in the brain and other parts of the central nervous system. In the Brachyura and Astacura, the source of the moult-accelerating hormone is unaffected by eyestalk removal which takes away only the moult-inhibitory hormone: consequently the Y-organs are stimulated and proecdysis is brought forward. But in those species which have the sensory pore X-organ as a neurohaemal organ in addition to the sinus gland, the medulla terminalis ganglionic X-organ (Fig. 2.20) is said to be an *additional* source of the moult-accelerating hormone. Eyestalk removal in these species therefore either accelerates or delays proecdysis according to the prevailing hormone balance at the time of the operation. It must be admitted that although such a complicated control over Y-organ activity is feasible, the evidence for its existence is not good. The insects have certainly developed a simpler, energy conserving system. There is better evidence for the existence of eyestalk neurosecretory hormones which affect metabolic processes independently of the moulting hormone (p. 219).

Mode of action of the moulting hormone

Proecdysis is initiated by the release of moulting hormone from the Y-organs. The crustacean moulting hormone is crustecdysone (p. 271) and injections of this hormone in the intermoult stage will initiate early proecdysis in a variety of species.[178a, 178b, 227a, 266h] The Y-organs must be present for the first part of proecdysis, but thereafter moulting can proceed in their absence (p. 213). This has led to the claim that an abrupt cessation of secretion of the neurosecretory moult-inhibiting hormone results in the increased production or release of crustecdysone which acts only as a trigger to start proecdysis. However, the titre of crustecdysone in the blood of the shore crab, *Carcinus maenas*, does not increase sharply just before proecdysis but builds up gradually to reach a peak towards the end of proecdysis.[1d, 1e] Similarly in the crab *Callinectes sapidus*, the titre of crustecdysone is greatest immediately after ecdysis.[86h] When crustecdysone is injected into the crayfish *Orconectes sanborni* at any stage of proecdysis, the animals enter the succeeding stage sooner than the controls.[251c] This evidence suggests that crustecdysone not only initiates proecdysis, but continues to act throughout the stage to stimulate the various processes involved in moulting.

A great many sequential and parallel changes are involved in proecdysis, and uncertainty about the primary action, or actions, of crustecdysone makes a clear analysis of the control of the moult cycle impossible. The situation is further complicated by evidence that neurosecretory hormones from the eyestalks have direct effects upon some metabolic processes. Moreover, the distinctively cyclic nature of the moulting process is reflected in cyclic metabolic changes, some of which are not directly concerned with moulting as such.

The evidence suggests, however, that crustecdysone has two major sites of action: the epidermal cells and the hepatopancreas. An accumulation of glycogen in the epidermal cells and an increased concentration of blood calcium, probably derived from the hepatopancreas and later from the old cuticle, mark the first appearance of crustecdysone in the blood. These early events can be slowed down or even reversed if external conditions become unsuitable, or if the Y-organs are removed. But once the epidermal cells enlarge and separate from the old cuticle, and use their accumulated glycogen to synthesize new cuticle, the progress of proecdysis is irreversible.

The epidermal cells

The many changes that occur in the epidermal cells during proecdysis are usually attributed to the action of crustecdysone. In the crayfish

Orconectes virilis, epidermal mitosis and DNA synthesis in the epidermal cells occur during proecdysis, and epidermal protein synthesis starts in early proecdysis, reaches a peak shortly after ecdysis, and then declines until cuticle growth is complete.[251b] All these events can be correlated with a rising titre of crustecdysone in the blood, and in *Orconectes limosus*, the synthesis of epidermal RNA follows the blood titre of the hormone during proecdysis.[167a] The injection of crustecdysone into intermoult *Orconectes* significantly increases the *in vivo* incorporation of amino acids into epidermal cells within 24 hours.[205] That such events can be correlated with the presence of crustecdysone in the blood, and do not occur in the absence of the hormone, does not necessarily mean that crustecdysone exerts direct control. Nor does it provide any evidence on the manner in which crustecdysone could exert its effects. All that can be said with certainty is that crustecdysone initiates and maintains at least some of the activities of the epidermal cells during proecdysis. The action of crustecdysone at the cellular and molecular levels is likely to be similar to that in insects—which is to say that in both groups the definitive mode of action of the hormone is still far from resolution.

The hepatopancreas

The crustacean hepatopancreas fulfils a variety of roles. It produces digestive enzymes, stores reserves, and is an organ of intermediary metabolism. It is in many ways analogous to the fatbody of insects and the liver of vertebrates. There is evidence that at least some aspects of hepatopancreas functions are controlled by hormones, but it is often uncertain whether crustecdysone or neurosecretory hormones are involved. During proecdysis, lipid is mobilized and there is an increase in both protein and RNA synthesis in the hepatopancreas: crustecdysone injected into intact intermoult *Orconectes* induces precocious moulting, and *in vitro* the hormone stimulates the incorporation of leucine into protein and uridine into RNA.[118a, 118b] This suggests that crustecdysone affects protein metabolism in the hepatopancreas, and this view is supported by the isolation from the hepatopancreas of two proteins which preferentially bind crustecdysone, and which are therefore likely to be specific receptor proteins for the hormone.[118c, 118d]

Removal of the eyestalks from the crab *Pachygrapsus crassipes* and the crayfish *Procambarus* increases lipid synthesis in the hepatopancreas. This could be the result of consequent crustecdysone release, but no effect is obtained after the injection of crustecdysone, and the increased lipid synthesis caused by eyestalk removal is unaffected by removal of the Y-organs.[21c, 214a, 214b] Thus the changes in lipid metabolism during early

proecdysis are not controlled by crustecdysone but instead are affected by the withdrawal of an eyestalk neurosecretory hormone.

The eyestalk neurosecretory system may also exercise control over the production of digestive enzymes by the hepatopancreas. When the eyestalks are removed from the crayfish *Procambarus clarkii*, the synthesis of amylase by the hepatopancreas is significantly reduced. Since amylase, like all enzymes, is a protein, this result could reflect an overall reduction in cellular protein synthesis by the hepatopancreas: in fact, during the 7 days after eyestalk removal, the RNA content of the hepatopancreas decreases to zero, to reappear when eyestalk extracts are subsequently injected.[91] The eyestalk hormone must clearly act upon some fundamental metabolic activities of the cells of the hepatopancreas. Since all RNA disappears from the hepatopancreas after eyestalk removal, then presumably all enzyme synthesis must cease, although those enzymes other than amylase have not yet been investigated. The endoplasmic reticulum of the hepatopancreatic cells shows degenerative changes after eyestalk removal which certainly suggests a general effect upon protein synthesis. The hormone may not affect RNA synthesis directly; instead, its removal could stimulate the production of RNAase. If this profound effect of eyestalk removal upon the hepatopancreas is confirmed, it would throw very considerable doubt upon the validity of many studies of the metabolism of the hepatopancreas after eyestalk removal.

Calcium metabolism

Calcium is an important constituent of the crustacean cuticle, and the extent of calcification varies with the stage of the moult cycle. The endocuticle, which is the principal calcified layer, is not formed until late metecdysis and so there is a net increase in the calcium content of the cuticle during late metecdysis and intermoult. During proecdysis, calcium is withdrawn from the old cuticle and is commonly hoarded in the hepatopancreas, gastroliths (calcium deposits in the stomach lining) or blood in terrestrial and freshwater species, but may be either stored or excreted in marine forms.

The control of calcium mobilization is complex, and in the crayfish *Orconectes virilis* neurosecretory hormones from the eyestalks and crustecdysone both appear to play a part. Eyestalk extracts prepared from animals in proecdysis reduce the calcium content of the cuticle when injected into intermoult animals, and extracts from animals in intermoult increase the cuticular calcium content when injected into animals in proecdysis.[204] Eyestalk hormones thus seem to be involved in the regulation of the normal changes in cuticular calcium content which occur during the moult cycle,

and there may be different hormones operating in proecdysis and inter-moult.

The injection of crustecdysone into intact intermoult *Orconectes* causes gastrolith formation, suggesting that this hormone is also involved in calcium mobilization. But when the amount of calcium/mg dry weight of cuticle is measured, crustecdysone is found to have no effect whereas the injection of proecdysial eyestalk extracts reduces the amount of calcium/mg dry weight of cuticle.[205] The most plausible interpretation of these results is that the eyestalk neurosecretory hormones regulate calcium mobilization alone, while crustecdysone influences the dissolution of the organic matrix of the cuticle, and in consequence its associated calcium content. Crustecdysone and the neurosecretory hormones thus act synergistically to control the calcium content of the cuticle.

Respiratory metabolism

It is sometimes claimed that crustecdysone primarily affects respiratory metabolism. As the epidermal cells secrete layers of protein and chitin in the new cuticle, their oxygen consumption rises from 0·49 to 0·85 µl O_2/mg dry weight/hour. After removal of the eyestalks in *Gecarcinus*, there is a prolonged increase in oxygen consumption indicative of proecdysial metabolism which continues until the next moult.[19] It is now thought more likely that these changes in respiratory metabolism are secondary to some other actions of crustecdysone. But in *Uca pugilator*, the eyestalks may produce hormones which regulate respiration independently of any effects of the hormones controlling the moult cycle. Removal of the eyestalks increases the oxygen consumption of warm-acclimatized crabs by 117%, and of cold-acclimatized crabs by 47%. Since cold-acclimatized crabs do not moult, it is assumed that crustecdysone is not produced even though the source of the moult inhibiting hormone is removed. That the oxygen consumption still increases in these animals suggests the existence of eyestalk hormones which regulate oxygen consumption independently.[245b] This is confirmed by the effects of sinus gland extracts on oxygen consumption. Extracts fom cold-acclimatized crabs increase oxygen consumption when injected into warm-acclimatized crabs, and extracts from warm-acclimatized crabs decrease oxygen consumption when injected into cold-acclimatized crabs. Two distinct eyestalk hormones might then affect respiratory metabolism, one reducing the metabolic rate of crabs under high temperature stress, and the other increasing the metabolic rate of crabs exposed to cold stress. These hormones would thus be adaptive in enabling the poikilothermic crab a greater metabolic independence of the ambient temperature.[245c]

Hyperglycaemic hormone

Aqueous extracts of eyestalks injected into crabs rapidly increase the concentration of reducing sugars in the blood (hyperglycaemia). As little as 0·001 'eyestalk equivalents' doubles the reducing sugar concentration in the blood of *Callinectes sapidus* one hour after injection. Eyestalk removal decreases the reducing sugar level of the blood in some species only, but injections of sinus gland extracts induce hyperglycaemia in all species tested.[238]

Stress of any kind also leads to hyperglycaemia, except after eyestalk removal or when the X-organ–sinus gland tracts have been cut. The release

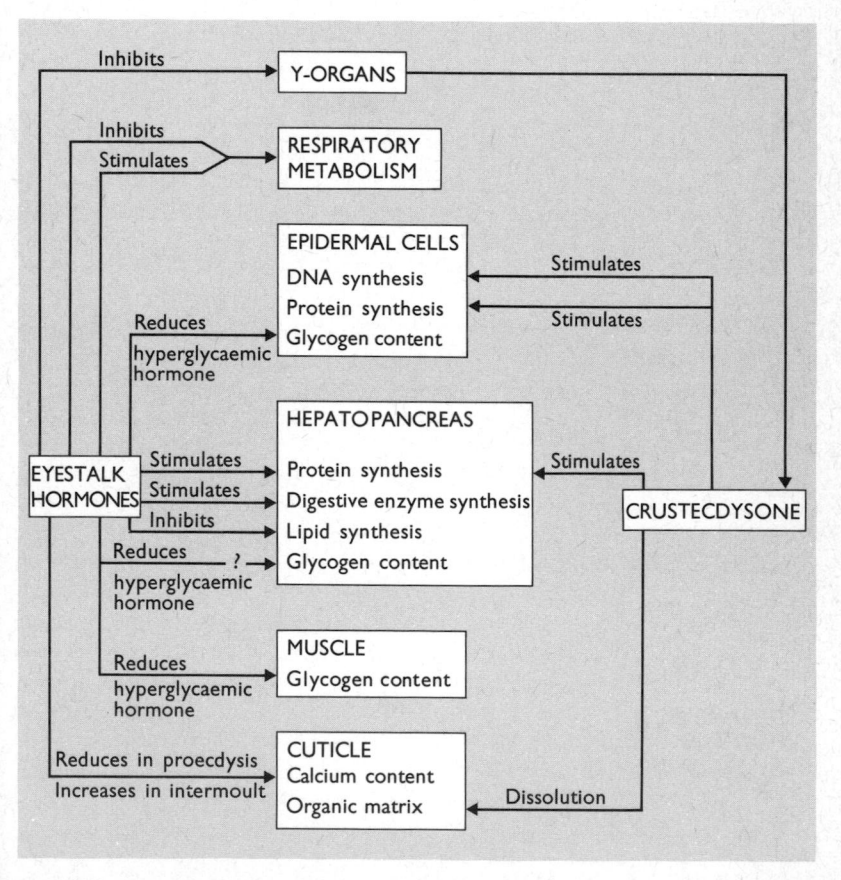

Fig. 9.5 The actions of crustecdysone and the eyestalk neurosecretory hormones during the crustacean moult cycle.

of neurosecretory hormones is a characteristic response to stress in many animals. It might be thought that the hepatopancreas would be the target organ for the hyperglycaemic hormone, but in *Uca* and *Orconectes* there is no decrease in the glycogen content of the hepatopancreas during the hyperglycaemic response. However, in *Orconectes* glycogen is reduced in the abdominal muscles, epidermal cells and gonads after injection of sinus gland extracts, and is also reduced in the epidermal cells and gills of *Uca*.[167b] This suggests that the epidermal cells and muscles are the more important target tissues for the hormone in these species, although the possibility must be considered that these tissues are reacting to other metabolic effects of the hormone, as is thought to be the situation in insects (p. 129).

Hypotheses about the hormonal control of metabolic processes in the Crustacea suffer from insufficient knowledge about the normal metabolism of the animals, and also from the lack of pure hormone preparations: the hyperglycaemic hormone alone has been isolated and partially purified (p. 265). The use of whole eyestalk, or even sinus gland, extracts means that several hormones are injected simultaneously; eyestalk removal similarly removes the source of not one but many hormones. In normal animals, the differential release of the separate hormones must occur: injections of eyestalk or sinus gland extracts may very well provide a mixture of hormones in proportions and concentrations that are never found naturally and in consequence could produce effects which are abnormal. Detailed and convincing studies on the control of metabolism in Crustacea must await the availability of pure hormone preparations. The influences of hormones upon various tissues during the moult cycle are summarized in Fig. 9.5.

HYDROMINERAL REGULATION

Water and ions in crustaceans are regulated with relation to two distinct processes. In many species, the uptake and retention of water during proecdysis is an integral part of the moulting cycle: the water is used for enlarging the new cuticle and the redistribution of the water can help in determining the shape of the new cuticle. In addition, fresh water and estuarine species need to regulate water uptake from, and ionic exchange, with, their environment.

There is evidence that water uptake during proecdysis is controlled by hormones. A normal *Carcinus* increases in volume by about 80% at each moult; the increase is nearer 180% in eyestalkless animals, due entirely to a greater water content. Sinus gland extracts injected into eyestalkless

animals counteract the abnormal water uptake, and when injected into normal *Carcinus*, less water than usual is taken up at the moult.[38]

The water-balance hormone is distinct from the moult inhibitory hormone. Sinus gland extracts taken from *Carcinus* at any stage of the moult cycle reduce water uptake in both normal and eyestalkless crabs, but the extracts delay proecdysis only when prepared from sinus glands taken during the intermoult period. Clearly, it is possible for the sinus glands to contain water balance hormone without any concomitant moult inhibitory hormone activity. Furthermore, decapods typically shed their exoskeletons at a time and place of their own choosing: ecdysis can be postponed if conditions are unsuitable. The moment of ecdysis is determined by the uptake of water to cause swelling and splitting of the old cuticle. Since this moment can be varied independently of the progress of proecdysis, it supplies further support for the independence of water balance and moult-inhibiting hormones.

The water balance hormone is clearly stored in and released from the sinus glands, but the neurosecretory centre responsible for its manufacture has not yet been determined. The way in which the hormone acts is also unclear: it could act by regulating nervous centres responsible for drinking, or by affecting directly the antennal and maxillary excretory organs.

Eyestalk removal in *Carcinus* thus has a dual effect: the onset of proecdysis is accelerated and water uptake at ecdysis is increased. Several moults may follow the operation, and the considerably shortened intermoult periods do not allow the complete replacement of water by new tissues. Consequently, the eyestalkless animal becomes more waterlogged at successive moults with the net result that the animal dies from a dilution of the tissues.

The control of water content during the moulting cycle is different in *Gecarcinus lateralis*, which is a land crab. *Gecarcinus* depends upon relatively infrequent rain showers and perhaps more frequent dews for its moisture, and can take up water from a damp substrate by means of setae and folds on its pericardial sacs. Thus water cannot be taken up from the environment during proecdysis, but must be acquired in advance.[19b] *Gecarcinus* increases in size at ecdysis by movements of water from the haemolymph into the gut and its diverticula. Removal of the eyestalks has no effect upon the water content of *Gecarcinus*, but if the fused thoracic ganglia are implanted into an eyestalkless crab several days before ecdysis, the water content of the gut is increased.[19b, 20] The fused thoracic ganglia could thus produce a hormone which controls the redistribution of water within the animal at ecdysis. The principal site of release of the hormone is probably the sinus gland, but peripheral release from the fused ganglia can also occur. Thus in the sense that water moves from the blood under

the influence of the hormone from the fused thoracic ganglia, the hormone could be called diuretic in its action. But to be a true diuretic hormone, the water would have to be passed out of the body under its influence; in *Gecarcinus*, it is likely that the water is eventually reabsorbed.

In normal *Gecarcinus*, the foregut is permeable to water and ions in both directions during intermoult and after ecdysis. But if the eyestalks are removed, the foregut becomes impermeable after ecdysis. If an extract of the fused ganglia is added to the 'blood side' of a foregut *in vitro*, there is a large and immediate increase in permeability. This is found both in normal crabs during intermoult, and in eyestalkless crabs after ecdysis. The 'diuretic' hormone of *Gecarcinus* thus seems to control the permeability of the foregut, and hence presumably the direction of the resulting net movement of ions and water.[201a] It is likely that the hormone controls water and ionic balance at all times, but its action becomes especially significant at ecdysis.

In the freshwater crayfish *Procambarus clarkii*, removal of the eyestalks causes an influx of water and an increased flow of urine, together with a reduction in the concentration of blood chloride ions. These changes are reversed by the injection of eyestalk or brain extracts; in normal animals, the injection of such extracts increases the concentration of chloride in the blood and also the uptake of sodium ions from the external medium. Brain extracts from other crustacean species have differing effects when injected into *Procambarus*, depending upon the environment from which the animals came. Brain extracts from a freshwater prawn have an effect similar to that of brain extracts from the crayfish itself, but brain extracts from marine crabs decrease both the blood chloride concentration and sodium influx.[158n] It is also possible that water influx can be controlled independently of sodium movement: in the estuarine crab *Thalamita crenata*, two active factors are present in the central nervous system which produce opposite effects on water movement when tested upon both marine and freshwater crustaceans. A water-soluble fraction reduces water influx, whilst an acetone-soluble fraction increases the rate of water influx.[265k]

The evidence thus suggests that hydromineral regulation in crustaceans is controlled by hormones which are probably neurosecretory in origin. But no uniform pattern of control has yet emerged, except that the mechanisms of control depend upon the environment of the animals and the particular osmotic problems which they face. The variety of osmotic behaviour exhibited by crustaceans probably necessitates a great diversity of endocrine controlling mechanisms.

IO

Endocrine Mechanisms in Crustacea—II

Endocrine Control of Reproduction

In the insects, sexual differentiation is held to be entirely genetic, both the primary and secondary sex characters developing under the direct influence of the genome without the intervention of circulating hormones. Recent work on the control of sexuality in the glow worm, *Lampyris noctiluca* (p. 131), indicates that such genetic control may not be of general occurrence. But in the Crustacea both sexual differentiation and gonadal activity can be influenced by hormones, to some extent resembling the situation in the vertebrates.

THE CONTROL OF SEXUAL DIFFERENTIATION

Most of the Malacostraca are bisexual, only a few groups showing protandric hermaphroditism. In the bisexual species, sex is determined genetically, but the morphological and functional expression of this genetic sex is largely influenced by hormones. The young malacostracan acquires its specific form either at hatching or at the end of a larval period: in either case, the sexes cannot usually be distinguished and this asexuality may persist for several moults. When sexual differentiation does begin, it progresses at successive moults, and in many species the

full development of secondary sexual characters is not complete until well after gonadal maturity. External differences between the sexes may be due not only to the presence of special sex organs, such as the male's penis or the female's oostegites, but also to variations in the morphology of structures common to both (Fig. 10.1).

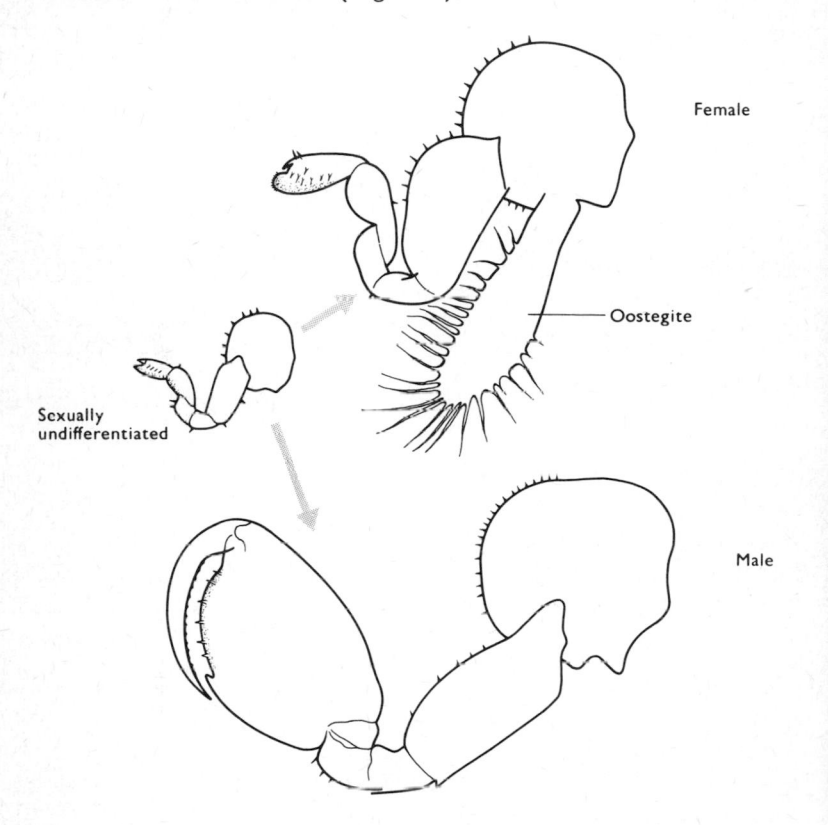

Fig. 10.1 Sexual dimorphism of the second gnathopod of *Orchestia gamma-rella*. The undifferentiated gnathopod develops into a powerful clawed structure in the male and a slender non-clawed limb in the female. The female gnathopods also develop oostegites—plates which form a brood pouch for the young. (After Charniaux-Cotton[48])

In undifferentiated young Crustacea, the primordia of the vasa deferentia are present in both sexes. Attached near the end of each vas deferens is a gland called, because of its function, the androgenic gland (see below), lying between the muscles of the coxopodites of the last pair of walking legs (Figs. 2.19, 10.3). In a genetic female, the androgenic glands fail to

develop, but in the male, the glands enlarge to form solid strands of cells, folded several times (Figs. 10.2, 10.3). The cells of the androgenic gland secrete in a holocrine manner, emptying their total contents into the blood. The androgenic glands were first discovered in the amphipod, *Orchestia gammarella*, in 1954, and have since been described in representatives of most of the malacostracan orders.[46, 47, 49] How are the androgenic glands involved in the control of sexual differentiation?

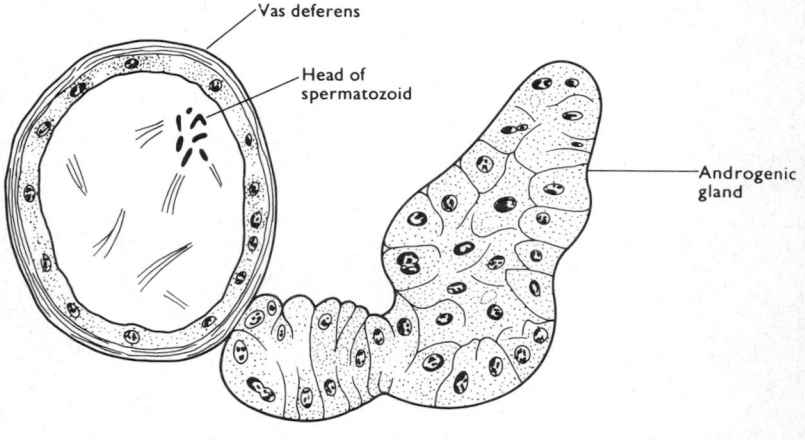

Fig. 10.2 Section through the vas deferens and androgenic gland of *Orchestia gammarella*. (After Charniaux-Cotton[48])

In *Orchestia*, when the androgenic glands are removed from a young male, the primary germ cells in the testis produce not spermatocytes but oocytes. Usually, this oogenesis is subsequently reversed because it is difficult to remove the glands completely from young animals and they regenerate. But it seems that a hormone from the androgenic glands is necessary for the normal development of the testes; in its absence, the primary germ cells develop spontaneously into oocytes.

In genetic females, the androgenic glands do not develop. But when androgenic glands from a young male are transplanted into a female, the ovaries rapidly transform into testes (Fig. 10.4), the primary germ cells giving rise to spermatocytes, which develop fully through spermatids to spermatozoa. The follicle cells turn into cells similar to those found in the normal testis, which secrete mucus around the spermatozoa. The rudimentary vasa deferentia in the female also grow and develop.[50] The implantation into females of testes, or of vasa deferentia without androgenic glands, has no effect upon ovarian development, but masculinization of the ovaries can be achieved by extracts of androgenic glands injected into females and by blood transfused from a male *Carcinus* into a

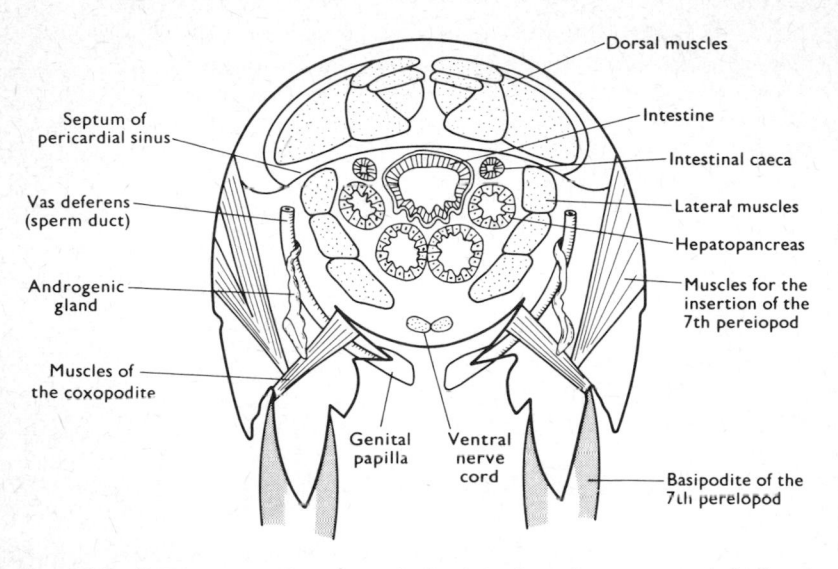

Fig. 10.3 Transverse section through the last thoracic segment of *Orchestia gammarella*, showing the position of the androgenic glands. (After Charniaux-Cotton[50])

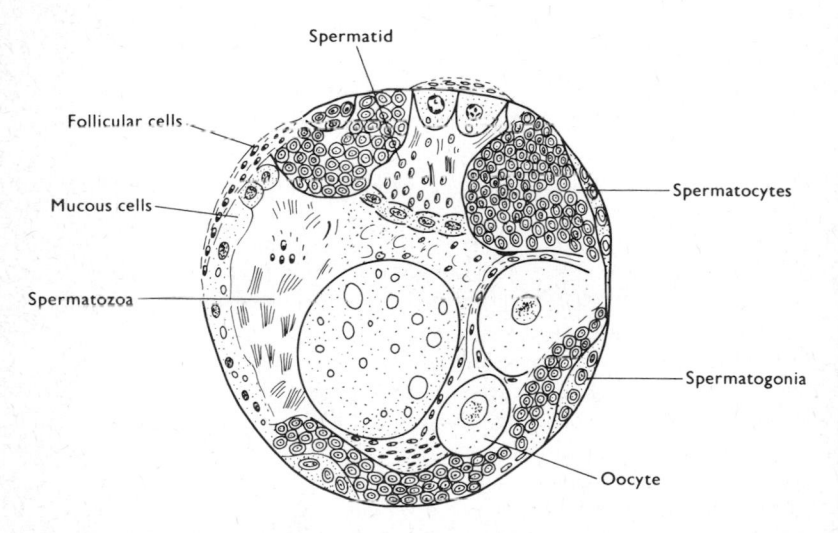

Fig. 10.4 Ovary of a mature female *Orchestia gammarella* which has been masculinized by the implantation of an androgenic gland. The primary germ cells have given rise to spermatocytes; spermatids and spermatozoa are also present. (After Charniaux-Cotton[50])

female *Orchestia*. The androgenic hormone can act in the female as it does in the male. Moreover, the hormone is not species-specific.

Thus the genetic determination of the primary sex characters—the testis and ovary—is expressed through the androgenic gland. In a genetic male, the glands develop and their androgenic hormone induces the gonads to become normal testes. In a genetic female, the androgenic

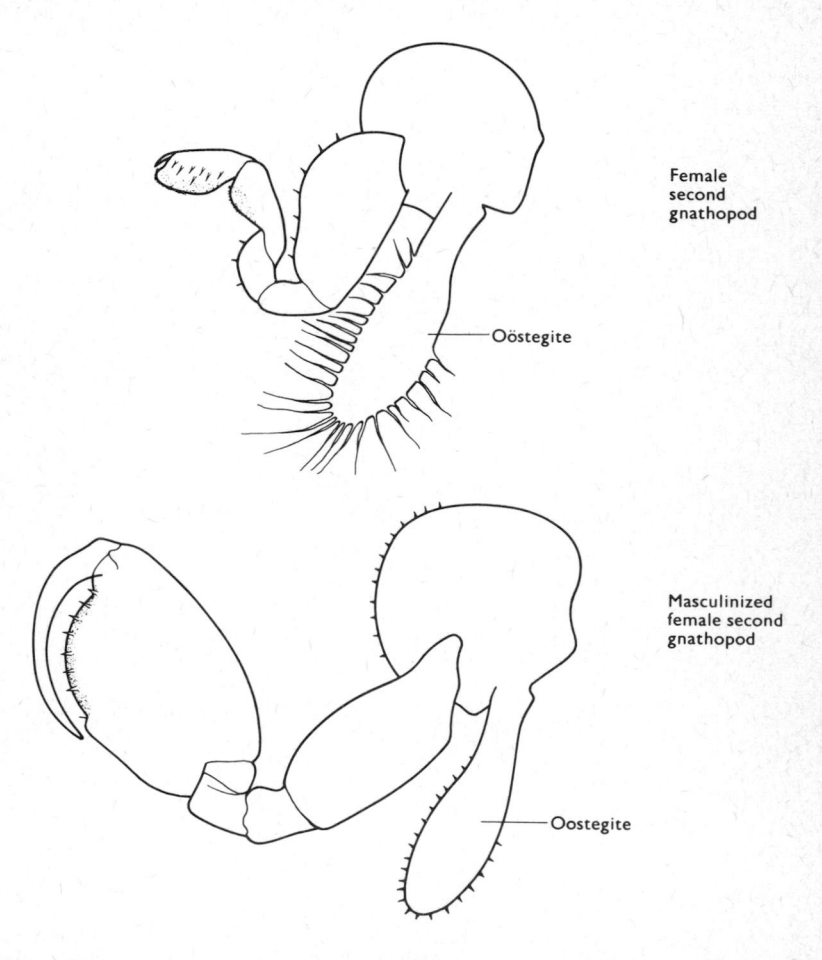

Female second gnathopod

Oöstegite

Masculinized female second gnathopod

Oostegite

Fig. 10.5 *Orchestia gammarella:* transformation of the female second gnathopod into the male form after implantation of an androgenic gland. The gnathopod progressively acquires the male form during the period following implantation of the gland. The only evidence of its original female structure is the presence of the oostegite. (After Charniaux-Cotton[48])

glands do not develop, and in the consequent absence of the androgenic hormone, the gonads develop into normal ovaries.

But how do the secondary sexual characteristics develop? Are they determined genetically, or are they also hormonally controlled? In *Orchestia*, the claw on the gnathopod is small and slender in the female, well developed and powerful in the male. In a female whose ovaries have been masculinized by the implantation of androgenic glands, the feminine gnathopod claw is transformed into the masculine form[48] (Fig. 10.5). Even in a female ovariectomized before the androgenic glands are implanted, the female claw transforms to the male type, showing that the effect is due directly to the action of androgenic hormone, and is not mediated through the masculinized gonad. Moreover, females masculinized by the implantation of androgenic gland *behave* exactly like males, and will even mate with normal females. Fertilization is impossible, however, because the newly induced sperm ducts do not have a lumen; but freshly laid eggs can be fertilized when sperm taken from masculinized ovaries is deposited upon them.

The male secondary sex characters are induced by the direct action of androgenic hormone upon the tissues. When the androgenic glands are removed from a male *Orchestia*, and at the same time the appendages are amputated, new appendages regenerate without the typical male form, as would be expected. But the new appendages do not have the female form either—they are undifferentiated rather than being male or female (Fig. 10.6). This can only mean that in the female, where the androgenic

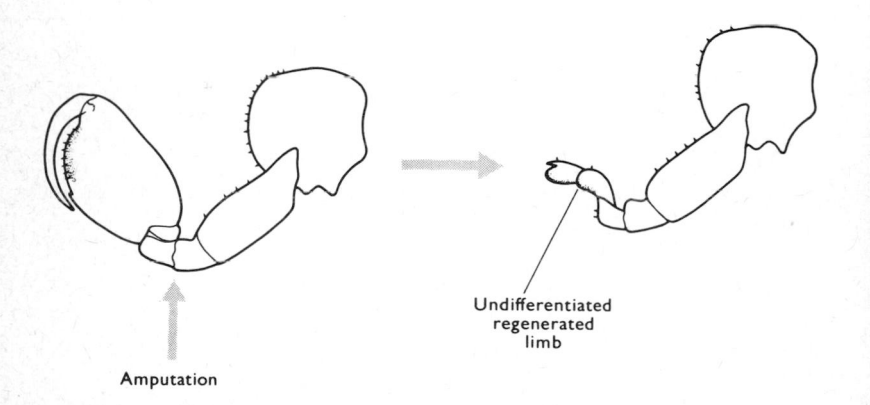

Amputation

Undifferentiated
regenerated
limb

Fig. 10.6 Regeneration of the second gnathopod of a male *Orchestia gammarella* when the androgenic gland is removed at the same time that the limb is amputated. The gnathopod regenerates as a sexually undifferentiated limb. (After Charniaux-Cotton[48])

glands are undeveloped, the secondary sex characters must be determined by something other than the mere absence of androgenic hormone.

The female *Orchestia* possesses oostegites—plates which form a brood
pouch for the young—which is a permanent secondary sex character.
But in addition, long ovigerous hairs appear on the oostegites during a
moult preceding the laying of eggs, replacing the short juvenile hairs of
immature forms, or of females in sexual repose[48] (Fig. 10.7).

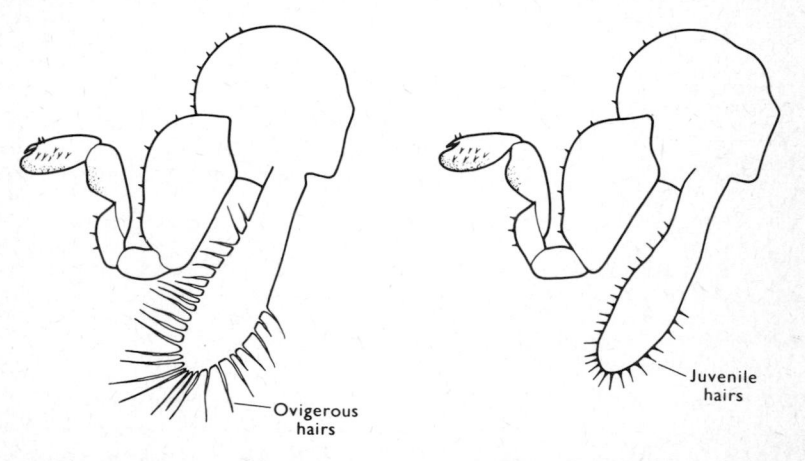

Fig. 10.7 Second gnathopods of female *Orchestia gammarella* showing the
juvenile hairs on the oostegites of immature animals or those in sexual repose;
and the *ovigerous* hairs which appear on the oostegites during the moult preceding egg laying.

When an ovary from a young or a maturing female is implanted into a
male whose androgenic glands have been removed, oostegites appear at
the first or second postoperative moult. This permanent female secondary
sex character is thus induced by an ovarian hormone. Similarly, total
ovariectomy of a reproductive female results in the replacement of
ovigerous by juvenile hairs at the first or second postoperative moult.
When an ovary, or any portion of one, is implanted into an ovariectomized female, yolk deposition in the oocytes begins after the first postoperative moult, and ovigerous hairs are developed after the second
postoperative moult. Thus the temporary female sex characters are also
controlled by an ovarian hormone.[45]

It is likely that in *Orchestia*, the ovarian hormones controlling oostegite
development and the formation of ovigerous hairs are distinct from each
other (Fig. 10.8). The latter is secreted only when vitellogenesis proceeds,
which accounts for the intermittency of ovigerous hair development, and
since all parts of the ovary produce the hormone the follicle cells are its

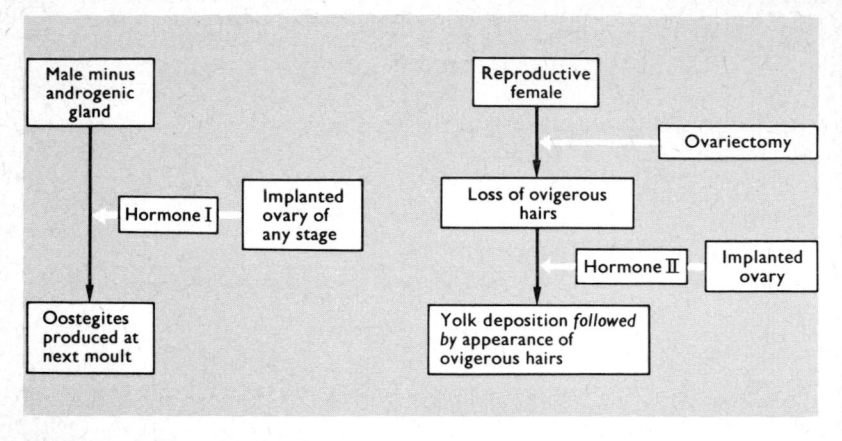

Fig. 10.8 Experiments which suggest that two hormones are produced by the ovary, controlling oostegite development and ovigerous hair formation respectively. An ovary implanted into a male whose androgenic glands have been removed will always induce oostegite formation at the next moult, implying a constant secretion of this hormone. The ovarian hormone inducing ovigerous hair formation is secreted *only* when vitellogenesis proceeds.

probable source. But the oostegite-controlling hormone must be secreted continuously, to explain the invariable effect of ovary transplantation into males without androgenic glands. The source of this hormone is probably the germinative zone or young follicles.

PARASITIC CASTRATION[51]

In the Crustacea, protandric hermaphroditism is readily explained by a hormonal mechanism of sexual differentiation. The regression of the androgenic glands terminates the male phase and the testes are transformed to ovaries in the absence of androgenic hormone. In the same way, the frequent occurrence of intersexes in the Malacostraca is understandable: a partial failure in the hormonal mechanism of control can lead to gonads which contain spermatozoa and oocytes, and to mixtures of male and female secondary characters. Testes containing some oocytes are common, due to failure of some germinative cells to get sufficient or any androgenic hormone.

The phenomenon of parasitic castration is also more readily explained now that it is known that sexual differentiation is hormonally controlled. Male crustacea parasitized by *Sacculina carcini* or other Rhizocephala, show changes in their secondary sex characters, which often resemble the female form. It was previously thought that destruction of the testes

brought about these changes, or that the parasite imposed metabolic demands upon the male which were similar to those caused by developing eggs in the female, and the male was feminized as a result.

In fact, there is no castration of the male by the parasite, although infrequently there may be slight atrophy of the testes. Moreover, the secondary sex characters of the female crustacean appear before vitellogenesis (see above) and femaleness therefore cannot be due to excessive metabolic demands. But very significantly, the testes of parasitized males frequently contain oocytes, and rarely the testes are transformed completely into ovaries. Consequently, parasitic castration must now be explained in terms of a reduction in the amount of androgenic hormone in the blood. It is likely that the parasite removes the hormone from the blood. This is followed by transient hypertrophy of the androgenic glands, in an attempt to produce the high concentration of hormone necessary to maintain the male characters. Later the glands atrophy, probably as the result of their previous unnatural hyperactivity together with the influence of the parasite. The male sex characters regress more or less completely at each moult as the androgenic hormone disappears; the appearance and development of female characters at subsequent moults would then depend upon the degree of metamorphosis of testes into ovaries.

THE CONTROL OF GONADAL ACTIVITY

Reproductive activity and growth in decapod Crustacea are antagonistic. Female reproduction alternates with moulting and occurs during the intermoult period. Some decapod crustaceans moult throughout the year or moult several times within a specific season, while others moult only annually. The moulting pattern thus imposes perennial, seasonal or annual reproductive cycles according to the species.

In the shrimp *Palaemon serratus*, the ovaries are quiescent during the summer, but vitellogenesis proceeds rapidly in October, and the first spawning occurs in November. Every moult is then followed by egg laying until the following May or June. But when the eyestalks are removed from *Palaemon* during the quiescent summer period, vitellogenesis begins at once. After the subsequent moult, the brooding characteristics appear and the eggs are laid. The implantation of eyestalk extracts into eyestalkless females prevents the precocious vitellogenesis.[215] Similar effects follow these operations in many other shrimps and crabs.[37, 116] They suggest that a hormone inhibiting vitellogenesis is produced by the X-organ–sinus gland system in decapod Crustacea. The X-organs show histological cycles associated with oocyte development,[4] and in some

isopods ovariectomy causes hypertrophy of the sinus glands; this provides supporting evidence for a relationship between the ovaries and the eyestalk neurosecretory system.

Accelerated vitellogenesis in maturing eyestalkless *Palaemon* is not merely a consequence of a reduced intermoult period and precocious proecdysis which would follow removal of the moult-inhibiting hormone (p. 213). In fact, premature vitellogenesis *prolongs* the intermoult period in the same manner as in a normal female. If the eyestalks are removed from an immature female *Palaemon*, the oocytes develop no more than they do in control animals, but somatic growth and ecdysis are accelerated. The ovarian inhibiting hormone thus appears not to be identical with the moult-inhibiting hormone. In juvenile forms, somatic growth and ecdysis take precedence over ovarian development; in mature individuals reproductive growth is of greater importance. Thus according to the developmental stage of an individual, eyestalk removal will accelerate whichever process is currently dominant. This view is supported by the situation in *Carcinus*, where the Y-organs are restrained by the moult-inhibiting hormone (p. 215) after the terminal ecdysis. If the same hormone were to inhibit both moulting and gonadal activity, it is difficult to envisage how reproduction could occur in such a permanently anecdysic species.

The ovarian-inhibiting hormone does not act alone to control the full course of ovarian development. When the Y-organs are removed from young female *Carcinus*, oogonial mitoses almost cease and no oocytes develop.[4a, 4b] Early oocyte growth must therefore be dependent upon crustecdysone, as in the insects (p. 133). Moreover, eyestalk removal in postpubertal females of some species at certain times of the year causes not vitellogenesis but moulting,[1f] suggesting either that the ovaries are only periodically or cyclically responsive to removal of the ovarian-inhibiting hormone, or that some other factor is involved in the control of vitellogenesis. The implantation of brains or thoracic ganglia into young females of some crab species results in the enlargement of the ovaries, and cycles of secretion of the neurosecretory cells within these organs can be correlated with vitellogenesis. This evidence suggests the presence of an ovarian stimulating hormone, although the sites of production of the hormone make extirpation experiments impossible.

Vitellogenesis may then be controlled by alternately fluctuating levels of ovarian-inhibiting and stimulating hormones produced by neurosecretory centres in different parts of the central nervous system. It is also possible that the moult and ovarian-inhibiting hormones act synergistically during postmoult phases, but antagonistically during intermoult phases when reproductive development occurs (Fig. 10.9).

The way in which the ovarian-inhibiting and stimulating hormones

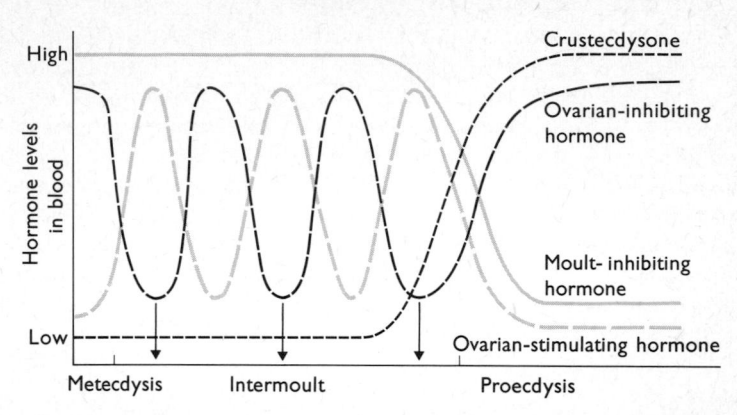

Fig. 10.9 The relationships between moult-inhibiting hormone and crustecdysone, and ovarian-inhibiting and stimulating hormones, during reproduction and moulting. The times of reproduction are shown by the black arrows. (After Adiyodi and Adiyodi[1f])

act upon the ovary is unknown. A major lipoprotein constituent of the yolk is synthesized outside the ovary, probably in the hepatopancreas, and is transported in the blood to the vitellogenic oocytes.[1f] The mobilization of this and other yolk proteins is prevented by the ovarian-inhibiting hormone. Female specific proteins have been found in a number of crustaceans, but it is not known whether their synthesis is hormonally controlled, nor whether the ovarian-inhibiting and stimulating hormones act gonadotrophically. The situation in crustaceans is as yet nowhere as clear as it is in insects.

An inhibitory hormone from the eyestalks is also present in male decapods.

In young male *Carcinus*, eyestalk removal causes precocious spermatogenesis, and growth of the vasa deferentia.[77] That these effects are produced by removing inhibition from the androgenic glands is very likely since the glands hypertrophy within a month after the operation. Eyestalk removal in mature crabs has no effect upon the activity of the testes or the androgenic glands. Inhibitory hormones thus control spermatogenesis in the male and vitellogenesis in the female, but the male hormone differs profoundly from that of the female in that it acts through epithelial endocrine organs, the androgenic glands, and at a quite different stage in the life-history. There is also circumstantial evidence that a stimulatory hormone acts upon the androgenic glands. These inhibitory and stimulatory hormones appear to be identical in both sexes.

In protandrous hermaphrodites, the change from the male to the female phase is accompanied by the degeneration of the androgenic glands

(p. 232). This is followed by the loss of the male morphology, the transformation of the gonad into an ovary, and the development of female secondary sexual characters. Sex reversal is delayed if an androgenic gland is implanted into a male about to change sex. In the prawn *Lysmata seticaudata*, eyestalk extracts injected into males of the size at which sex reversal normally takes place considerably reduce the degree of change, whereas eyestalk removal in such animals enhances the degree of change.[37] Thus the eyestalk neurosecretory system may maintain the male phase in some animals but only at the time during which sex normally changes. The exact timing of sex reversal could be correlated with external conditions through such a mechanism.

II

Endocrine Mechanisms in Crustacea—III

In the Crustacea, neurosecretory hormones are of the utmost importance in controlling growth and development, both through their trophic effects upon epithelial endocrine glands (Y-organs, androgenic glands) and also by a direct intervention in metabolic processes. But neurosecretory hormones are also concerned in the regulation of much more temporary physiological events, such as heart beat, colour change and the movement of retinal pigments, which are related to short-term environmental fluctuations (Chapter 14).

HORMONAL CONTROL OF HEART BEAT

The crustacean heart is essentially a single-chambered sac of striated muscle, pierced by afferent and efferent openings. In most Crustacea, the heart is neurogenic, with the elongated dorsal cardiac ganglion as pacemaker. Inhibitory and acceleratory nerve fibres can affect the beating of the heart, acting directly upon the neurones of the cardiac ganglion. In the decapods, one inhibitory and two acceleratory fibres enter the nerve plexus, which forms part of the pericardial organ (Fig. 11.2) on each side of the heart, before passing to the cardiac ganglion as a single dorsal nerve[203] (Fig. 11.1).

In decapod Crustacea, major portions of the pericardial organs lie across the openings of the branchio-cardiac veins (Fig. 11.2), and

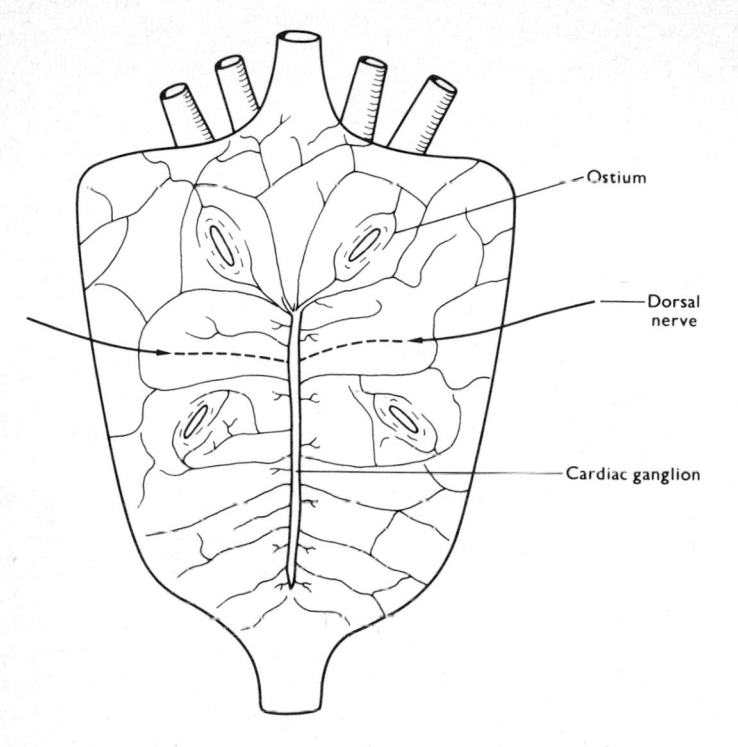

Fig. 11.1 Dorsal view of the heart of the lobster showing its innervation. The single dorsal nerve which pierces the wall of the heart on each side and enters the cardiac ganglion consists of both inhibitory and acceleratory nerve fibres. (After Maynard[203])

elementary neurosecretory granules, about 120 nm in diameter, are present within the nerves making up the organs. The position of the pericardial organs across the flow of venous blood from the gills, together with their contents of neurosecretion, suggest a role in regulating the heart beat.[12]

When isolated hearts have aqueous extracts of pericardial organs added to the perfusion medium, the amplitude of beat is always increased, and often the frequency of beating increases also. Since the heart is isolated from its nervous connections, the increased amplitude and frequency of beating cannot be due to any effect upon the inhibitory or excitatory fibres which run to the cardiac ganglion. But when nerve fibres from the cardiac ganglion are monitored to record their nervous impulses, both the frequency and duration of bursts of impulses increase when the

ganglion is treated with aqueous extracts of pericardial organs[64] (Fig. 11.3).

Electrophoresis and chromatography of extracts of pericardial organs of *Cancer* separate two peptides with cardio-excitor activity. Both peptides have a similar amino acid composition. Whether these represent two distinct hormones, or are merely the degradation products of a single hormone, is conjectural.[12] It had previously been thought that the decapod cardio-excitor was 5-hydroxytryptamine or 5:6-dihydroxytryptamine. But responses of the heart to 5-HT can be blocked without affecting its reaction to aqueous extracts of pericardial organs or to the separated peptides[64] (Fig. 11.4). This suggests that the pericardial organs

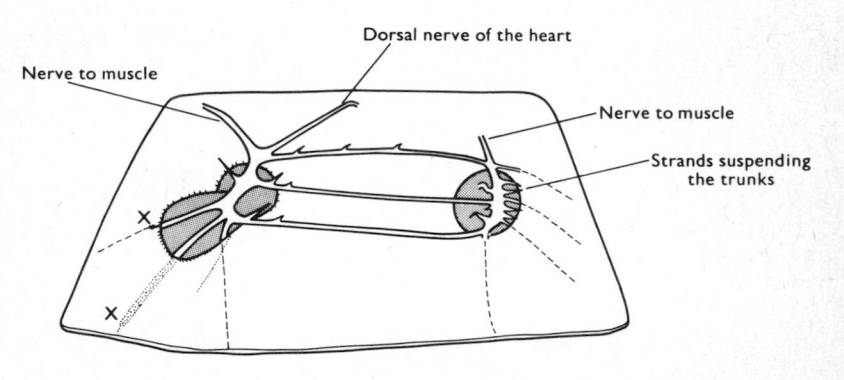

Fig. 11.2 Pericardial organs of the right side of the spider crab, *Maia squinado*. These are neurohaemal organs lying within the pericardial cavity and bathed in blood moving towards the heart. The organs are shown in place on the inside of the lateral pericardial wall. The nerves from the central nervous system which enter the pericardial organs are shown as dotted lines, and the points at which the two anterior nerves enter the lumen of the veins are shown as crosses. (After Carlisle and Knowles[41])

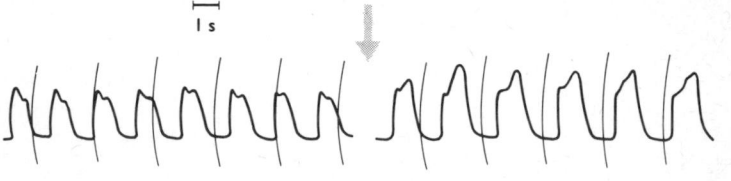

Fig. 11.3 A record of the electrical activity from an isolated cardiac ganglion of the lobster. The ganglion is held in the perfusion fluid by forceps, anteriorly and posteriorly, which also serve as recording electrodes. When extract of pericardial organs is added to the perfusion fluid, the burst *duration* increases. The burst *rate* may increase or decrease when extract is added. (After Cooke[64])

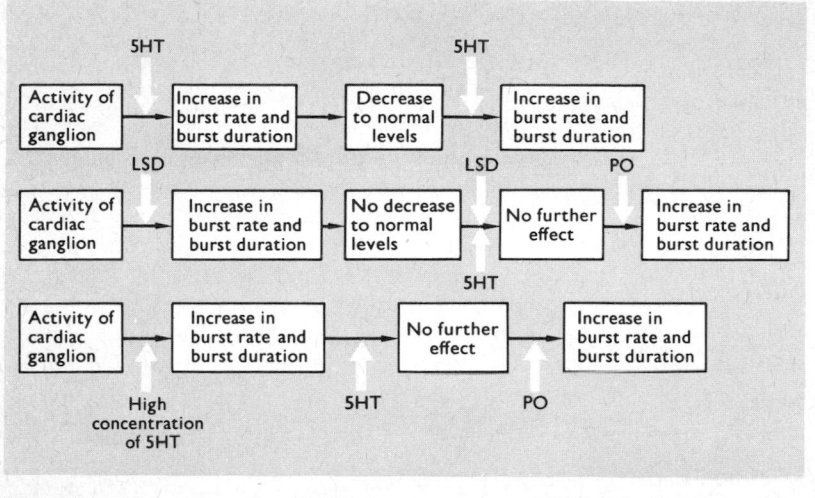

Fig. 11.4 The action of pericardial organ extract (PO), 5-hydroxytryptamine (5HT) and D-lysergic acid diethylamide (LSD) upon the electrical activity of the isolated, perfused cardiac ganglion of the lobster. LSD blocks the responsiveness of the ganglion to 5HT but not to PO. This suggests that the cardio-excitor substance from the pericardial organs is not 5-hydroxytryptamine.

release a peptide neurosecretory product, designed to initiate the action of a neurotransmitter acting upon the neurones of the cardiac ganglion.

It might be wondered why hormonal cardio-excitors are necessary when the crustacean heart can be influenced by *nervous* inhibitors and accelerators. The answer probably lies in the different natures of the two control mechanisms: the effects of the nervous regulators are essentially transitory, while that of the hormone can be very much more prolonged.

HORMONAL CONTROL OF COLOUR CHANGE[5, 88]

For more than a hundred years biologists have attempted to explain the spectacular changes in colour shown by crustaceans. Colour change may be slow and predictable, synchronized with cyclic environmental events such as season, the ebb and flow of tides, or the daily progression of night and day; or it may be quickly and reversibly associated with local and transient fluctuations in illumination or background colour.

The effectors which bring about colour changes in an individual are the chromatophores, situated directly beneath the epidermis or scattered in the deeper tissues of the body. The crustacean chromatophore is a much branched cell, whose overall shape remains unchanged during

colour changes. Instead, pigment contained within each chromatophore disperses or concentrates to alter its shade or tint (Fig. 11.5). When dark pigment is dispersed through each of many chromatophores, an overall dark coloration of the animal results; when the pigment is concentrated, the animal appears blanched.

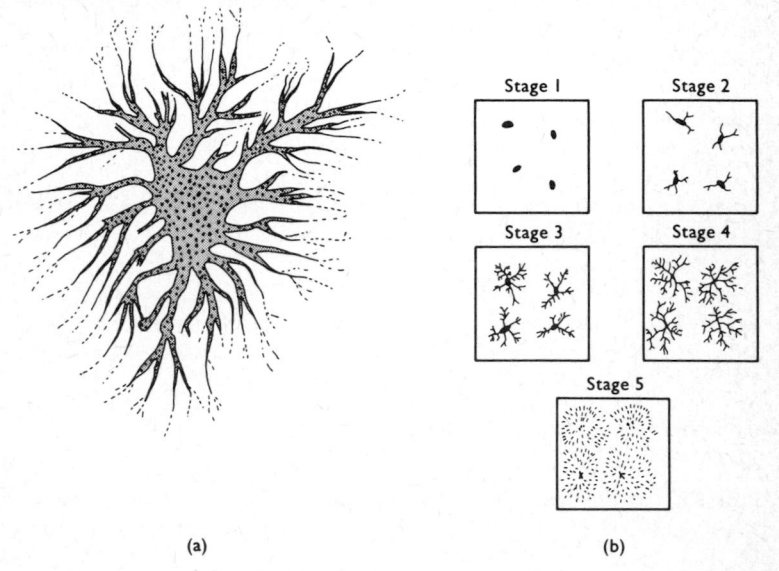

(a)   (b)

Fig. 11.5 (a) Diagram of a chromatophore from the uropod of the prawn, *Leander serratus*; the pigment is shown in its expanded state. (b) The arbitrary scale by which measurements of pigment expansion in chromatophores are made. In Stage 1 the pigment is fully contracted; Stages 2, 3 and 4 indicate progressive pigment expansion; Stage 5 is a chromatophore with pigment fully expanded.

Crustacean chromatophores may contain one (monochromatic), two (dichromatic) or several (polychromatic) pigments. Some chromatophores are multinucleate, and seem to be groups of anastomosing cells. A polychromatic chromatophore might be a syncytium of physiologically distinct cells, each containing a single pigment. Ultrastructural examination has shown that in *Crangon septemspinosa*, the chromatophores previously considered to be tetrachromatic are four separate chromatophores in close contact with each other.[51a, 84a] Red, blue, yellow, white and black pigments are commonly found in crustacean chromatophores. Polychromatic chromatophores are characteristic of the prawns and shrimps (Decapoda Natantia); mono- and dichromatic chromatophores are found in the decapod Astacura, Anomura and Brachyura, and also in the Isopoda and

Stomatopoda. Most experimental work on colour change has been limited to the Decapoda.

The shrimp, *Crangon crangon*, has a typical polychromatic chromatophore pattern with four types of pigment cell: monochromatic black, dichromatic black-red, trichromatic brown-yellow-red, and tetrachromatic black-white-yellow red. The fiddler crab, *Uca pugilator*, a brachyuran, has three kinds of monochromatic chromatophores, containing red, white or black pigment.

The control of polychromatic chromatophore systems is likely to be complex, for potentially the *independent* movement of several pigments is involved. At one time, pigment movements were thought to be under nervous control in spite of the failure of fifty years of experimentation to provide evidence for the hypothesis. But in 1928, it was found that the blood of the shrimp *Crangon crangon*, and the prawn *Palaemonetes vulgaris*, contained substances which affected the movements of chromatophores.[222, 176] This provided the first indisputable evidence for the existence of a hormone in any invertebrate, and the endocrine control of colour changes in the Crustacea has been intensively studied ever since.

The source of chromactivating hormones

After the discovery of chromactivating hormones in the blood, a systematic search suggested that tissues near the basement membrane of the eyes were a potent source of the hormones. Histologically, two possible secretory organs, the sinus gland and the X-organs (Fig. 2.20) were found in this region. Soon after, chromactivating substances were also found in the central nervous system outside the eyestalks, the circumoesophageal connectives and the postoesophageal commissures being particularly potent sources for the materials.

It was only in 1951, when the concept of neurosecretion had become firmly established, that the full significance of chromactivating hormones originating in the central nervous system was realized. It is now known that the crustacean eyestalk X-organ-sinus gland system is the source of many neurosecretory hormones, some of which control development and reproduction (Chapters 9, 10). But the chromactivating hormones in the circumoesophageal and postoesophageal commissures are also neurosecretory hormones in transit from other neurosecretory centres to neurohaemal organs in the epineurium of the posterior commissure nerves—the postcommissural organs (Fig. 2.19, and p. 27).

The control of pigment movement

Once eyestalks and other parts of the central nervous system had been

identified as sources of chromactivating hormones—now called *chromatophorotrophins*—it was quickly shown that individual chromatophores would react to more than one hormone. In *Crangon*, one hormone, soluble in alcohol, from the postcommissural organs concentrates the pigment in black chromatophores on the body, causing an overall blanching; another, insoluble in alcohol, also from the postcommissural organs, disperses the pigment in the black chromatophores on the body *and on the tail*, making the individual an overall dark shade. But in addition, the eyestalks of *Crangon* also contain two hormones, one of which blanches the body only, the other blanching the body and the tail, by concentrating the pigment in the black chromatophores. In *Crangon*, therefore, as many as four hormones are involved in controlling the movement of pigment in one chromatophore type[28] (Fig. 11.6). But

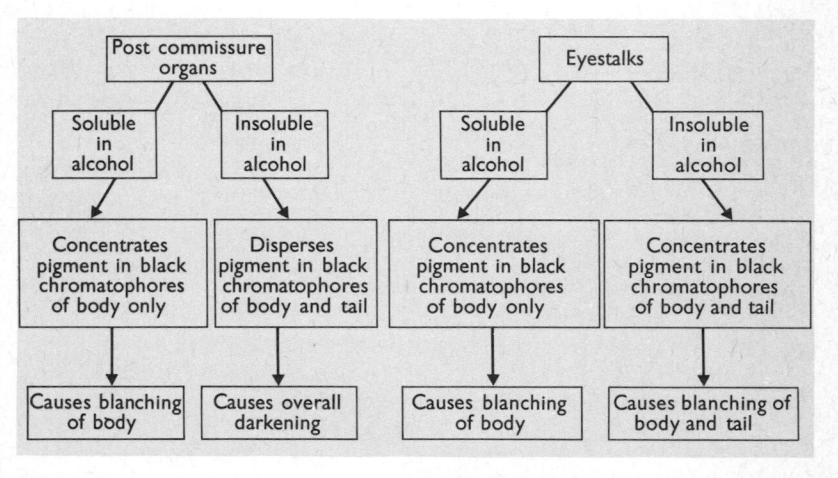

Fig. 11.6 Hormones controlling the movement of the black pigment in the chromatophores of *Crangon crangon*. Alcoholic and aqueous extracts of the postcommissure organs and the eyestalks suggest the presence of four different chromatophorotrophins controlling pigment movement in this one chromatophore type.

one very important principle emerges: when a chromatophorotrophin induces either dispersion or concentration of chromatophore pigment, its *absence* alone does not bring about the opposite movement. This is caused by a second hormone which acts antagonistically to the first. Such antagonistic dispersing and concentrating chromatophorotrophins have now been identified in at least eight crustacean genera, and the 'multiple hormone hypothesis' for chromatophore control is well estab-

lished, although the *complete* control system for all chromatophores in any one species has not yet been fully worked out.

In the prawn, *Palaemonetes*, the sinus glands contain a hormone which concentrates pigment in the red chromatophores, and another, probably identical, in the tritocerebral commissures, and also in the circumoesophageal commissures. An antagonistic red pigment dispersing chromatophorotrophin is present also in the circumoesophageal commissures and in the abdominal nerve cord.[26] The presence of the dispersing chromatophorotrophin is difficult to detect. If normal prawns are kept on a white background, the red pigment is fully contracted, but injections of nerve cord extracts containing dispersing hormone have only a slight and transient effect. But when eyestalkless animals (thus with a major source of concentrating hormone removed) are kept on a dark background, the

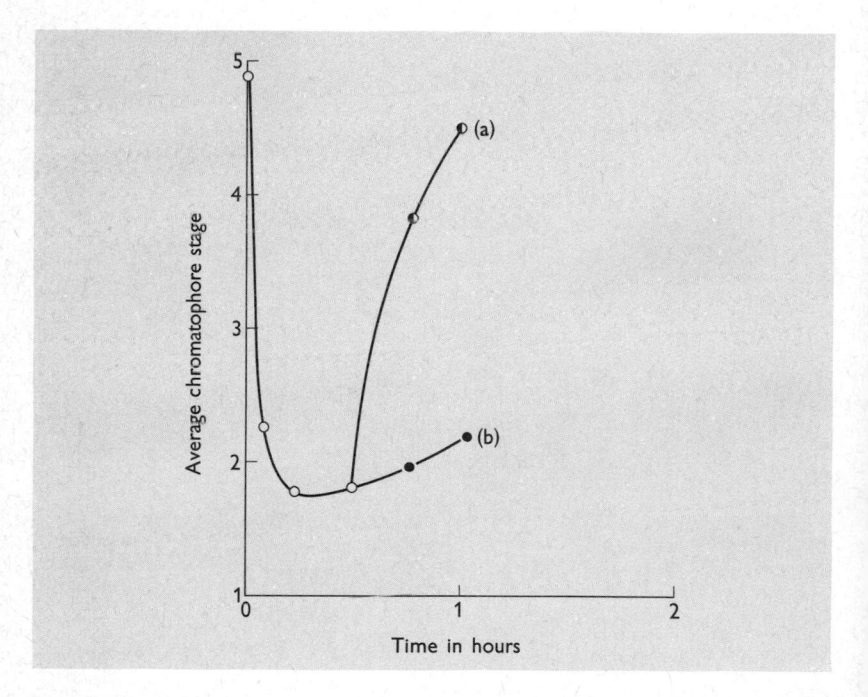

Fig. 11.7 Dual hormonal control of the red chromatophores of *Palaemonetes*. In an eyestalkless animal the pigment is fully dispersed (stage 5) ; when sinus gland or postcommissural organ extracts are injected (time zero), the pigment quickly concentrates. At the end of 30 minutes, the injection of an extract of the abdominal nerve cord (a) results in the rapid dispersal of the pigment, whereas in controls injected with sea water (b) a slow dispersal continues. (After Brown, Webb and Sandeen[26])

red chromatophores are fully expanded and the injection of sinus gland extracts (containing concentrating hormone) into these animals causes rapid and complete concentration. After a little time, the effect of the hormone wears off, and the pigment starts to expand: injection of extracts of dispersing hormone at this time markedly accelerates the rate of pigment dispersion (Fig. 11.7).

In recent years, success has been achieved in the isolation of pigment dispersing and concentrating factors from the eyestalks of various crustacean species. Three distinct chromatophorotrophins can be separated from aqueous extracts of the eyestalks of the crab *Rhithropanopeus harrisi* and the shrimp *Crangon crangon* by gel filtration on Sephadex G-25: these are a black pigment dispersing hormone, a white pigment concentrating hormone and a red pigment concentrating hormone.[246b, 246c] A different result is obtained when eyestalk extracts of *Crangon septemspinosa* are separated by gel filtration: separation on a column of Bio-gel P6 produces two distinct peaks of activity: one which disperses, and one which concentrates, the pigment in red, white and black chromatophores (Fig. 11.8). Although pigment dispersing and pigment concentrating activities separate completely from each other, the three individual dispersing and the three concentrating activities do not separate (Fig. 11.8).[89a] However, when the extracts are separated on a column of Sephadex G-25, an extra peak of black pigment concentrating activity can be isolated (Fig. 11.9). The presence of a separate black pigment concentrating hormone in the eyestalk is unique, and might be explained by the unusual chromatic response of *Crangon* to eyestalk removal. All other stalk-eyed crustaceans respond to eyestalk removal by either permanent blanching (all brachyurans) or permanent darkening, but *Crangon* first darkens the telson and uropods and blanches the rest of the body, but an hour later the telson and uropods blanch and the rest of the body shows an intermediate darkening. In *Crangon*, the black chromatophores are the most numerous, and a chromatophorotophin that concentrates the black pigment alone together with another that concentrates all three pigments would provide a finely modulated chromatophore response to different backgrounds.[89a]

Pigment concentrating factors have also been separated from the dispersing factors in extracts of the eyestalks of *Uca pugilator*.[91a, 91c] The pigment concentrating factors all separate as a single peak. This raises the possibility that different chromatophore types all respond to a single hormone. But observations on chromatic adaptations argue against this concept, as black, white and red chromatophores in *Uca* can vary independently of each other. Moreover, the synthetic red pigment concentrating hormone (p. 265) will concentrate the red pigment in the chromatophores of *Uca* and the crayfish *Cambarellus shufeldtii*, but has no effect upon the white chromatophores or the black chromatophores of *Uca*

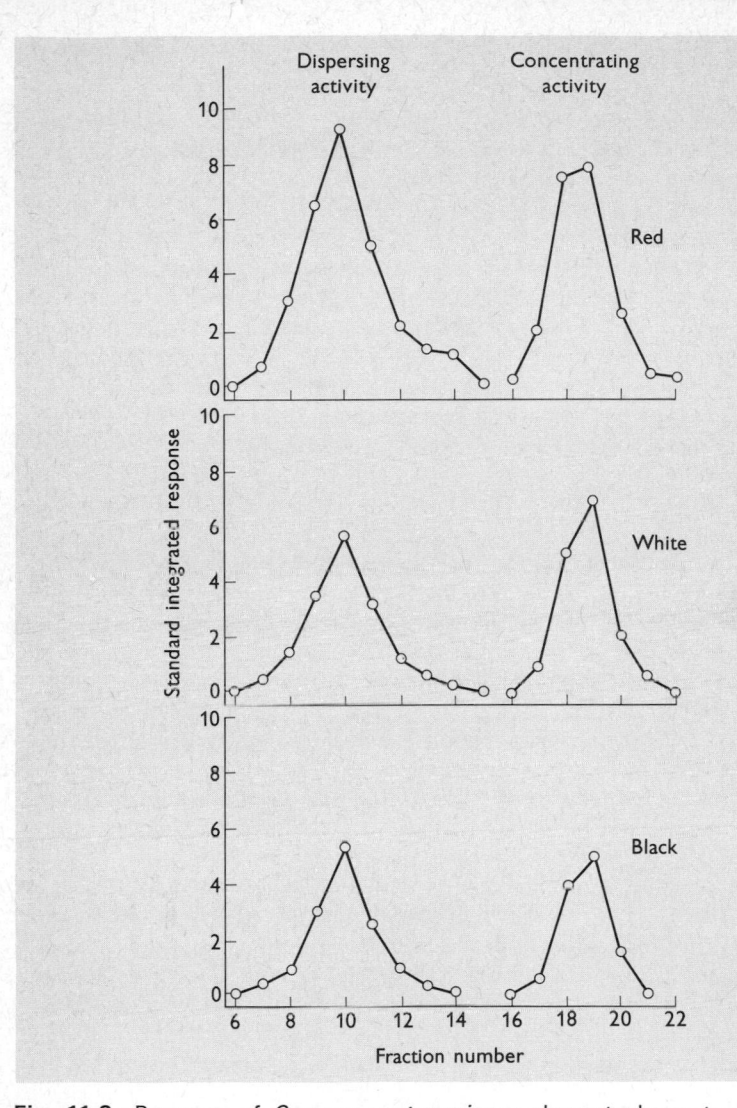

Fig. 11.8 Response of *Crangon septemspinosa* chromatophores to partially purified eyestalk extracts. The extracts were purified by gel filtration on Bio-gel P6. The extract is passed down a tube containing the gel and molecules of different molecular weights travel at different rates. The liquid emerging from the tube is collected into fractions, each fraction therefore containing molecules of a particular size range. The chromatophorotrophic activity of each fraction can then be tested. The results are expressed as a standard integrated response—a measure that includes both the amplitude and the duration of the response. Separation on this gel reveals that the dispersing activity for all three pigments is contained in fractions different from those with concentrating activity, but that there is no separation of different concentrating or dispersing activities. (After Fingerman[89a])

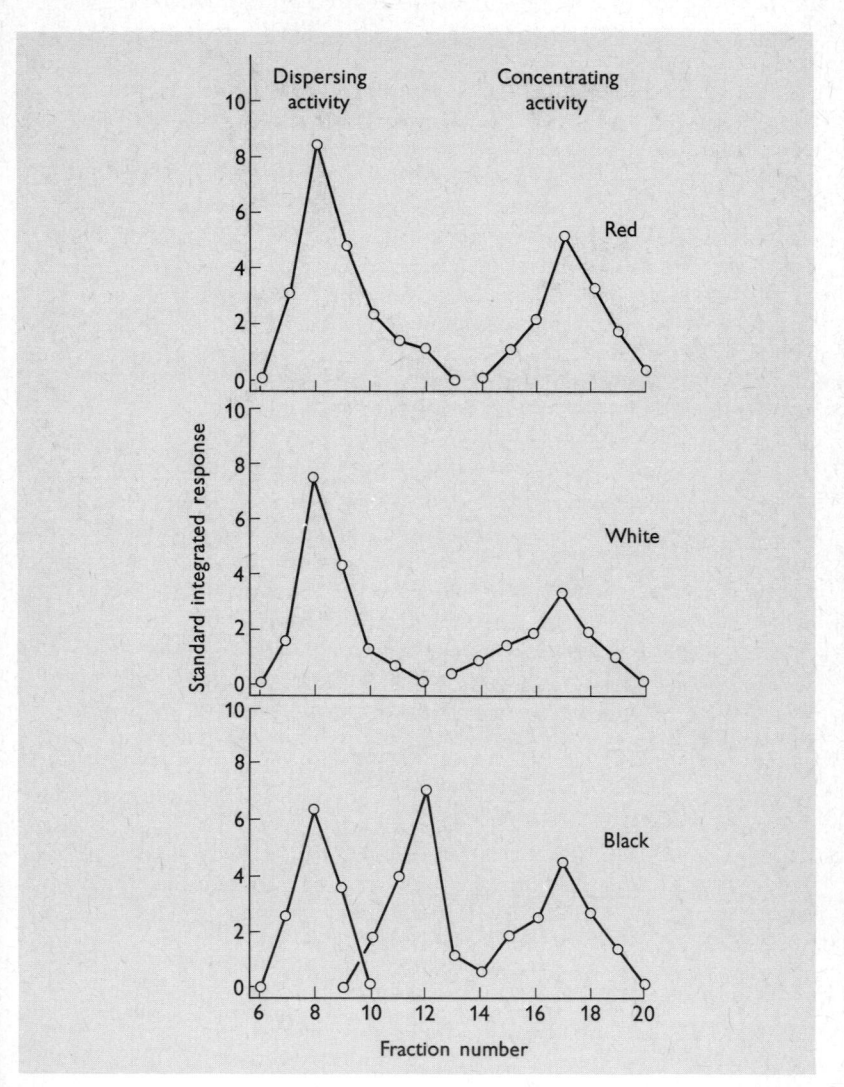

Fig. 11.9 Response of *Crangon septemspinosa* chromatophores to partially purified eyestalk extracts. The extracts were separated by gel filtration on Sephadex G-25 (compare Fig. 11.8). A separate peak of black pigment concentrating activity is revealed, indicating the presence of two distinct black pigment concentrating chromatophorotrophins. (After Fingerman[89a])

(*Cambarellus* has no black chromatophores).[89b] It must therefore be concluded that each chromatophore type is controlled by separate and distinct pigment concentrating and dispersing hormones. The black pigment dispersing hormone, and the red and white pigment concentrating hormones are not species specific, and these three chromatophorotrophins at least are probably common to all decapods.[171a, 246d]

The presence of chromatophorotrophins throughout the central nervous system of Crustacea poses the problem of whether they can properly be called hormones, since one half of the essential endocrinological experiment—removal of the source of the hormone—is impossible, for the animals die as soon as large parts of the nervous system are removed. Only chromatophorotrophins from the eyestalk neurosecretory system would be true hormones if the strict letter of the definition were followed. But the injection of extracts of the nervous system has a hormonal effect, and it is possible to show, even in normal animals, that their chromatophorotrophins are released into the blood. This has been done in *Palaemonetes* by keeping the animals for long periods on either a white or a black background. When transferred to the opposite background, the chromatophores take longer to change to the opposite condition the longer they have been maintained under the previous conditions (Fig. 11.10). This result is interpreted to mean that a greater concentration of dispersing or concentrating hormone is maintained *in the blood* when particular conditions are prolonged. But even more significantly, after 14 days on a white background, the central nervous system contains very much more dispersing hormone when assayed on eyestalkless animals than it does after only 2 hours. Similarly, the sinus glands, circumoesophageal and tritocerebral commissures contain *less* concentrating hormone after 14 days on a white background than after 2 hours, when similarly assayed. Exactly opposite effects are found when the animals are kept on a black background for 14 days.[92] These results can only mean that the hormone which is not in use during these prolonged periods accumulates within the nervous system. Together with the evidence for high blood levels of the antagonistic hormone, this provides extremely good evidence for the origin of chromatophorotrophins within the nervous system, *and* their release into the blood. Although circumstantially, all the chromatophorotrophins are therefore hormones.

Diurnal colour change

Uca pugilator, the fiddler crab, changes colour in a daily rhythm, becoming dark by day and pale by night. The rhythm persists for a long time when the crab is kept in constant darkness; and when the crabs are

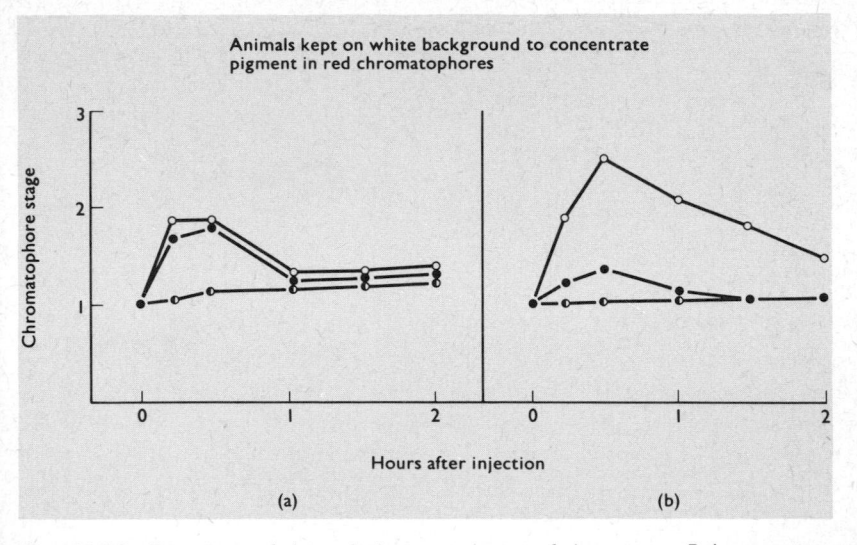

Fig. 11.10 Response of the red chromatophores of the prawn, *Palaemonetes vulgaris*, to extracts of circumoesophageal connectives. (a) Animals kept on a white background received injections of extract of circumoesophageal connectives from animals that had been kept on a white background (○) or a black background (●) for two hours. Control animals (◐) received injections of sea water. (b) Animals again on a white background received injections of extract of circumoesophageal connectives from animals kept on a white background (○) or a black background (●) for two weeks. Control animals (◐) received injections of sea water.

The results show that the amount of red pigment dispersing hormone in the circumoesophageal connectives increases significantly when animals are kept for long periods on a white background. So when the red pigment in the chromatophores is maintained in a concentrated state, dispersing hormone is not released and accumulates in the circumoesophageal connectives. (After Fingerman, Sandeen and Lowe[92])

kept at 0–3°C for several hours, the rhythm is delayed by an interval closely approximating the period of chilling (Fig. 11.11). The rhythm must therefore be controlled internally, and does not merely reflect changes in environmental illumination. The rhythm is expressed mainly through pigment movements in the monochromatic black chromatophores which are predominant in *Uca*, although red, white and yellow chromatophores are also present.[23]

When the eyestalks of *Uca* are removed, the animals become permanently pale due to concentration of pigment in the black chromatophores. The eyestalks must be the source of a black pigment dispersing hormone. When sinus glands are implanted into eyestalkless *Uca*, the rhythm is

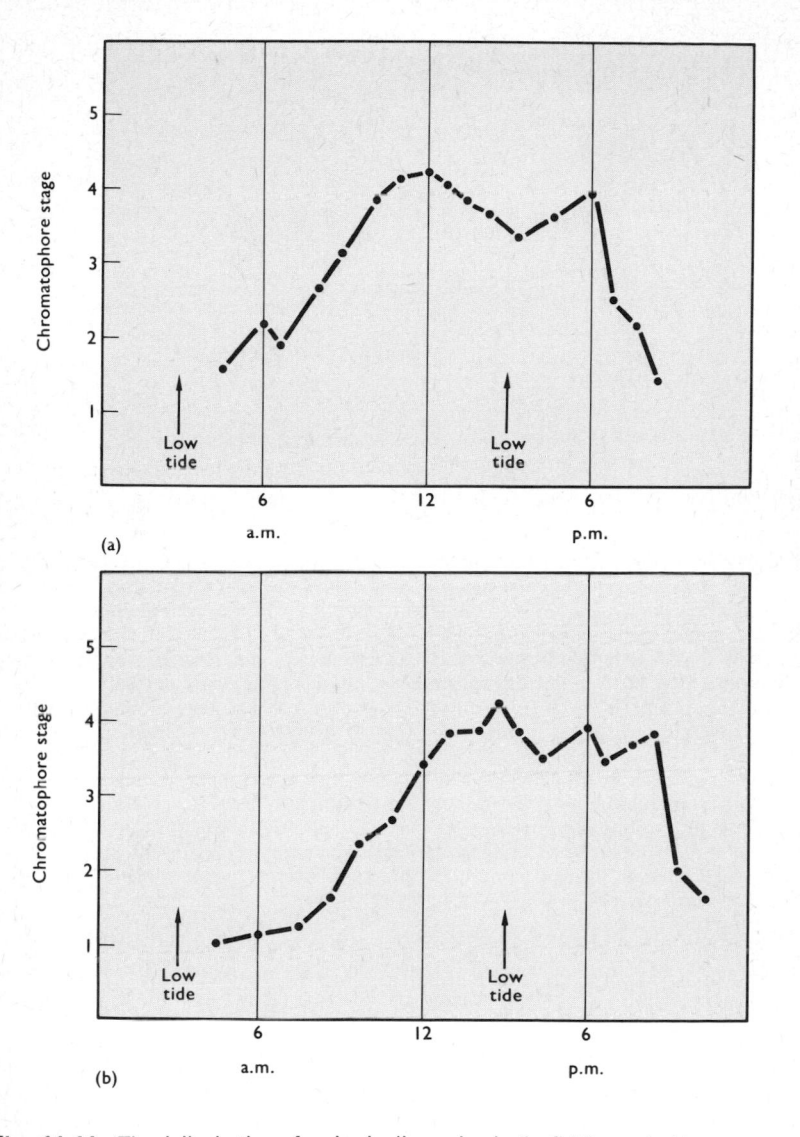

Fig. 11.11   The daily rhythm of melanin dispersion in the fiddler crab, *Uca pugnax*. (a) the normal rhythm which is maintained in the laboratory after the crabs are collected. (b) The rhythm in crabs collected at the same time as those in (a), but placed at 0–3°C for three hours. The rhythm is interrupted by the low temperature period, and the maximum dispersion within the chromatophores is delayed by three hours. (After Brown, Fingerman, Sandeen and Webb[23])

restored. It must be supposed that the pigment dispersing hormone from an isolated implanted sinus gland diffuses constantly out of the gland into the blood. The restoration of the rhythm must therefore result from the nightly release of an antagonistic pigment concentrating hormone from a source other than the sinus gland (Fig. 11.12). This is supported by the

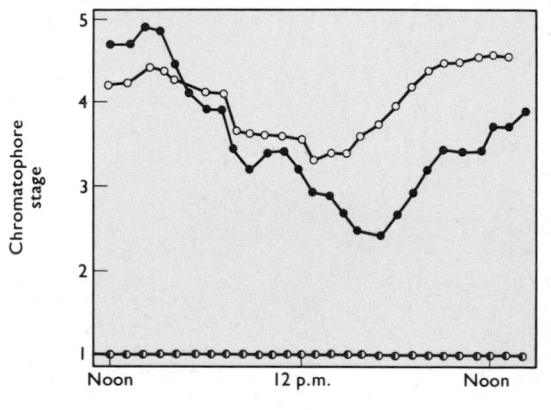

Fig. 11.12   Daily rhythm of pigment movement in the black chromatophores of normal and operated *Uca pugnax*. ○—normal crabs. ●—eyestalkless crabs that have received an implant of three sinus glands. ◐—eyestalkless crabs. For further details see text. (After Fingerman [89]).

fact that extracts of sinus glands are less effective in dispersing the black pigment in eyestalkless *Uca* when injected at night than when injected during the day.[89]

Uca shows a tidal rhythm of pigment movement superimposed upon the diurnal rhythm. Supplementary dispersion of melanin in the chromatophore occurs between 1–3 hours prior to the time of low tide. Thus when low tide occurs near midday, the peak of diurnal chromatophore pigment movement is reinforced, whereas low tides in the early morning and evening cause a double peak of pigment dispersion (Fig. 11.13). The two rhythms are functionally related: when the diurnal rhythm is experimentally shifted by exposing *Uca* to light during the normal dark period, the tidal rhythm of pigment dispersion is also shifted. Rather similar rhythms of colour change occur also in other crabs.

Significance of colour change

In general, chromatophores are of considerable importance for protective coloration, thermoregulation and displays associated with mating and parental behaviour. Because its colour merges with the sea bottom,

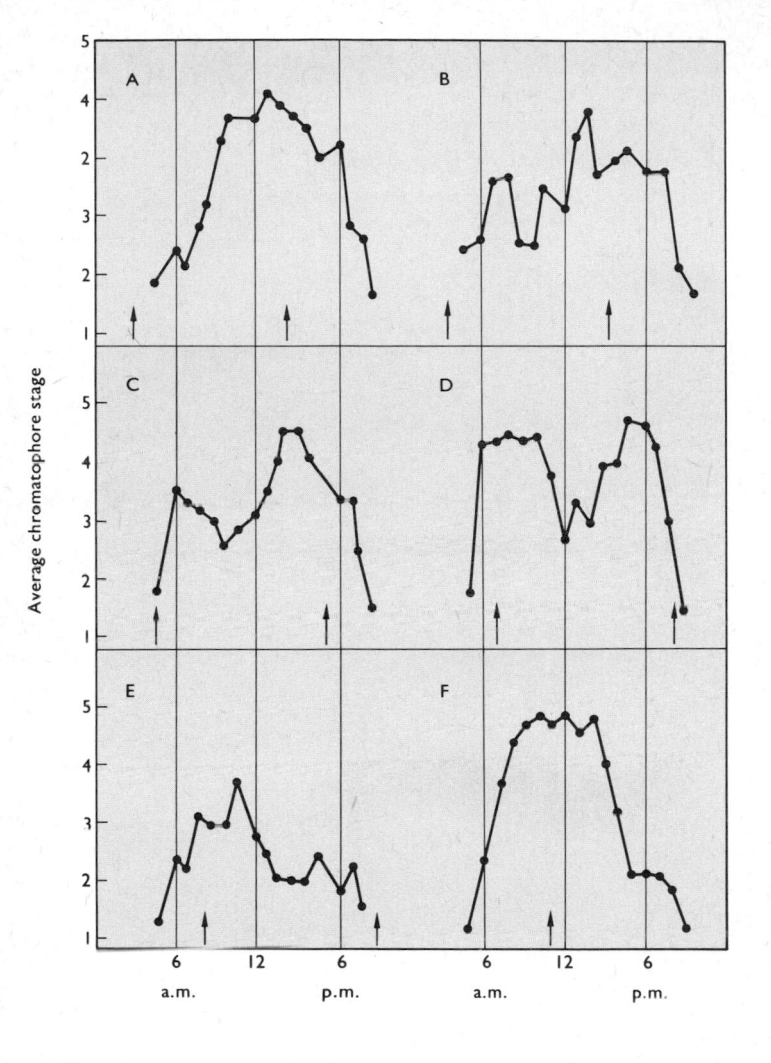

Fig. 11.13 Changes in melanin dispersion on the day following collection in *Uca* kept in darkness. The arrows indicate the times of low tides on the days the observations were made. There are clearly two cycles of melanin dispersion: one is a diurnal rhythm, and the other a tidal rhythm with greatest dispersion of melanin occurring 1–3 hours prior to the time of low tide. Thus when low tide occurs near midday, the peak of melanin dispersion is reinforced, whereas low tide in the early morning or evening results in a double peak of melanin dispersion because the two rhythms are out of phase. (After Brown, Fingerman, Sandeen and Webb[23])

an individual is protected against predation from above; blanching or transparency camouflages with the sea's surface and protects against predation from below. It is not surprising, therefore, that direct light and

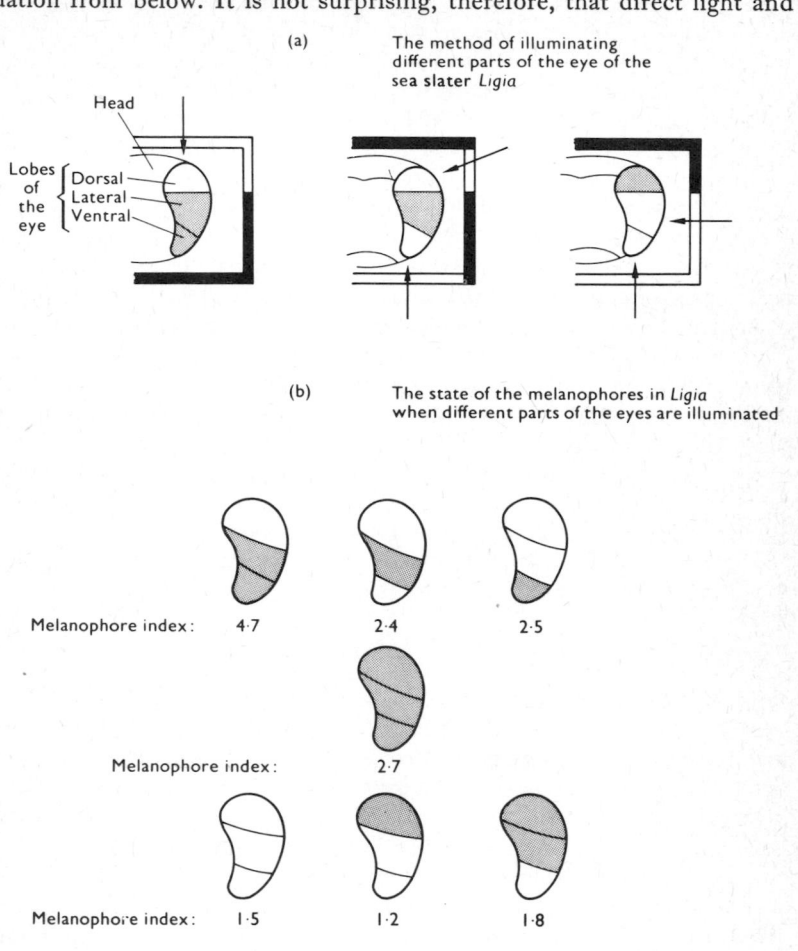

(a) The method of illuminating different parts of the eye of the sea slater *Ligia*

Head

Lobes of the eye { Dorsal — Lateral — Ventral — }

(b) The state of the melanophores in *Ligia* when different parts of the eyes are illuminated

Melanophore index: 4·7   2·4   2·5

Melanophore index: 2·7

Melanophore index: 1·5   1·2   1·8

Fig. 11.14 Effects of illuminating different parts of the compound eyes upon the state of the chromatophores in *Ligia*. (a) Method used to illuminate different parts of the eye. (b) Relationship between illumination and the chromatophore index. When the dorsal part of the eye is illuminated, as would happen naturally with direct light upon a dark background, the chromatophore pigment disperses fully, darkening the animal to tone with its surroundings. When the eye is illuminated from the ventral surface, or from several directions, as would happen with direct light together with light reflected from the background, the chromatophore pigment concentrates, toning the animal with a light background.

reflected light together control the particular shade of the individual. It is likely that the ommatidia in particular parts of the compound eyes, acting through the central nervous system, control the release of dispersing and concentrating hormones (Fig. 11.14). The presence of antagonistic pairs of chromatophorotrophins in Crustacea may thus be related ultimately to integration centres in the central nervous system which effect the release of the appropriate hormones.

In addition, pigmentation of the body can be used for thermoregulation. In *Uca*, the black pigment in the chromatophores tends to concentrate when the temperature rises above 15°C. The white pigment tends to disperse at a temperature above 20°C.[25]. The concentration of the black pigment and expansion of the white with increasing temperature has the effect of reducing the dark areas which absorb light and heat, and increasing the reflective white areas. If the intensity of illumination increases, the white chromatophores of three species of *Palaemonetes* expand even when the animals are on a black background. But in *Palaemonetes*, unlike *Uca*, the white pigment concentrates with increasing temperature. Heat and intense light usually occur together in nature, as in direct sunlight, and the antagonistic responses of the white pigment in *Palaemonetes* to the two factors may be a device to keep the chromatophores in a steady state.[93] But only the combination of detailed ecological and physiological studies is likely to reveal the manifold value of chromatophores to the animal in its natural habitat.

Control of retinal pigment movement

The crustacean compound eye, like that of the insect, consists of a number of units or ommatidia. Each ommatidium is surrounded by two sets of pigment cells—distal and proximal—and in the vicinity of their basement membranes, cells containing 'reflecting' pigment are present.

In the light adapted eye, the distal and proximal pigments prevent the sensitive rhabdome from receiving too much light, and the reflecting pigment cells move beneath the basement membrane (Fig. 11.15). In the dark adapted eye, the distal and proximal segments uncover the rhabdome, and the reflecting pigment moves externally to the basement membrane (Fig. 11.15). The rhabdomes thus receive light from adjacent as well as their own corneas.[170]

When extracts of eyestalk are injected into *Palaemonetes*, the distal pigment migrates towards the light-adapted position; sinus gland and brain extracts have the same effect.[171]

Sinus gland extracts injected into an animal maintained in a light intensity such that the distal pigment is midway between the fully dark adapted and light adapted positions, cause a light adapting response

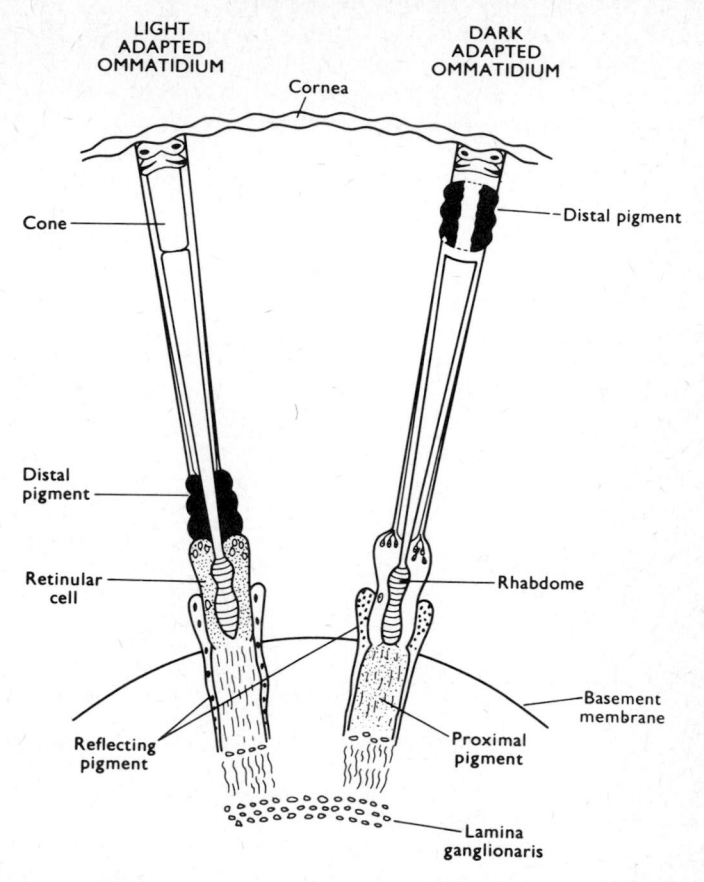

Fig. 11.15   Position of pigments around an ommatidium of a crustacean eye in the light adapted (left) and the dark adapted (right) condition. The compound eyes, are composed of large numbers of such ommatidia. (After Carlisle and Knowles[41])

which lasts for 2–3½ hours, but this is followed by a dark adapting response which lasts for 3½–6 hours. Does the sinus gland therefore contain two antagonistic hormones or is the dark adapting response merely over-compensation following the disappearance of the light adapting hormone?

The injection of brain extracts produces a light adapting response equal to that produced by the sinus glands, but there is no subsequent dark adapting response. Electrophoretic separation of sinus gland extracts

produces a fraction which induces a light adapting response after injection, and another with a dark adapting effect. The movement of the distal retinal pigment is thus controlled by two antagonistic hormones (Fig. 11.16). Moreover, when *Palaemonetes* is kept in the dark for long

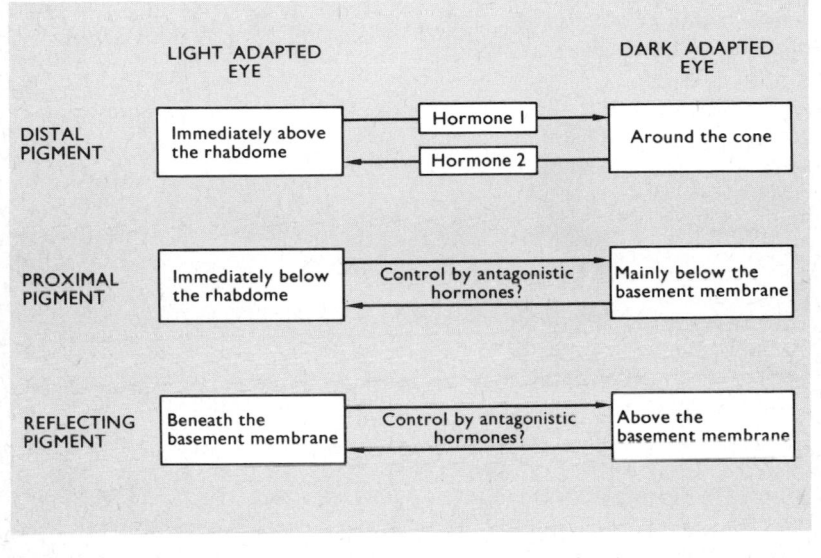

Fig. 11.16 Hormonal control of pigment movement in the crustacean eye. Movement of the distal pigment is controlled by two antagonistic hormones. It is not yet certain that movement of the proximal and reflecting pigments is hormonally controlled. For further details see text.

periods of time, extracts of brain can have up to four times the normal concentration of light adapting hormone, suggesting its accumulation when conditions do not call for its release.[24, 22, 27]

The proximal and reflecting pigments also appear to be hormonally controlled, but the evidence is meagre and confused, and will not be discussed further here.

Electrophoretic separation of eyestalk extracts of *Cambarellus shufeldtii*, produces distinct fractions with light adapting, red pigment dispersing and red pigment concentrating effects.[87] In *Palaemonetes*, the distal retinal pigment light- and dark-adapting hormones can be separated by gel filtration; both evoke graded responses directly proportional to the injected dose.[91c] The retinal pigment light-adapting fraction can also be separated from the pigment concentrating chromatophorotrophins, but not from the pigment dispersing hormones. However, differences in response show

that the hormones causing pigment dispersion in the red and white chromatophores (*Palaemonetes* has no black chromatophores) are not the same as the retinal pigment light-adapting hormone.[90] Thus it seems very likely that the hormones controlling retinal pigment migration are distinct from those controlling the chromatophores. However, the *Palaemonetes* retinal pigment light-adapting hormone will cause pigment dispersion in the black chromatophores of *Uca*; this adds to the evidence that the hormones involved in controlling both retinal pigment migrations and chromatophore changes must be rather similar in their molecular structures and configurations. Recent advances in the isolation and purification of eyestalk neurosecretory hormones (p. 265) will undoubtedly clarify the situation in the near future.

12

The Chemistry of Invertebrate Hormones

Before a hormone can be analysed chemically it must be extracted from the animal in pure form and in sufficient quantity. This entails its separation from other materials by partition through a variety of solvents and its eventual crystallization. At every stage of the separation and purification procedure, the biological activity of the extracts has to be tested to ensure that the hormone has passed into one solvent and has been removed from another. The extraction and purification of a hormone therefore involves the development of a suitable bioassay, which for preference should be simple, quick and easily repeatable so that accurate statistical estimates of activity can be made. The assay used for the eventual identification of ecdysone and juvenile hormone in insects both come into this category; the crustacean chromatophorotrophins, which can often be assayed on isolated pieces of cuticle, are another good example.

THE STARFISH SPAWNING HORMONES

The shedding hormone

Using extracts of radial nerves from about 7400 *Asterias amurensis*, a yield of about 1·3 mg of purified shedding hormone has been obtained.

The hormone is heat stable and insoluble in organic solvents; its activity is destroyed by treatment with trypsin, pepsin and pronase. Shedding hormone is a polypeptide, probably consisting of 23 amino acids and with a molecular weight of 2200.[158s, 158t] Its amino acid composition is:

aspartic acid (2), threonine (1), serine (7), glutamic acid (1), proline (1), glycine (4), alanine (2), valine (1), isoleucine (1), leucine (1), histidine (1) and ornithine (1).

Since the hormone contains no cysteine, it is likely to be a chainlike molecule. It is of interest that the shedding hormones of some starfish species have no effect when tested upon other species (p. 48). The polypeptide hormone from *Asterias amurensis* is therefore unlikely to be common to all species.

The maturation inducing substance

Starting with 20 kg of fresh ovarian fragments of *Asterias amurensis* incubated with 100 litres of artificial sea water containing radial nerve

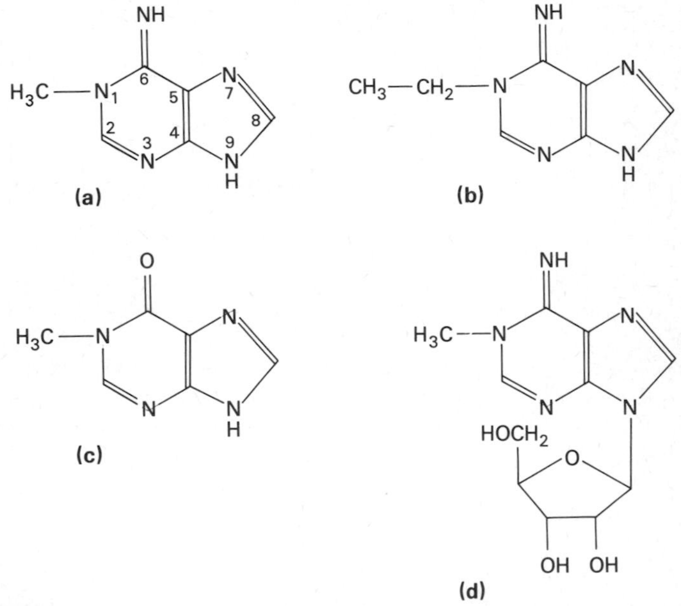

Fig. 12.1 (a) 1-methyladenine, the maturation inducing substance of starfish : the atoms are numbered in the standard manner. (b) 1-ethyladenine, which is less effective than (a) in inducing oocyte maturation. (c) 1-methylhypoxanthine, which is ineffective in inducing oocyte maturation. (d) 1-methyladenosine, the precursor for (a), present in the follicle cells.

shedding hormone, 8·5 mg of purified maturation-inducing substance has been isolated after concentration of the extract and separation by gel filtration and ion exchange column chromatography.[158w] The purified material induced 100% maturation of isolated oocytes of *Asterina pectinifera* at a concentration of 0·2 μg/ml, and induced maturation in 97% of the oocytes at a concentration of 0·012 μg/ml. Unlike the radial nerve shedding hormone, the maturation-inducing substance shows no species specificity.[158o, 251a]

The ultraviolet and infrared absorption spectra of the purified material, together with its mass spectra, suggested that the compound was 1-methyladenine (Fig. 12.1). Synthetic 1-methyladenine has the same melting point as the purified substance, and exactly similar minimal concentrations of the synthetic compound will induce oocyte maturation. It is therefore concluded that the maturation-inducing substance is, in fact, 1-methyladenine. 1-Ethyladenine (Fig. 12.1) is less effective than 1-methyladenine in inducing oocyte maturation (1×10^{-6} M compared with 3×10^{-7} M respectively as the minimal doses to effect maturation of 100% of the test oocytes) and 1-methyl hypoxanthine (Fig. 12.1) is totally ineffective.[158v] So the presence of a short alkyl radical at N-1 and of an imino radical at C-6 of the purine nucleus are necessary for the biological activity of the compound.

NEUROSECRETORY HORMONES IN INSECTS

The thoracotrophic hormone

The starting material for the extraction of thoracotrophic hormone should naturally be insect brains, and since hormones are active in minute quantities, very large numbers of brains must be used initially. The silk industry, based upon the silkmoth *Bombyx mori*, breeds enormous numbers of the insects every year, and it is not surprising that *Bombyx* has been the starting material for the extraction not only of thoracotrophic hormone but also of ecdysone (p. 266ff). Equally, the economic importance of *Bombyx* has led Japanese biologists to an intensive study of the insect.

Kobayashi and Kirimura in 1958 extracted thoracotrophic hormone from 8500 pupal brains of *Bombyx*, assaying their preparations upon brainless pupae—the positive result being the initiation of adult development. They eventually produced an oily material, soluble in organic solvents, and apparently lipid in nature. This result was surprising, since it had been expected that insect neurosecretory hormones, like those in the vertebrates, would prove to be polypeptides or low molecular weight

proteins. But a few years later, 4 mg of the crystalline hormone was isolated from 220 000 *Bombyx* brains, dissected free from corpora cardiaca
and allata.[174] This quantity of material was sufficient for its chemical
characterization: it turned out to be identical with cholesterol (Fig. 12.2).

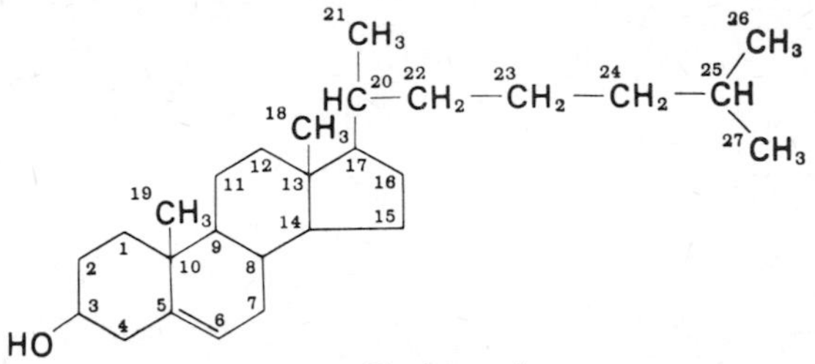

Cholesterol

Fig. 12.2 The formula for cholesterol. The carbon atoms are numbered in the
standard manner.

Since the thoracic gland hormone is a steroid (p. 267), it was thought
significant that the thoracotrophic hormone could be the precursor for
steroid biosynthesis.

When this identification of thoracotrophic hormone was published,
many laboratories attempted to confirm it. Although in some instances
very small quantities of commercial cholesterol would induce development in brainless insects, a normal dose-response relationship (increasing
percentage development with increasing cholesterol concentration) could
not be obtained[243] (Fig. 12.3). Very highly purified cholesterol often
produced no result. Moreover, many insects normally contain large
quantities of cholesterol, so why the further addition of a minute amount
of the substance should activate the thoracic glands was bewildering.

At about the same time that these conflicting results appeared,
Ichikawa and Ishizaki[153] reported that they had extracted from *Bombyx*
brains a water soluble hormone, stable within only a narrow range of pH,
non-dialysable, precipitated by ammonium sulphate and trichloroacetic
acid, and which was inactivated by incubation with bacterial proteases.
The preparation was later purified 8000 fold, and was found to consist of
a mixture of hormonally active polypeptides or small proteins with
molecular weights ranging from 9000 to 31 000.[153d]

It is difficult to reconcile the chemical natures of these two brain
hormones from *Bombyx*. It is possible that the brainless pupae used for
the bioassay of the lipoidal extract were very sensitively poised on the
brink of development. The addition of minute amounts of true brain
hormone in the extracts could then initiate development; or perhaps

contaminating steroids—present in both extracts and commercial cholesterol—activated the thoracic glands. Ecdysone itself will activate thoracic glands.[287]

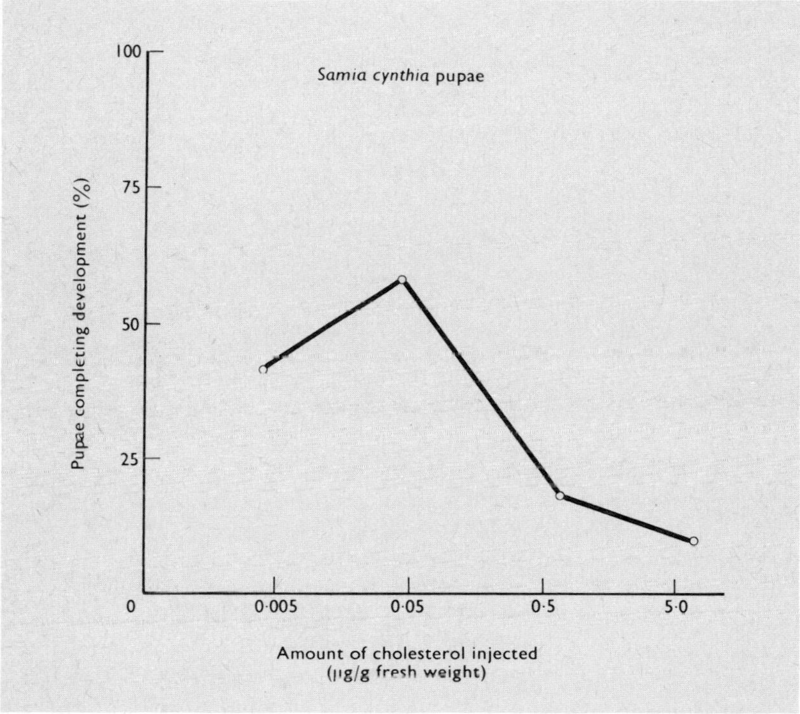

Fig. 12.3 Percentage development of *Samia* pupae with increasing dosage of cholesterol. Note the absence of a typical dose-response effect. Logarithmic scale on abscissa. (After Schneiderman and Gilbert[243])

Although not abandoning the concept of a cholesterol (or steroidal) brain hormone, Kobayashi and his colleagues subsequently extracted another brain hormone from the lipoidal fraction of *Bombyx* brains which they concluded was a glycoprotein (with carbohydrate equivalent to about 15% glucose) with a molecular weight of about 20 000.[174, 175a, 293b] A protein brain hormone from the cockroach *Periplaneta americana* also has a molecular weight of about 20 000.[105] The brain hormone from the saturniid moth *Antheraea pernyi*, because of its heat stability, its lack of specific absorption at 280 nm, its insensitivity to pepsin, trypsin and chymotrypsin, together with its insensitivity to classical methods of removing proteins

during preparative procedures, may, it is suggested, possibly be a muco-polysaccharide.[289a]

Some of the confusion relating to the nature of the brain hormone arises from the use of different assay insects: Kobayashi and his colleagues used decerebrate pupae of *Bombyx mori* and Ishizaki and Ichikawa used decerebrate pupae of *Samia cynthia*. A later isolation and purification of the *Bombyx* brain hormone overcomes this difficulty by using both species as assay material, and also by adhering strictly to a particular period after removal of the brain from the pupae.[213a] The preparative technique used methods which removed cholesterol, lipids, nucleic acids, glycoproteins, lipoproteins and polysaccharides. The final preparation was resistant to trypsin and chymotrypsin, but its activity was destroyed by nagase and pronase. Of particular interest was the discovery that the active principle was dialysable: taken in conjunction with the results of Sephadex separation, it is concluded that the brain hormone, or an active core of the hormone, or even perhaps one component of the hormone in a heterogeneous mixture, is a polypeptide with a molecular weight of less than 5000.[213a]

Many of the preparations of brain hormone can be resolved into two or more active fractions of different molecular weights. It is possible that the hormone could exist in different molecular forms, or perhaps a single hormone is associated with proteins of different molecular weights.[115j] In any event, the bulk of the evidence points to the brain hormone being a small molecular weight protein or a polypeptide, which accords with what is known of neurosecretory hormones in animals other than insects.

Other insect neurosecretory hormones

Purified extracts of **bursicon** are stable up to temperatures of 55°C, but are inactivated by repeated freezing and thawing. High salt concentrations, extremes of pH and short-chain alcohols all inhibit bursicon activity. Bursicon is inactivated by proteolytic enzymes and is non-dialysable. The hormone is not species-specific and bursicon from blow-flies has chemical properties and a molecular weight similar to that from cockroaches. There is little doubt that bursicon is a protein hormone; its molecular weight is about 40 000.[207, 95b]

The **hyperglycaemic hormones** of cockroaches and locusts are soluble in methanol and water, can be separated chromatographically in solvent systems typically used for amino acids and peptides, and are inactivated by the enzyme chymotrypsin. They are thus likely to be polypeptides. The hyperglycaemic peptide of the cockroach contains one or two dithio groups and is made up largely of aromatic amino acids.[104c] In the locust, the less potent hyperglycaemic hormone is present in the neuro-

secretory storage lobes of the corpora cardiaca and also in the pars inter-cerebralis of the brain: it is undoubtedly neurosecretory in origin. The more potent hormones can be extracted only from the glandular lobes of the corpora cardiaca: its origin is therefore not *immediately* neurosecretory, although it is likely that cerebral neurosecretion is concerned indirectly with its production. The situation recalls that of the glandular anterior pituitary producing polypeptide hormones under the control of neuro-secretory releasing factors from the hypothalamus.

The locust *diuretic hormone* has the same physical properties as the neurosecretory hyperglycaemic hormone and is similarly inactivated by chymotrypsin. This hormone, too, is therefore likely to be a polypep-tide. However, in the cockroach the diuretic hormone has a molecular weight greater than 30 000.[113a] In *Rhodnius*, extracts containing diuretic hormone are inactivated by trypsin, chymotrypsin and pronase, suggesting that the hormone is a high molecular weight polypeptide. However, the hormone present in the haemolymph, and also separable by gel filtration of thoracic ganglion extracts, appears to be a low molecular weight compound.[4c]

The locust *adipokinetic hormone* unlike the cockroach hyper-glycaemic hormone, does not contain a large proportion of aromatic aminoacids. It probably contains an imino amino acid in the NH_2 ter-minal.[115j]

The *diapause hormone* of *Bombyx mori* is present in the sub-oeso-phageal ganglion of the male moth.[123b] Since the Japanese silk industry discards large numbers of adult males after mating, these are a convenient source of the hormone. Using an extraction procedure suitable for complex lipids, 30 μg of the diapause hormone was obtained from 2·18 million heads (4 kg). This represents a yield of less than 1%, determined by com-parison with the effects of sub-oesophageal ganglia in inducing diapause egg production.[153e] Gel filtration suggests the molecular weight of the hormone to be between 2000 and 4000. Although preferentially soluble in lipid solvents, the hormone is unaffected by lipid hydrolases, but is completely inactivated by the proteinases papain and bromelain. The hormone is therefore probably a protein, perhaps a lipoprotein.[123c, 153e]

NEUROSECRETORY HORMONES IN CRUSTACEA

Presumably because of the ease of assaying chromatophorotrophic hormones in the crustaceans, the extraction, purification and identifica-tion of these hormones has proceeded more rapidly than that of the developmental hormones. All the available evidence suggests that the chromatophorotrophins and the retinal distal pigment-adapting hormones

are peptides. They are inactivated by proteases and behave as peptides when separated by electrophoresis and gel filtration. Two hormones have been sufficiently purified to allow an analysis of their amino acid composition. The *retinal distal pigment light-adapting hormone* is an octadecapeptide with a molecular weight of about 2000 and the following amino acid residues: arginine (1), aspartic acid (2), threonine (1), serine (2), glutamic acid (1), proline (1), glycine (2), alanine (1), valine (1), methionine (2), isoleucine (3), and leucine (1).[86j] The *red pigment concentrating hormone* of *Pandalus* has been further characterized: the hormone is an octapeptide with the amino acid sequence: pyroglutamic acid-leucine-asparagine-phenylalanine-serine-proline-glycine-tryptophan amide.   The molecule has been synthesized and the synthetic hormone has the same biological activity as the purified eyestalk hormone.[86k] The synthetic *Pandalus* hormone cannot be separated by gel filtration from the red pigment-concentrating hormones from the eyestalks of *Uca* and *Palaemonetes*, indicating that the three hormones have the same or nearly the same molecular size.[91b] Similarly, the retinal distal pigment light-adapting hormone cannot be separated by gel filtration from the chromatophore pigment-dispersing hormones, and these hormones are likely to be peptides of similar sizes.

There is some evidence that the *black pigment dispersing hormone* in *Uca* is associated with a lipoprotein carrier both within the neurosecretory cells and in the blood.[5a]

The crustacean *hyperglycaemic hormone* has been partially purified, and is a protein of relatively small molecular size. Comparative studies suggest that the molecular structure of this hormone differs amongst species.[172]

The *cardioaccelerator hormone* released from the pericardial organs of crabs is also a peptide. In *Cancer borealis*, two cardioaccelerator peptides with molecular weights of about 1000 have been extracted: whether these are distinct hormones or are the breakdown products of a single hormone is uncertain.[12] Experiments designed to reveal the presence of a carrier protein in association with the cardioaccelerator hormone in the pericardial organs of the spider crab *Libinia dubia* failed to reveal any binding of the hormone to a protein, and completely eliminated the possibility of the existence of a binding protein with properties analogous to those of the neurophysins of vertebrates.[12d]

Evidence from both the insects and the crustaceans strongly supports the notion that the neurosecretory hormones are protein or peptide compounds. The arthropod neurosecretory mechanisms thus fall into line with those of the vertebrate groups, where neurosecretory hormones are of a similar chemical nature. A great deal of histochemical evidence exists for the presence of protein or lipoprotein substances within neuro-

secretory cells of many animals. This evidence has been deliberately disregarded because neurosecretory material is not necessarily related to the hormonally active products which are released into the blood. In fact, in the vertebrates the histologically and histochemically demonstrable neurophysin appears to be a carrier substance for the biologically active peptides and proteins. It is conceivable that in other animals, carrier neurosecretion could be associated instead with other active molecules. So far, the evidence in arthropods is generally against such an idea: but the possibility should always be borne in mind.

STEROID HORMONES IN ARTHROPODA

The insect moulting hormone: ecdysone

As early as 1935, a bioassay for the estimation of moulting hormone extracts was devised. This consists of ligaturing blowfly larvae behind Weismann's ring before pupation: puparium formation proceeds normally anterior to the ligature, but is prevented posteriorly if the ligature is applied before hormone is released from the ring gland[94] (Fig. 12.4). Extracts containing moulting hormone injected into the abdomen of the larva will then induce pupation. With the exception that two ligatures are now applied, so that injections can be made through the second

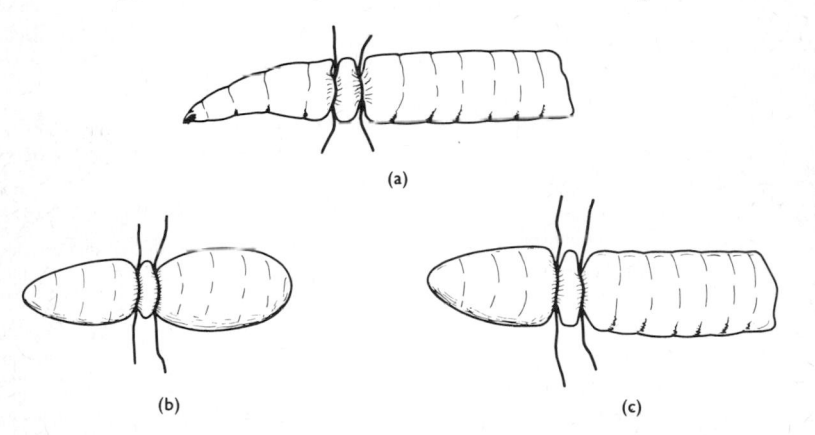

(a)

(b)   (c)

Fig. 12.4 Double ligature technique in the *Calliphora* larva bioassay for ecdysone (a). If the ligatures were applied after endogenous ecdysone were released, both parts of the larva would pupate (b). If the ligatures are applied earlier, only the anterior part of the body pupates (c) and the posterior region can be injected with test solutions from a needle inserted between the two ligatures.

ligature into the abdomen without loss of fluid, the bioassay has remained virtually unchanged for almost 40 years.

In 1954, Butenandt and Karlson extracted 25 mg of crystalline moulting hormone, which they called ecdysone, from 500 kg of *Bombyx* pupae.[33] The crystalline ecdysone induces pupation when as little as 0·0075 μg is injected into ligatured blowfly larvae. With most batches of larvae, 0·010 μg of ecdysone is effective; and this quantity of hormone is known as one *Calliphora unit* (C.U.).

Although ecdysone was isolated in crystalline form in 1954, the quantity extracted was too small for accurate chemical analysis with the means then available. But some 10 years later, starting with 1000 kg (dryweight) of Bombyx pupae (equivalent to about 4 tons fresh weight) 250 mg of ecdysone were prepared and the hormone was proved to be indisputably a steroid,[162] with the structure shown in Fig. 12.5. That one of the most important of the insect developmental hormones is a member of a class of compounds well known and widespread in the vertebrate classes is of exceptional interest.

In fact, two ecdysones can be extracted from *Bombyx* pupae: that which has just been described is sometimes called α-ecdysone; a second steroid present in very much smaller quantity is called β-ecdysone. The same two ecdysones have now been isolated from the Moroccan locust, *Dociostaurus maroccanus*, and it is interesting that in this species β-ecdysone is the predominant steroid.[250] In the tobacco hornworm, *Manduca sexta*, α-ecdysone is also present in smaller quantity than a second ecdysone, which appears to be 20-hydroxyecdysone[159] (Fig. 12.6). A third ecdysone, called ecdysterone, has been isolated from *Bombyx* extracts.[149] Despite reported differences in biological activity, which could be the result of using different insects for assay, or the same insect at slightly different times in its development, or even differences in chemical purity, 20-hydroxyecdysone, β-ecdysone from different sources, and ecdysterone are now known to be the same compound.[100a, 148d] A further ecdysone

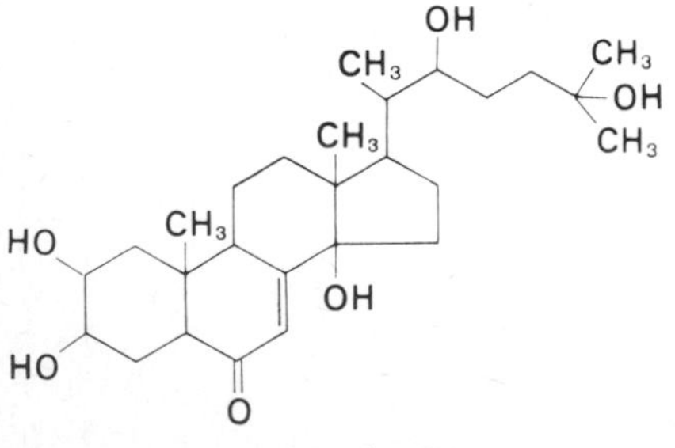

Fig. 12.5 Formula for α-ecdysone (compare with Fig. 12.2).

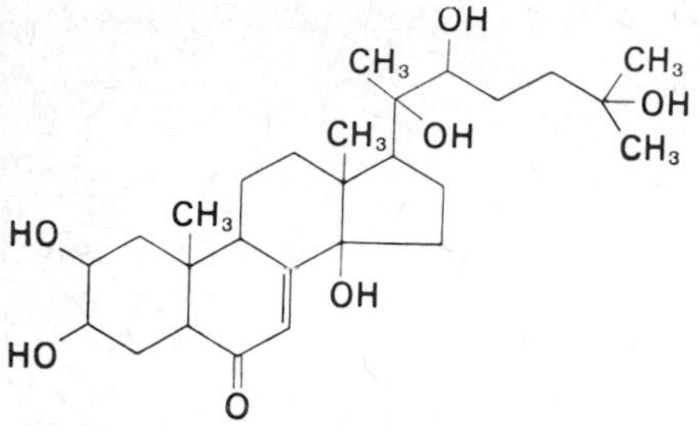

Fig. 12.6 The formula for 20-hydroxyecdysone. Note the additional -OH group on carbon atom 20.

20-hydroxyecdysone

has been isolated from *Manduca sexta*, which has four -OH groups on the side chain (Fig. 12.7) and can thus be called 20, 26-dihydroxy ecdysone.[259a]

Large yields of ecdysones, with marked moulting hormone activities,[150] have been extracted from the roots and leaves of plants such as *Achyranthes*,[175] *Podocarpus, Taxus, Polypodium*, etc. These usually have the same structure as the insect ecdysones except that the side chain is modified to a greater or less degree (Fig. 12.8). One exception is Polypodine B, which has an -OH on C-5,[126d] which is unknown in the insect ecdysones. Both α- and β-ecdysone can be extracted from plant material.[126c, 153b, 158x] The plant steroids can have greater moulting hormone activity than the natural insect ecdysones, probably because they are less easily metabolized by the assay insects, and consequently have a longer biological half-life in which to act.[214f, 214g, 214h, 289b]

Ecdysone has the same carbon skeleton as cholesterol (Fig. 12.1). When radioactively labelled cholesterol is administered to *Calliphora* larvae, a great deal of the radioactivity is incorporated into ecdysone.[160, 161] Cholesterol is therefore very likely to be the precursor of

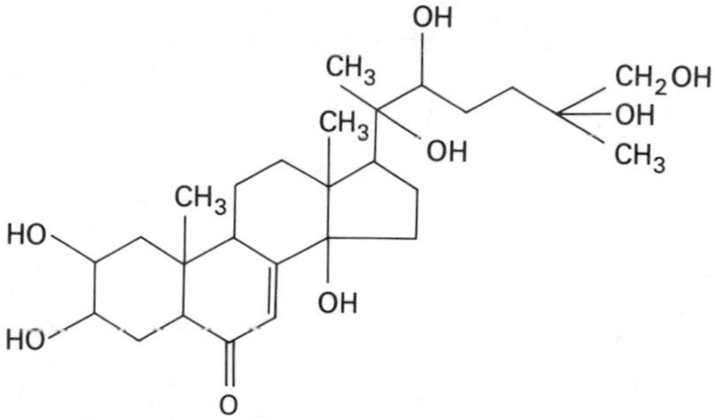

Fig. 12.7 The formula for 20,26-dihydroxyecdysone (26-hydroxy-β-ecdysone) a third moulting hormone from *Manduca sexta*.

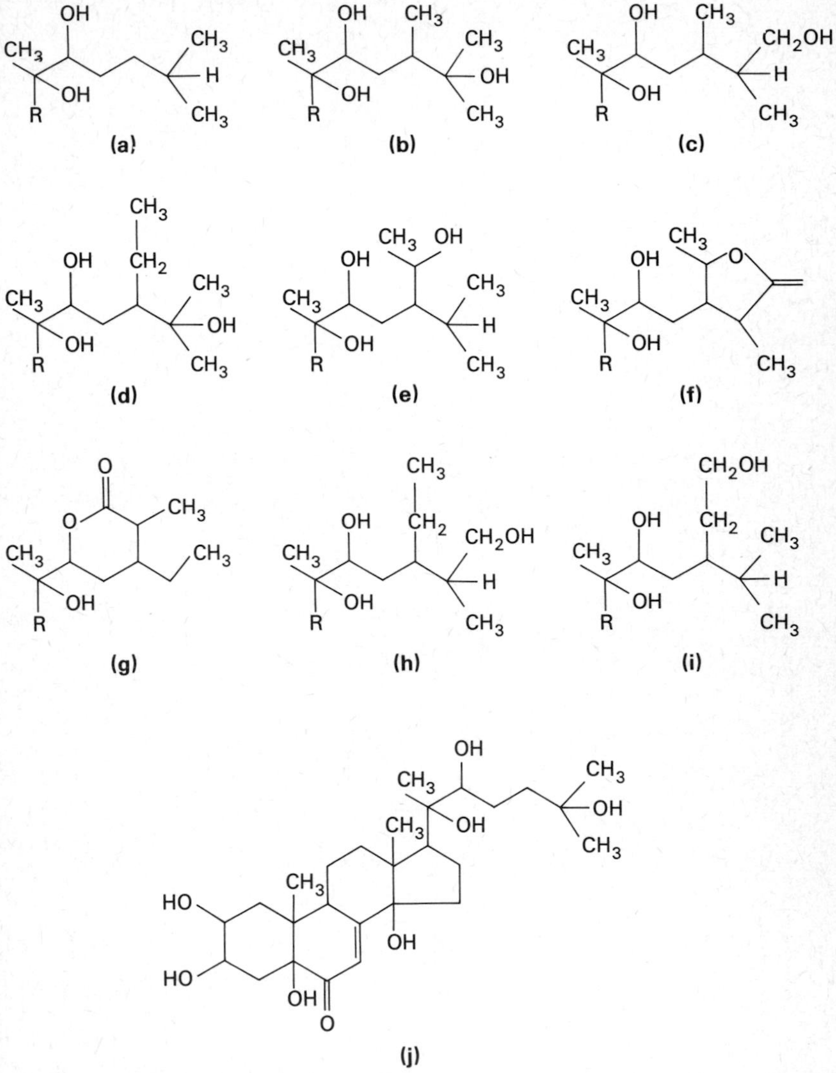

Fig. 12.8 Ecdysones from plants : (a) ponasterone A, (b) makisterone A, (c) makisterone B, (d) makisterone C, (e) makisterone D, (f) cyasterone, (g) capitasterone, (h) amarasterone A, (i) amarasterone B, (j) polypodine B. R=the steroid nucleus of ecdysone.

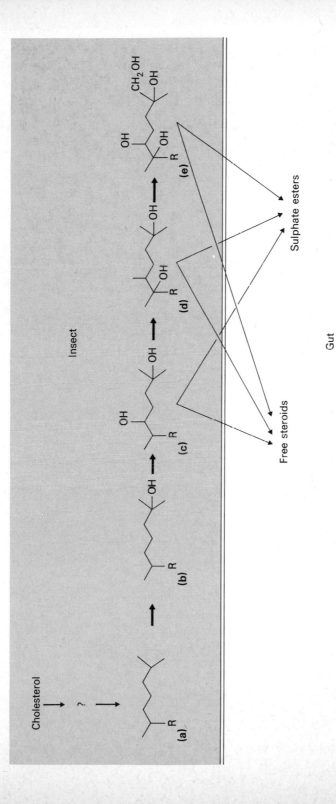

Fig. 12.9 Metabolism of ecdysones in *Manduca sexta*: progressive hydroxylation of the side chain occurs, from 22,25-dideoxy-α-ecdysone (a), through 22-deoxy-α-ecdysone (b), α-, β-ecdysone and 20,26-dihydroxyecdysone (c, d and e respectively). The last three hormones can be secreted into the gut either as free steroids or as sulphate conjugates. R=the steroid nucleus of α-ecdysone. (After King⁻⁶⁹ᵇ)

ecdysone in insects. Insects cannot synthesize cholesterol from simpler precursors: however, phytophagous insects can convert plant sterols to cholesterol, and carnivorous insects will obtain cholesterol from their prey.

The metabolic pathway between cholesterol and ecdysone is not known, and is difficult to elucidate because of the large pool of endogenous cholesterol in insects which dilutes the injected radioactive precursor. However, when tritiated 22-deoxy-α-ecdysone and 22,25-dideoxy-α-ecdysone are injected into prepupae of the tobacco hornworm, *Manduca sexta*, radioactively labelled α-ecdysone is one of the metabolic products.[169b, 169c] It is thus possible that these compounds are intermediates in the metabolism of cholesterol to α-ecdysone in *Manduca sexta* (Fig. 12.9), although this appears not to be so in the blowfly *Sarcophaga bullata* and the grasshopper *Gastrimargus africanus*.[169b]

There is better evidence that in both holometabolous and hemimetabolous insects, α-ecdysone is rapidly converted into β-ecdysone.[51b, 101a, 118e, 148e, 169b, 169c, 210a] In *Manduca sexta*, β-ecdysone is further hydroxylated to 20,26-dihydroxyecdysone (26-hydroxy-β-ecdysone).[169b] In larvae of *Locusta migratoria*, α-ecdysone and β-ecdysone are dehydrogenated to inactivate the hormones before excretion (Fig. 12.10)[148e] and both α- and β-ecdysone, together with the 3-dehydro compounds, are also metabolized to glucosides, glucuronides, and sulphate esters as part of the inactivation process (Fig. 12.10).[148e, 169b] Some of the difficulties of relating steroid metabolism in different insects may be resolved when particular developmental stages are compared. In fifth instar larvae of *Locusta migratoria*, α- and β-ecdysone titres are low early in the instar, rise sharply at the time of apolysis and new cuticle synthesis, then decrease rapidly to a low level at ecdysis. When endogenous ecdysone titres are high, excretion of the hormones and their metabolic products is remarkably low and *vice versa*.[148e] Moreover, the relative importance of the metabolic pathways of α-ecdysone varies during the course of the larval instar (Fig. 12.10). These changes may reflect a very potent system for regulating the titres of ecdysones.

Crustacean moulting hormones

Ecdysone is able to accelerate moulting in Crustacea, and crayfish homogenates, extracted for steroids, have some activity in the *Calliphora* larva test. These observations led to the idea of a crustacean steroid moulting hormone, which was called **crustecdysone**. This steroid has now been isolated and identified: it has the same structure as β-ecdysone (20-hydroxyecdysone, ecdysterone).[121] A second ecdysone has been isolated from the marine crayfish *Jasus lalendei*, in which the hydroxyl group on

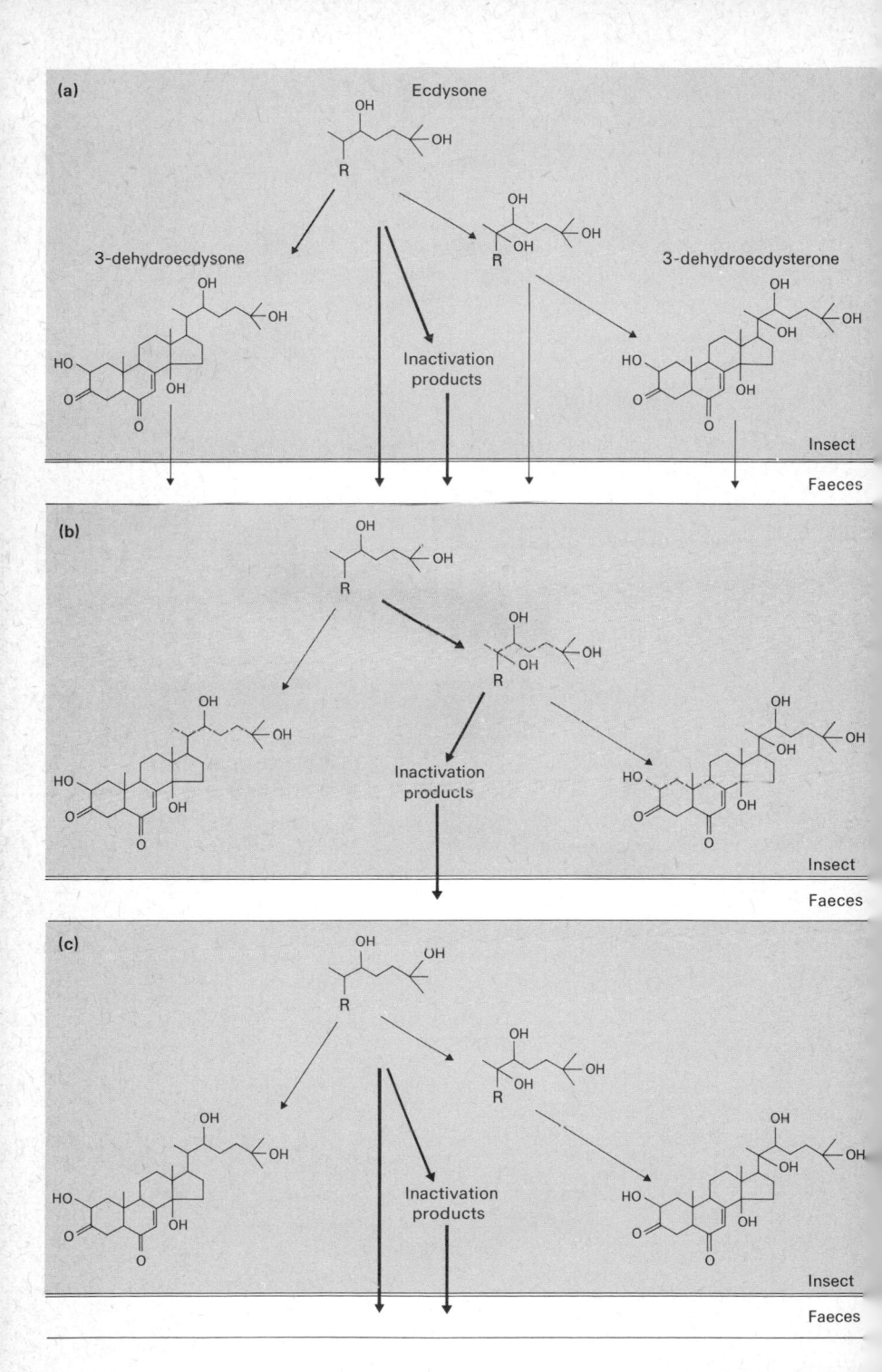

(a) Ecdysone

3-dehydroecdysone

Inactivation products

3-dehydroecdysterone

Insect

Faeces

(b)

Inactivation products

Insect

Faeces

(c)

Inactivation products

Insect

Faeces

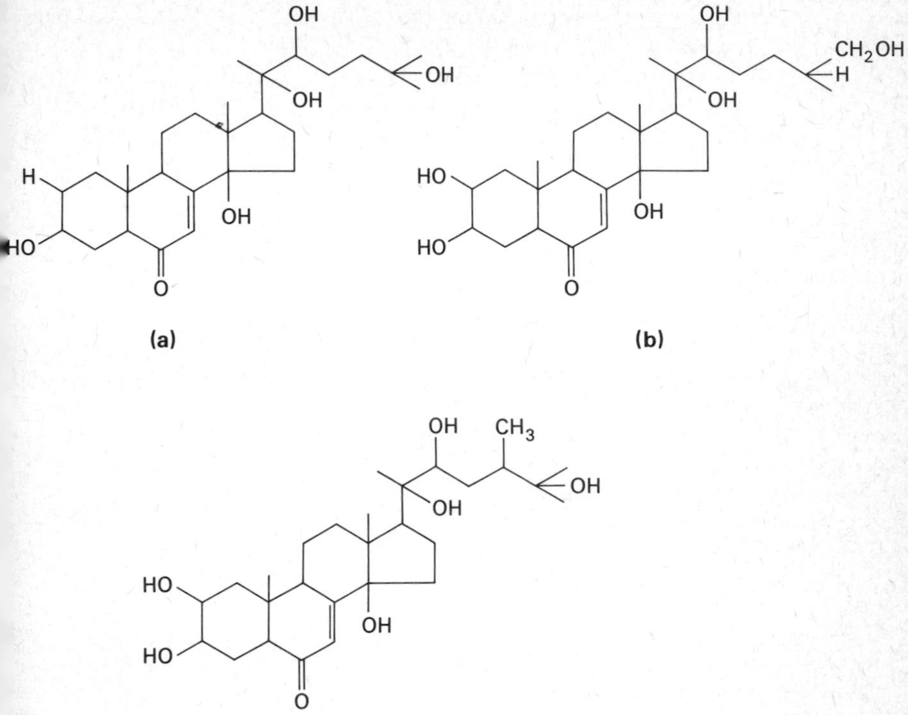

(a)   (b)

(c)

Fig. 12.11 Moulting hormones from crustaceans, additional to crustecdysone. (a) 2-deoxycrustecdysone from *Jasus lalendei*, possibly a deactivation product of crustecdysone. (b) callinecdysone A, and (c) callinecdysone B, both from *Callinectes sapidus*.

C-2 is replaced by hydrogen: the compound is therefore called 2-deoxy-crustecdysone[101] (Fig. 12.11). In the crab, *Callinectes sapidus*, two ecyd-sones in addition to crustecydone have been identified. One, called callinecdysone A, is an isomer of crustecdysone with a hydroxyl group on C-27 instead of C-25 (Fig. 12.11) and is probably identical with the plant steroid inokosterone; the other callinecdysone B, has an additional methyl group on C-24 compared with crustecdysone[86h] and thus resembles the plant steroid makisterone A (Fig. 12.8) extracted from *Podocarpus macrophyllus*.[153c]

Fig. 12.10 Metabolism of injected ecdysone in *Locusta migratoria*: (a) in a 1 day old fifth instar larva; (b) in a 7 days old fifth instar larva; and (c) in a 5 days old adult. The thickness of the arrows is proportional to the importance of the pathway. (After Hoffman *et al.*[148e])

The metabolic relationships (if any) between the crustacean ecdysones are unclear. When α-ecdysone is injected into crustaceans it is rapidly converted to crustecdysone[169c] and β-ecdysone will induce moulting in crustaceans.[178a, 178b, 193c] However, when the hepatopancreas of the crayfish *Orconectes virilis* is incubated with α-ecdysone, a steroid can be extracted associated with two binding proteins which is neither α- nor β-ecdysone, but appears to be a metabolite of these compounds.[118c] The same compound can be extracted from the hepatopancreas after the injection of β-ecdysone *in vivo* and since the hepatopancreas is one of the target organs for β-ecdysone (p. 218), it is suggested that the metabolite is a biologically active compound responsible for the particular reactions of the hepatopancreas. It is known that certain steroid hormones in vertebrates are converted to active forms in target organs, and a similar process in invertebrates is not unlikely. The hypothesis is attractive, since it implies that only the specific targets for ecdysone will have the enzymes necessary for conversion to the active form; non-target tissues will be unaffected by the precursor ecdysones.[118c]

In *Callinectes sapidus*, callinecdysone A first appears in the early premoult stage, rises to a maximum at late premoult, and is absent after ecdysis. Crustecdysone, on the contrary, can be detected first at the late premoult stage, and is present in high concentrations after ecdysis, when callinecdysone B is also detectable.[86h] A metabolic relationship between these compounds is unlikely to be close: it is possible that in any species, the different ecdysones could each play a specific role in the complicated processes associated with moulting. Because of the alkyl group on the side chain, it has been suggested that callinecdysone B may be synthesized directly from marine phytosterols.[86h]

The crustacean androgenic gland hormone

Preliminary work suggested that the hormone from the androgenic gland was protein in nature. However, when the androgenic glands of *Ocypoda platyaris* are extracted for sterols, the non-saponifiable lipid fraction mimics the effects of androgenic gland transplants when injected into female crabs. Moreover, injected testosterone will convert ovarian to testis tissue, although it will not by itself induce the secondary male characters.[230]

In the lobster, *Homarus americanus*, testosterone is produced when androgenic glands are incubated with a precursor known to produce testosterone in vertebrate systems. The glands seem therefore to contain an enzyme capable of producing testosterone,[111] although this steroid has not yet been identified in the blood. The significance of mechanisms

in invertebrate tissues which duplicate the enzymatic conversion found in vertebrates must always be viewed with caution until the biological function of the reactions are known. But the evidence so far suggests that the crustacean androgenic gland could produce a sex steroid, perhaps in combination with a protein hormone to fulfil the complete range of androgenic gland activities.

Source of moulting hormone in insects and crustaceans

Long standing experimental studies have shown the central importance of the insect thoracic glands and the crustacean Y-organs in the control of moulting. But there is no direct evidence that these glands produce the definitive moulting hormone. In some insects, an abdominal factor is required in addition to the thoracic glands before moulting can be induced,[266i] and it has been suggested that the oenocytes are the source of this factor.[191b, 191c, 266i] Isolated abdomens can convert α-ecdysone to β-ecdysone,[210a] and can even synthesize both steroids from cholesterol.[106, 211a]

There is good evidence, however, that the insect thoracic glands can synthesize ecdysone. Ecdysone stimulates spermatogenesis in the isolated testes of diapausing silkworm pupae incubated in cell free haemolymph; the same effect is obtained when the testes are incubated with activated thoracic glands or with inactive glands plus active brains, but inactive glands alone or active brains alone produce no response.[158l] These results indicate that the activated thoracic glands produce ecdysone, although they do not rule out the possibility of simultaneous synthesis in other tissues, nor the peripheral activation of an inactive form of the hormone produced by the glands.

In crustaceans, there is no comparable evidence that the Y-organs secrete crustecdysone. However, the Y-organs of the crab *Hemigrapsus nudus* take up 650 times more radioactively labelled cholesterol per unit weight of tissue than the general body mass, and 100 times more cholesterol is taken up by the Y-organs in late intermoult than in pro- or postecdysis. No absolute identification of the fate of the labelled cholesterol has been made, but substances containing label and co-chromatographing with α-ecdysone and crustecdysone have been isolated from the Y-organs; 60 and 260 times more labelled 'crustecdysone' is present in the Y-organs of crabs in proecdysis than in intermoult and postecdysis glands respectively.[249b] These results indicate an accelerated uptake of cholesterol by the Y-organs in late intermoult, and the accumulation of ecdysone-like metabolites in the Y-organs in proecdysis: reasonably good evidence that the Y-organs synthesize a steroid hormone.

The trophic relationship between the brain hormone and the thoracic

glands is well established experimentally, and the *in vitro* studies described above provide more support for the concept. Moreover, brain hormone extracts stimulate significantly the synthesis of RNA by the thoracic glands *in vitro*.[104d] That the targets for the ecdysones may convert the steroids to more active molecules has already been described (p. 274). Therefore, although other tissues in the body may convert α- to β-ecdysone, or even synthesize the hormones from cholesterol, it must be concluded that the thoracic glands of insects and the Y-organs of crustaceans play a significant role somewhere along the metabolic pathway from cholesterol to the— perhaps various—steroids which will exert their biological effects upon their target tissues.

THE INSECT JUVENILE HORMONES

The first active extracts of juvenile hormone (neotenin) were made by Carroll Williams, from the abdomens of male *Hyalophora cecropia* moths where it accumulates in very large quantities (Fig. 6.10) (for an unknown reason).[289] A cold ether extract of a single moth abdomen yields about 200 mg of a yellow oil, which contains virtually all the lipids in the body, together with juvenile hormone. When 50 mg of the extract are injected into a pupa the individual moults into a second pupa—just as if active corpora allata had been implanted.[109] An even more sensitive bioassay has been developed for both hemimetabolous and holometabolous insects. When extracts containing juvenile hormone activity are applied locally to the cuticle of last larval hemimetabolous or pupal holometabolous insects, the epidermal cells are induced to secrete another larval or pupal cuticle at the next moult, whereas the surrounding cells produce a normal adult cuticle.[109]

Although very highly purified, the juvenile hormone from *Hyalophora cecropia* proved very difficult to characterize. In the meantime, a large number of compounds, and extracts from a multitude of organisms were assayed for juvenile hormone activity with a surprisingly large number of positive results[244] (Fig. 12.12). Of particular interest was the discovery in 1959 by Karlson and Schmialek that extracts of insect faeces often possessed considerable juvenile hormone activity.[164] In 1961 Schmialek extracted 80 kg of the faeces of the mealworm, *Tenebrio molitor*, and identified the active agent as the terpene farnesol and its oxidation product farnesal[241] (Fig. 12.12a). In *Tenebrio*, *Rhodnius*, the cockroach and several other insects, farnesol and particularly its methyl ether and its diethylamine derivative (Fig. 12.12a), are able to reproduce completely all the effects of juvenile hormone: delayed metamorphosis in

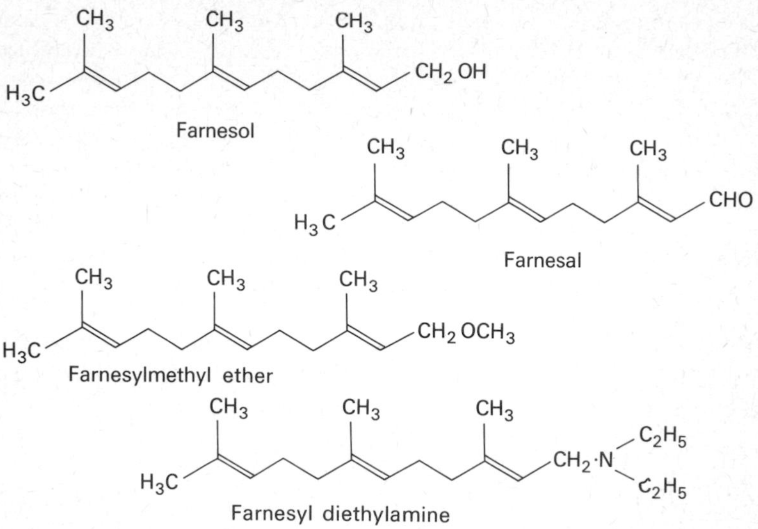

Fig. 12.12 (a)   The structures of farnesol, farnesal, farnesyl methyl ether and farnesyl diethylamine : compounds with juvenile hormone activity.

Substance	Juvenile hormone activity (*Galleria* units/g)
Farnesol	16
Farnesal	32
Farnesyl methyl ether	1000–1500
Farnesyl diethylamine	1000–1500
Crude *H. cecropia* extract	1000
Concentrated *H. cecropia* extract	300 000 000

Fig. 12.12(b)   The relative juvenile hormone activity of farnesol and its derivatives compared with extracts from adult males of *Hyalophora*. (From Gilbert and Goodfellow[107])

Rhodnius and reversal of adult characters in a moulting adult[281, 282]; gonadotrophic activity in several insects; prolongation of diapause in insects such as the rice stem borer, where the corpora allata are important in its control[96]; and in many insects farnesol replicated the cuticle bioassay in a manner exactly similar to extracts of juvenile hormone. It seemed conclusive that the insect juvenile hormone was farnesol, or some very closely related terpenoid compound. When Schmialek[242] isolated farnesol from extracts of male *Hyalophora cecropia* in 1963 the issue seemed to be beyond doubt.

But farnesol and its derivatives are many times less effective than highly purified extracts from *Hyalophora cecropia*[243, 107] (Fig. 12.12b).

Moreover, farnesol is present in larger amounts in chilled pupae of *Hyalophora* than in adult males, although the pupae possess no extractable juvenile hormone, while adult males contain enormous amounts of juvenile hormone (Fig. 6.10). Finally, removal of the corpora allata from *Hyalophora* pupae prevents completely the appearance of juvenile hormone in the adult males, but their content of farnesol is unaltered.[117, 107] These results contradict completely the idea that farnesol actually is the juvenile hormone. Farnesol and its derivatives *mimic* the effects of juvenile hormone, more so in some insects than in others, but they are not themselves the true juvenile hormone.

In 1967, Röller and his colleagues in the University of Wisconsin published the definitive structure of insect juvenile hormone.[228] It is methyl 10,11-epoxy-7-ethyl-3,11-dimethyl-2,6-tridecadienoate (Fig. 12.13).

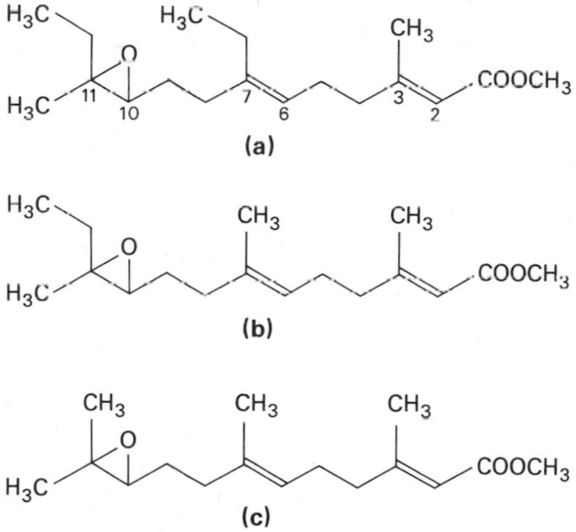

Fig. 12.13 Three naturally occurring juvenile hormones: (a) the C_{18} JH from *Hyalophora cecropia* (methyl *cis*-10,11-epoxy-3, dimethyl-7-ethyl-*trans, trans*-2,6-tridecadienoate); (b) the C_{17} JH, also from *H. cecropia* (methyl *cis*-10,11-epoxy-3,7,11-trimethyl-*trans, trans*-2,6-tridecadienoate); and (c) the C_{16} JH from *Schistocerca vaga* and *S. gregaria* (methyl *cis*-10,11-epoxy-3,7,11-trimethyl-*trans, trans*-2,6-dodecadienoate).

With two double bonds and its cyclic oxirane component, the molecule can exist in 16 different stereoisomeric forms, only one of which can be the authentic hormone.[65] A second juvenile hormone, with a methyl group in place of the ethyl on C-7 has also been identified from *Hyalophora*,[206d] and a third hormone from *Schistocerca vaga* has been characterized as

methyl-10,11-epoxy-3,7,11-trimethyl-2,6-dodecadienoate.[158c] The compounds are sometimes called the C_{18}, C_{17} and C_{16} juvenile hormones: they have been synthesized from isoprene and farnesol derivatives.[203d, 265b]

The biosynthesis of the sesquiterpenoid skeleton of the juvenile hormones involves a complex pathway *via* mevalonate and homoisoprenoid precursors. *In vitro*, the corpora allata of *Schistocerca vaga* will manufacture the C_{16} juvenile hormone from acetate, while the glands of *S. gregaria* will produce both the C_{16} and C_{18} juvenile hormones from the corresponding farnesenic acids, esterification preceding epoxidation, and the hormones are released from the glands in simple form and not as prepackaged hormone-protein complexes.[225c]

Since the juvenile hormones are highly lipophilic substances which are poorly soluble in water, it was postulated that the hormones would circulate in the insect haemolymph bound to a hydrophilic carrier such as a protein or lipoprotein.[284a] Such carriers have now been identified in a number of species. In *Hyalophora*, there is a 'yellow high density lipoprotein', consisting of 52% protein and 48% lipid, with a molecular weight of about 200 000 which alone of the six haemolymph proteins will bind juvenile hormone both *in vivo* and *in vitro*.[268g] About 1 mg of the lipoprotein will bind 60 μg of juvenile hormone, although this large amount is never present in the insect and the lipoprotein probably transports many lipids for various purposes. In *Locusta migratoria*, six lipoproteins are also found in the haemolymph, and one of these (lipoprotein VI), with a molecular weight of 220 000, is specialized for juvenile hormone transport.[84c] In *Tenebrio*, analogues of juvenile hormone are also bound to haemolymph proteins,[265a] but it is not known whether the true hormone also binds to these proteins.

The juvenile hormones can be rapidly metabolized to inactive compounds, and excreted. The early steps consist of hydrolysis of the ester and hydration of the epoxide, followed by conjugation to glucuronides and glucosides or sulphation[1h, 246e, 268f, 268i] (Fig. 12.14). It is also likely that the 6, 7 double bond can undergo oxidative attack to form a bisepoxide, which is then hydrated to form the JH-tetrol (Fig. 12.14).

It is of considerable interest that juvenile hormone injected into the pupal stage (which normally does not contain the hormone) of *Hyalophora cecropia* and *H. gloveri* will induce the production, probably by the fatbody, of carboxylesterases which pass into the haemolymph.[268h, 268i] These enzymes hydrolyse the methyl ester of juvenile hormone to the corresponding acid (Fig. 12.14), which is biologically inactive. It is suggested that the esterases are present in the fatbody in inactive 'protein bodies', and that the juvenile hormone induces an enzyme which breaks down the protein bodies to release the carboxylesterases.[268h] Juvenile hormone is protected against enzymatic attack when bound to its transporting lipo-

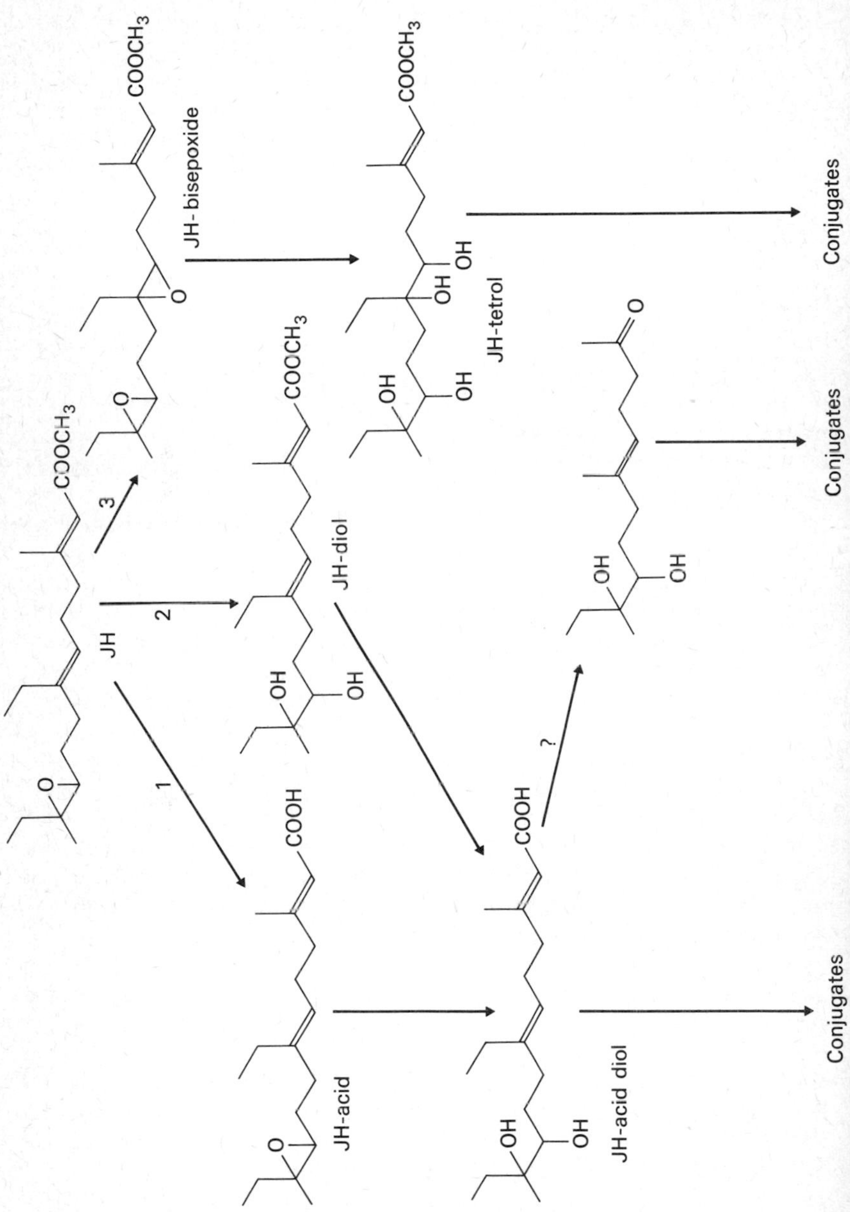
COOCH3
JH
1
2
3
COOCH3
JH-bisepoxide
COOCH3
JH-diol
COOH
JH-acid
COOCH3
JH-tetrol
COOH
JH-acid diol
?
Conjugates
Conjugates
Conjugates

protein,[84c, 268h] so the carboxylesterases will only degrade unbound juvenile hormone in the haemolymph. This is clearly an important method of regulating the concentration of juvenile hormone in the haemolymph, and can explain the rapid deactivation and excretion of exogenous juvenile hormone—and some of its analogues—when injected into insects at the 'wrong' times in their developmental histories. The mechanism can also have important consequences if analogues and mimics of juvenile hormone are used in pest control. It is also probable that a different group of carboxylesterases can degrade protein bound juvenile hormone, enabling a rapid diminution of hormone titre to be achieved at the appropriate time.[229f]

Hormonomimetic insecticides

During normal insect development, juvenile hormone is necessary during precisely determined periods (see p. 110); its continuing presence, for example, during the last larval instar, can prevent metamorphosis and lead to the death of the insect without its becoming adult. In particular, juvenile hormone must be absent from the eggs in many species or embryonic development is abnormal and the larvae do not hatch.

It has long been envisaged that juvenile hormone, and its mimics, could be used as insecticides by administering them at the wrong times in insect life-histories. Such compounds would have a tremendous advantage over insecticides at present used, since it was thought that the insects would be unlikely to develop immunity to substances which play a natural role in their own lives. Some juvenile hormone mimics are more effective in some species than in others, so that the development of *specific* insect pesticides could also be possible—in contrast to compounds like DDT which kill good and bad insects alike.

The feasibility of such developments was confirmed in a startling way when Dr. Karel Slama of the Charles University in Prague spent a year working at Harvard with Professor Carroll Williams. Slama took a stock of *Pyrrhocoris apterus* with him, but found that in Harvard the bugs would not develop into normal adults. Instead, at the end of the 5th larval instar, the insects all moulted into a supernumerary 6th larval instar, or into adultoids which preserved many larval characters. A few

Fig. 12.14 Metabolism of juvenile hormone. The molecule can be converted to the JH-acid by esterase activity at the methyl ester function (1), followed by epoxide hydration by hydrolase activity. Alternatively, epoxide hydration can precede the esterase action (2). The double bond may also be attacked, forming the bisepoxide and tetrol (3). Other metabolites of uncertain origin have also been found. The conjugated polar metabolites are glucosides or glucuronides. (After Ajami and Riddiford[1h], Slade and Zibitt[246e])

even grew and moulted into 7th instar larvae (Fig. 12.15). By analogy
with the effects of implanting corpora allata into last instar larvae (see
Chapter 6), it seemed likely that the *Pyrrhocoris* were being unintentionally exposed to a source of juvenile hormone, or a substance with juvenile
hormone activity.

A systematic comparison of the culture conditions in Harvard with
those in Prague eventually revealed that paper towels in the rearing jars
were the source of the substance.[248, 249] The chemical was not added
during manufacture because all papers, including newspapers, of
American origin had the same effect upon *Pyrrhocoris*. Papers of
European or Japanese manufacture had no effect. The major source of
wood pulp for paper manufacture in the United States is the balsam fir
(*Abies balsamea*), and extracts from this tree proved to contain high
juvenile hormone activity when assayed on *Pyrrhocoris*. Hemlock and
yew also showed high activity, but red spruce, American larch and southern pine had little activity. The substance was called **paper factor** by
Slama and Williams; it has now been identified (Fig. 12.16). It is the
methyl ester of todomatuic acid, and is now known as juvabione.[21d] The
dehydro derivative (Fig. 12.16) is additionally present in Czechoslovakian
balsam fir,[43c] and also shows juvenile hormone activity.

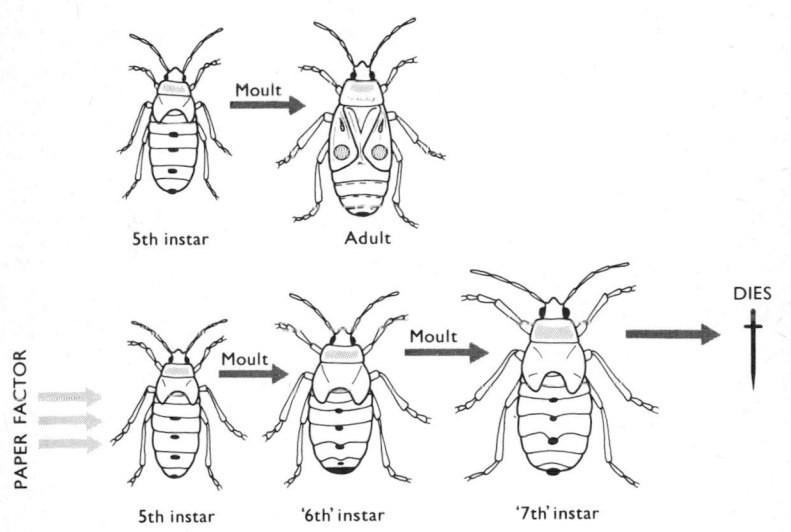

Fig. 12.15 The action of paper factor upon the development of *Pyrrhocoris
apterus.*
Upper row : normal moult from last instar larva to adult.
Lower row : paper factor induces the development of two giant supernumerary
larvae followed by death during the adult moult. (After Slama and Williams[249])

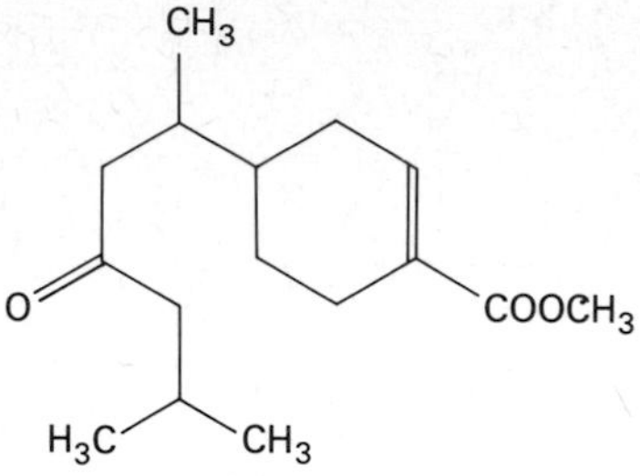

Juvabione ('paper factor')

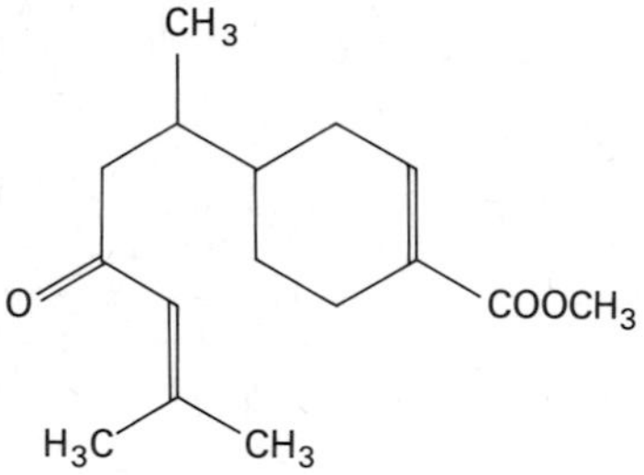

Dehydrojuvabione

Fig. 12.16 The structure of paper factors from American and European firs.

The most significant feature of this investigation, however, is that the paper factor proved to be inactive when tested upon *Hyalophora cecropia*, *Telea polyphemus*, and even upon *Oncopeltus fasciatus* and *Rhodnius prolixus*—bugs belonging to the same Order as *Pyrrhocoris*. This immediately raises the possibility of using such chemicals as specific insect pesticides. Figure 12.17 illustrates a few of the many thousands of compounds which have been tested for juvenile hormone activity; some of these are effective on whole families of insects, others will affect some genera but not all in a family, while some of the compounds are species specific in their juvenilizing effects.[247a, 257a] Chemicals are also known which potentiate the activity of the natural juvenile hormones by interfering with the breakdown of the hormones[227d] (Fig. 12.18).

Such insecticides can have two general applications: control against pests of stored products, and against those of field crops and animals. Laboratory experiments in which a juvenile hormone analogue is mixed with grain have already shown that infestations of *Tribolium castaneum* and other insect pests can be reduced or eliminated.[259] This method of control is likely to be widely successful for stored products, where usually *all* the insects present are undesirable. But in addition a technique has

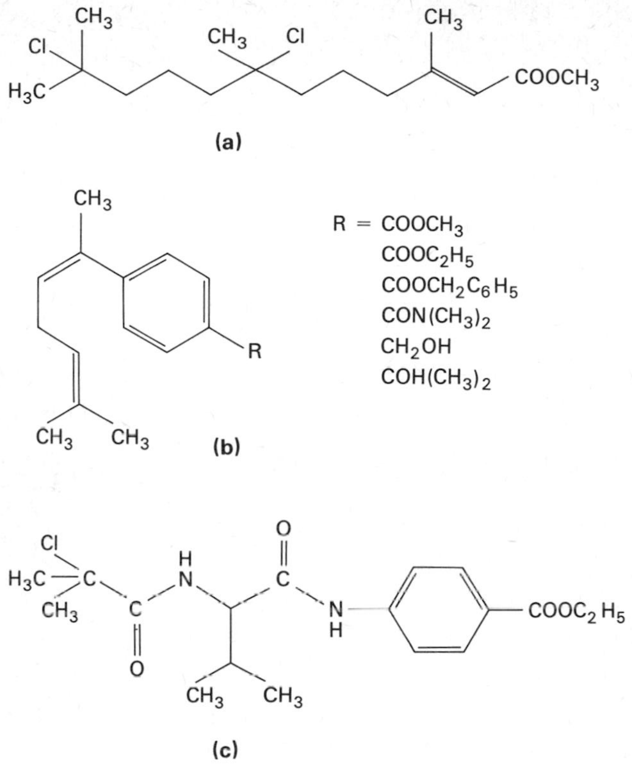

Fig. 12.17 A few of the very large number of compounds with juvenile hormone activity; (a) methyl 3,7,11-trimethyl-7,11-dichloro-2-dodecanoate; (b) derivatives of p-(1,5-dimethylhexyl) benzoic acid; the peptide derivative (c) is perhaps the most active juvenile hormone analogue yet known. (After Slama et al.[247b])

been developed which could provide a quite specific control of insect pests in the field.[202]

The method is effectively an extension of the sterile male technique so successfully used against the screw-worm, *Cochliomyia americana*, in America. This involves rearing large numbers of flies, irradiating the males to sterilize them, and releasing them in large numbers at the appropriate time. If an area is saturated with sterile males, the females will mate with them and lay infertile eggs. The female screw-worm mates only once: clearly this method of control is much less effective in insect species in which the female mates many times. The population of sterile males would have to be kept up for such extended periods that the

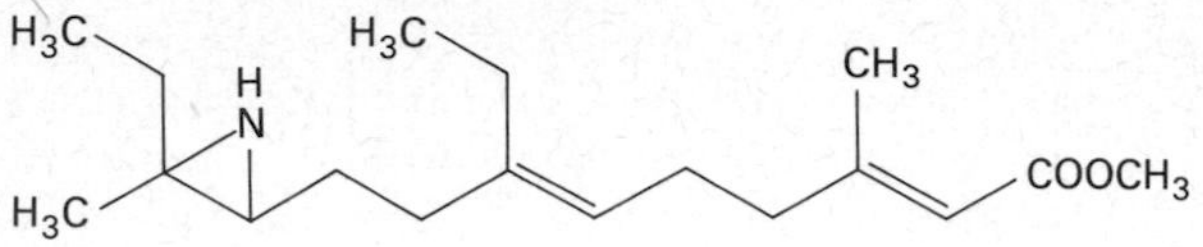

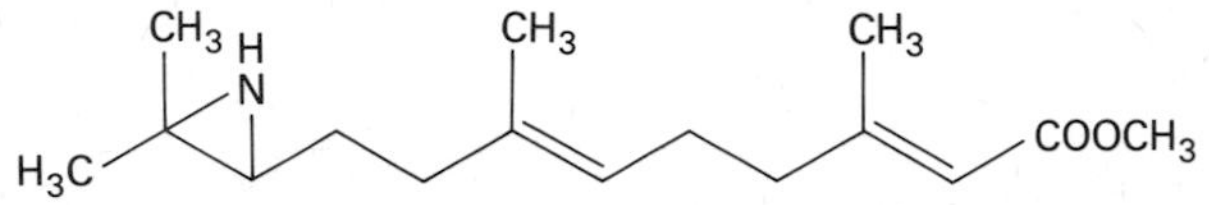

Fig. 12.18 Synthetic imino derivatives of the C_{16} and C_{18} juvenile hormones. These potentiate the activity of the hormones, probably by interfering with their natural deactivation mechanisms. (After Riddiford *et al.*[227d])

numbers involved would become practically and economically impossible.

But in *Pyrrhocoris apterus*, for example, a juvenile hormone analogue such as the dihydrochloride of methyl farnesoate (DMF) seriously disturbs embryonic development when applied to the eggs, and less than 1 μg applied to a mature female causes her to lay inviable eggs for the rest of her life. The males of *Pyrrhocoris* will tolerate up to 1000 μg of DMF applied topically—and they will pass sufficient DMF to females during mating to sterilize them for life. Males with as little as 100 μg each of DMF will pass 1–5 μg to females when kept with them for only 4 hours. With the much more effective compounds now being developed, even 10 μg per male will be sufficient.[202]

This method has the great advantage that a female once mated with a treated male will be sterile *for life*, no matter if she subsequently mates with any number of untreated, normal males: it extends the exact specificity of the screw-worm kind of pest control to species with polygamic females.

The stability of juvenile hormone and many of its mimics when applied under field conditions is low. This implies that frequent applications of the probably expensive chemicals would have to be made if the compounds are to be effective against a large proportion of a particular pest population. Moreover, the catabolizing enzymes of juvenile hormone (p. 279) can also act upon many of the juvenile hormone analogues and mimics. In fact, many of the enzymes which detoxify current insecticides, conferring immunity to large numbers of species, could also be effective against chemicals with juvenile hormone activity. In addition, much more detailed knowledge of the life-histories of insect pests will become necessary so that

the most appropriate times for application of the hormonomimetic compounds can be ascertained. And since the chemicals do not have the dramatic 'knock down' properties of current insecticides, they are likely not to be especially popular with farmers and other agriculturalists. However, in spite of these reservations, it is very probable that these compounds will play an important part in the integrated control of insect pests, involving cultural, biological and chemical methods in combination.

13

Pheromones

Pheromones are defined as substances produced by one individual of a species which cause particular behavioural and/or developmental reactions when received by another individual of the same species.[163] They are effective in minute amounts and act as chemical messengers *between* individuals in much the same way as hormones *within* an individual; in fact, they are sometimes called ectohormones. Pheromones may generally be subdivided into those which act upon the nervous system to evoke immediate and reversible behaviour changes (releaser pheromones), and those which cause permanent physiological changes in the recipient (primer pheromones).[291b] Releaser pheromones include the alarm substances of fish, ants and earthworms, trail markers, territorial and aggregation chemicals, sex attractants and aphrodisiacs. Primer pheromones include substances which affect developmental morphology, maturation, sexual cycles, blastocyst implantation in mammals, and many other facets of growth and reproduction in animals. The distinction between releaser and primer pheromones is not always clear cut: the latter often have initial releaser effects, and the behavioural reactions to the former, particularly those involved in reproduction, can have subsequent developmental consequences.

Pheromones are ubiquitous in the animal kingdom, and are found also in some algae and fungi. This account will be restricted to pheromones which are associated with particular developmental changes, and in which the endocrine system is involved. Only in the insects among the invertebrates have both pheromones and hormones been studied in detail, so this group inevitably predominates. For further information on insect pheromones, reference should be made to recent reviews.[16a, 171, 34, 154a]

Sex pheromones in insects

Many insects recruit their mates by means of chemicals, which are consequently often called sex attractants. Pheromones which play a role in the mating behaviour itself are sometimes called aphrodisiacs. The term 'sex attractant' has been severely criticized, since it originates from the final result of the behaviour and clearly implies that the behaviour is aimed towards that goal.[169a] Behavioural response to the pheromone, including orientation and locomotion, often depends upon many factors in addition to the presence of the chemical itself (p. 290), and can also vary dramatically with the concentration of the substance. Moreover, when the sexes are in proximity, stimuli other than the pheromone may serve to initiate and maintain the act of mating. These reservations should always be borne in mind, but 'sex attractant' is now so firmly embedded in the literature that little purpose can be achieved in attempting to coin a new, although more exact, terminology.

The first sex attractant to be identified was that of the female silkmoth, *Bombyx mori*, and consequently called bombycol[32] (Fig. 13.1). A rather similar compound from the female gypsy moth, *Porthetria dispar* (gyptol), was first suggested to be the sex attractant of this species,[155] but inconsistencies in its effects under laboratory and field conditions prompted a reinvestigation of the substance. The pheromone has now been identified (Fig. 13.1) and is called disparlure.[17h] Sex attractants have now been demonstrated in almost 200 species of Lepidoptera alone, and many have been chemically identified.

Female cockroaches produce sex attractants from one or more anal, sternal and tergal glands. In the Cuban cockroach, *Byrsotria fumigata*, removal of the corpora allata from the female shortly after the imaginal moult prevents the production of sex pheromones, and the allatectomized females remain unattractive to the males and do not elicit the masculine

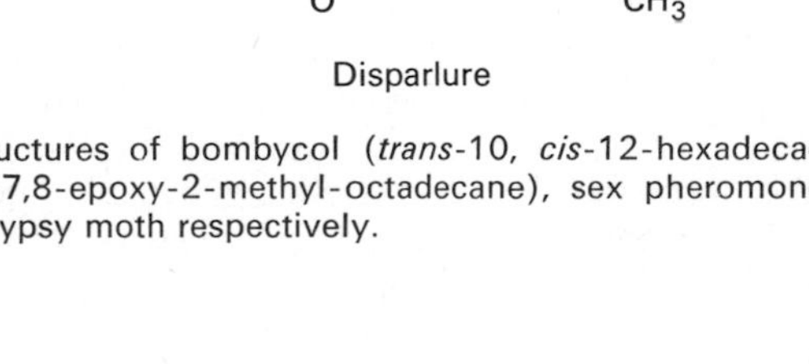

$$CH_3 - (CH_2)_2 - \overset{H}{C} = \overset{H}{C} - \overset{H}{C} = \overset{}{\underset{H}{C}} - (CH_2)_8 - CH_2OH$$

Bombycol

$$CH_3 - (CH_2)_{10} - \overset{H}{\underset{}{C}} \overset{H}{\underset{O}{-}} \overset{}{C} - (CH_2)_4 - CH \overset{CH_3}{\underset{CH_3}{<}}$$

Disparlure

Fig. 13.1 Structures of bombycol (*trans*-10, *cis*-12-hexadecadien-1-ol) and disparlure (*cis*-7,8-epoxy-2-methyl-octadecane), sex pheromones of the silkmoth and the gypsy moth respectively.

precopulatory behaviour. Reimplantation of corpora allata is followed by pheromone production within 10–16 days.[6] Moreover, in *Periplaneta americana* the production of sex pheromone ceases during the time the ootheca is being formed and extruded—when mating is impossible. In these cockroaches, oocyte development is controlled by the corpora allata, so that the concomitant endocrine control of pheromone production provides a mating signal during the correct period. Support for this idea comes from experiments with the cockroach *Pycnoscelus surinamensis*, which has bisexual and parthenogenetic strains. Allatectomy of females of bisexual *Pycnoscelus* prevents the production of sex pheromones as it does in *Byrsotria*; but allatectomy of the parthenogenetic females has no effect upon sex pheromone production. Further, females of the bisexual strain do not produce pheromone when they are carrying an egg case: those of the parthenogenetic strain do. Oocyte development is controlled by the corpora allata in both bisexual and parthenogenetic *Pycnoscelus*; but when, as in the parthenogenetic females, a mating signal is no longer required, the endocrine co-ordination of mating with oocyte development is lost.[7]

Male cockroaches may also produce sex pheromones, but their control differs from that of the female. When a male *Nauphoeta cinerea* recognizes a receptive female, he releases a volatile sex pheromone (seducin) which attracts the female to his tergum which she mounts and palpates. This mounting and feeding behaviour places her in the correct position for copulation. Seducin may therefore be called an aphrodisiac pheromone. Tergal feeding by the female ceases within a few seconds after mating. Allatectomy of males has no effect upon seducin production or mating. It is therefore proposed that the pheromone in males is released under direct nervous control,[123a] and its production is not hormonally controlled. This would economize on pheromone production by the male, where cyclic periods of readiness to mate are absent.

In many Orthoptera, either the male or the female (and sometimes both) produces a mating song by rubbing one part of the body against another, frequently the legs against the elytra, or the elytra against each other. This stridulatory song is an auditory communication mechanism comparable with the olfactory ones described above. The corpus allatum hormone co-ordinates the onset and termination of female stridulation with oocyte development.[193] Although affecting central nervous activity in the one instance, and glandular biosynthesis in the other, a signal is sent out at the appropriate time. The importance of factors other than the sex attractant in influencing behavioural responses has been well documented.

In the honey bee, *Apis mellifera*, the drones are attracted by a queen when she is flying in the afternoon of a warm, sunny day and only when

she is at a certain height relative to the wind-velocity.[34] The queen's sex
pheromone is only stimulatory when other sensory impulses are also passing to the drones' central nervous systems. Drones are not attracted to a
queen in the hive or on the alighting board. This could be a device to
prevent colony inbreeding. The honey bee's sex attractant is 9-oxodecenoic
acid[34] (Fig. 13.2): but in addition to its sex pheromone activity, stimulating

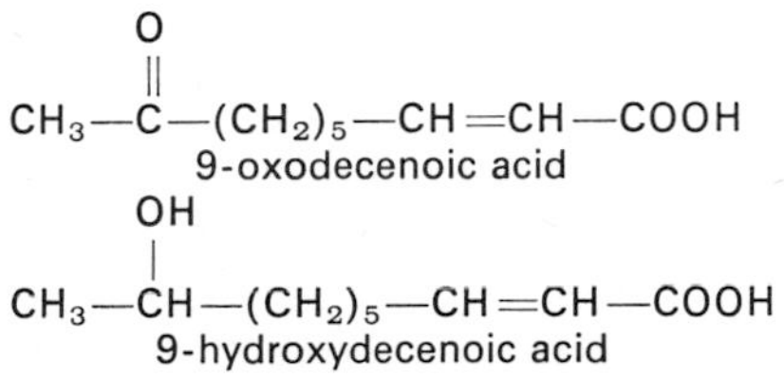

$$CH_3-\overset{\displaystyle O}{\overset{\displaystyle \|}{C}}-(CH_2)_5-CH=CH-COOH$$
9-oxodecenoic acid

$$CH_3-\overset{\displaystyle OH}{\overset{\displaystyle |}{CH}}-(CH_2)_5-CH=CH-COOH$$
9-hydroxydecenoic acid

Fig. 13.2   9-oxodecenoic and 9-hydroxydecenoic acids: honey bee pheromones.

the drone's olfactory receptors, the same substance plays important
developmental and behavioural roles *within* the hive, primarily affecting
the (female) worker bees.

In the cabbage looper moth, *Trichoplusia ni*, the males respond be-
haviourally to the female sex attractant only during that period of the day
when they are actively flying,[218b] although their antennal receptors will
respond physiologically throughout the day. Coordination within the
central nervous system to produce orientation responses to the pheromone
clearly depends upon more than the receipt of stimuli from the attractant
alone. A circadian rhythm of flight activity is common in many insects,
and it is not surprising that sex pheromone communication is linked with
this rhythm. Thus in several species of saturniid moths, sex pheromone
release by the female ('calling') occurs at different times in the 24 hour
cycle, but corresponds in each instance with the flight activity of the male
of the species.[227c] In these moths, calling involves a particular kind of
postural behaviour in the female which is mediated through the central
nervous system in response to a hormone released from the corpora
cardiaca.[227g] In *Hyalophora cecropia*, hormone release is governed by the
environmental light: dark cycle, and calling is consequently determined at
a particular time; in *Antheraea pernyi*, hormone release additionally requires the presence of the host plant (Oak), or more particularly a volatile
substance (*trans*-2-hexenal) from oak leaves.[227f]

In insects such as the cockroaches which are long-lived as adults, the
corpora allata coordinate both oocyte development and pheromone production. In insects which have short adult lives, such as the moths described above, high priority must be given to early reproduction since the

female emerges with almost the full complement of mature eggs, and a different endocrine mechanism has been developed to control pheromone production and release. In the saturniid moths, the behaviour of the female changes after mating: she ceases to call, flight activity is enhanced and oviposition is stimulated. This behavioural change is initiated by an interaction between the sperm and the bursa copulatrix: the bursa now produces a humoral substance which prevents the release of the calling hormone from the corpora cardiaca and stimulates instead the release of an oviposition hormone,[227e, 265i] which may act in the same way as the neuro-secretory myotropin in *Rhodnius* (p. 149). Thus although the number and kind of the hormones involved are different, endocrine mediated changes in behaviour after mating are characteristic of both long-lived and short-lived adult females.

Maturation pheromones

The queen is the only reproductive female in the honey bee hive. In the remaining females—the workers—oogenesis is inhibited. When the queen is removed from the hive, three consequences result. First, within a few minutes fanning activity—the rapid wing-beating which ventilates the hive and possibly also distributes pheromones—increases greatly, and the workers become 'agitated' with remarkably enhanced locomotor activity. Secondly, within 48 hours without the queen, the workers modify the cells of larvae less than three days old into queen cells and provision them lavishly, leading ultimately to the production of new queens. Finally, the ovaries of the workers begin to develop. Thus in the absence of the queen, both behavioural and developmental responses occur in the workers: the pheromones involved have both releaser and primer effects.

The inhibition of both oogenesis and queen rearing in workers is normally caused by queen substance produced in the queen's mandibular glands, spread over the body during grooming, and licked off by attendant workers who share it with their colleagues in regurgitated food.[34] Queen substance is largely composed of 9-oxodecenoic acid, but this substance alone does not completely inhibit oogenesis. It has to be combined with inhibitory scent, which has been identified as the odour of 9-hydroxydecenoic acid[35] (Fig. 13.2).

How these substances act in inhibiting oogenesis is unknown. 9-Oxo-decenoic acid can be identified in the crops of worker bees. But whether it stimulates specific chemoreceptors during feeding, thereby affecting the endocrine system, or whether it inhibits oogenesis directly after passing into the blood is conjectural. The initial growth of the corpora allata in workers is inhibited by 9-oxodecenoic acid.[103a] The behavioural

response to the two substances is more likely to be due to specific sensory stimulation affecting central nervous activity.

9-Hydroxydecenoic acid and 9-oxodecenoic acid together in a queenless hive are not quite as effective in inhibiting queen-rearing as a live mated queen. They are less attractive than a queen, and fewer individuals are consequently influenced by the substances. The queen, *in the hive*, presumably secretes another attractive substance.[34] But the two acids together enable *swarming* bees to find and remain with their queen.[34]

Old queens become deficient in the two acids and are superseded by new queens, when the workers react to the decreasing concentrations of the substances. Similarly, when the hive becomes overcrowded, workers may obtain less queen substance and swarm. In *Apis mellifera*, the versatility of 9-oxodecenoic acid, alone or in combination with other compounds, in influencing different kinds of behaviour, and development, must be acknowledged.

In the dry wood termite, *Kalotermes flavicollis*, there is a pair of royal reproductives, the king and the queen. The worker caste is actually composed of juvenile forms, consequently called pseudoworkers or pseudergates. These can develop into soldiers, or in the absence of the royal pair into secondary reproductive forms (Fig. 13.3).

In a normal colony of *Kalotermes*, the development of secondary reproductives from pseudergates is inhibited by a pheromone present in the excrement of the queen and distributed among the workers.[196] For the complete inhibition of secondary reproductives, the king must also be present. It seems that each member of the royal pair inhibits the development of reproductives of its own sex, and that each stimulates the production of pheromone by the other (Fig. 13.4). Moreover, the king produces another pheromone which stimulates the production of female reproductives in the absence of the queen.[196] Other pheromones determine the soldier caste, and still more are involved in the elimination of excess numbers of any one caste.

Although the evidence so far is slight, it is very likely that the pheromones work through the endocrine systems of the recipients. For example, the development of pseudergates to first and second stage nymphs is reversible, and hormones must be involved in such morphological plasticity. Soldiers can be developed from pseudergates by implanting the corpora allata of reproductives.[196] Moreover, competent pseudergates moult very soon after the removal of the royal pair, implying a rather direct effect upon the brain—thoracic gland system.[196] *Kalotermes flavicollis* forms small colonies with a limited number of castes. The pheromone-hormone interactions in species with large colonies and many more castes must be extremely complicated.

In the desert locust, *Schistocerca gregaria*, mature males produce a

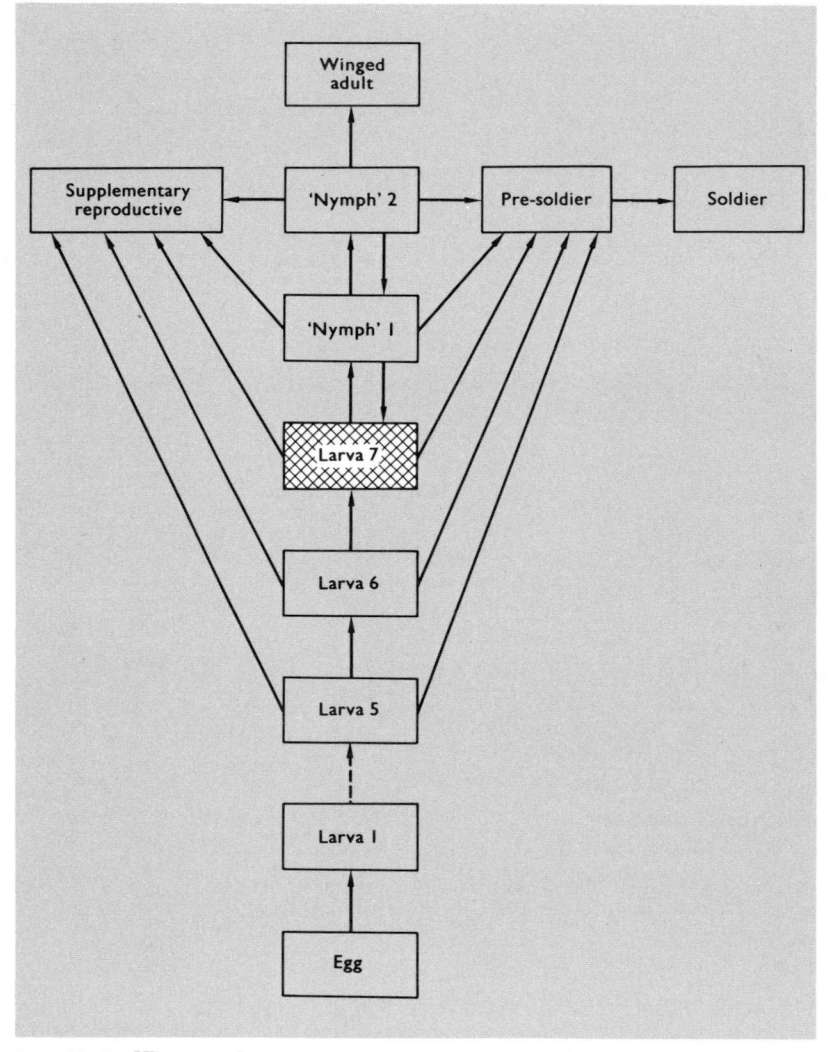

Fig. 13.3 Diagram of the caste structure of the termite *Kalotermes flavicollis*. The late larvae are 'false workers' or pseudergates. Note that regressive moults from nymphs to larvae can occur.

pheromone which accelerates maturation in other males and in females. The pheromone is not produced by allatectomized males, but the subsequent reimplantation of corpora allata reinitiates pheromone production[192] (Fig. 13.5). Maturation itself is controlled by the corpora allata

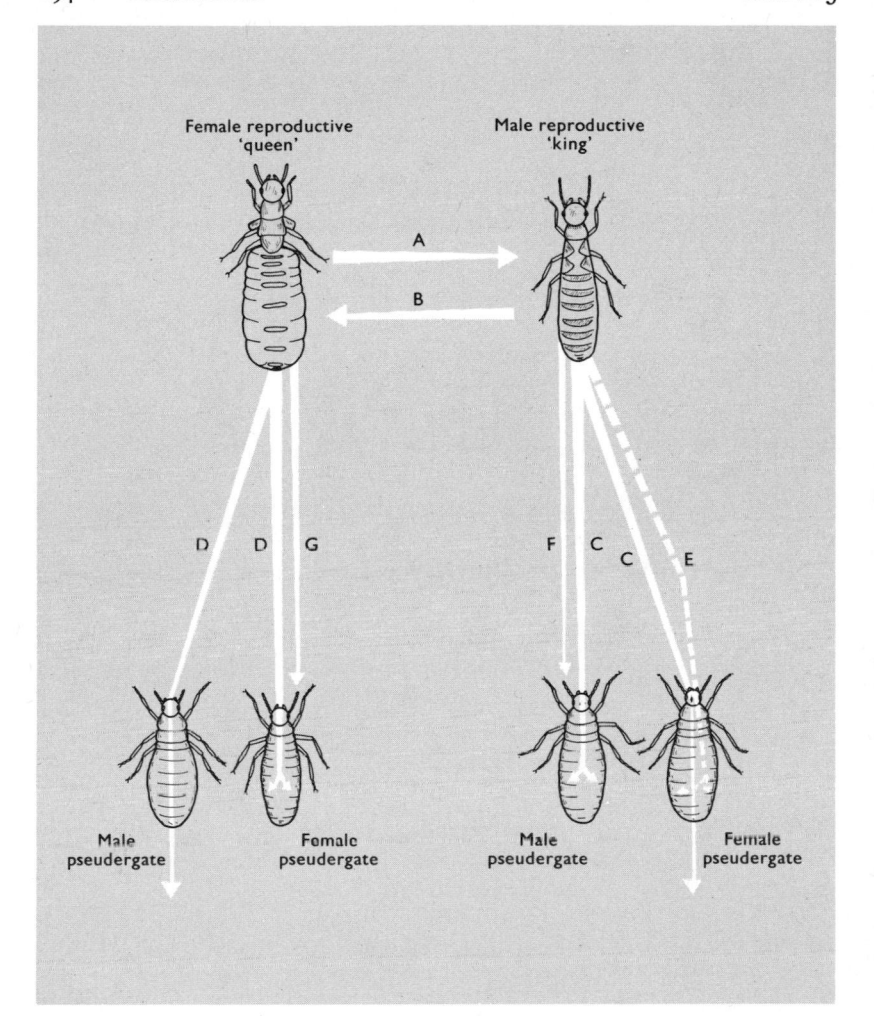

Fig. 13.4. Pheromones and the development of supplementary reproductives in *Kalotermes flavicollis*. The male reproductive ('king') produces a pheromone (C) which prevents reproductive development in male pseudergates but is without effect upon the female pseudergate. The female reproductive ('queen') similarly produces a pheromone (D) which inhibits reproductive development in female pseudergates, leaving the males unaffected. In the absence of the queen, the king produces a pheromone (E) which stimulates reproductive development in female pseudergates. The royal pair are reciprocally stimulated to produce their pheromones C and D by other pheromones (A and B). Yet other pheromones (F and G) affect the *behaviour* of pseudergates to eliminate an excess of supplementary reproductives if these develop. (After Luscher[196])

and the cerebral neurosecretory systems (see p. 157) so the maturation pheromone in some way stimulates the activity of the neuroendocrine system. It is not known whether this activation is direct, the pheromone passing through the recipient's cuticle into the blood, or indirect by stimulating specific chemoreceptors whose impulses eventually affect the

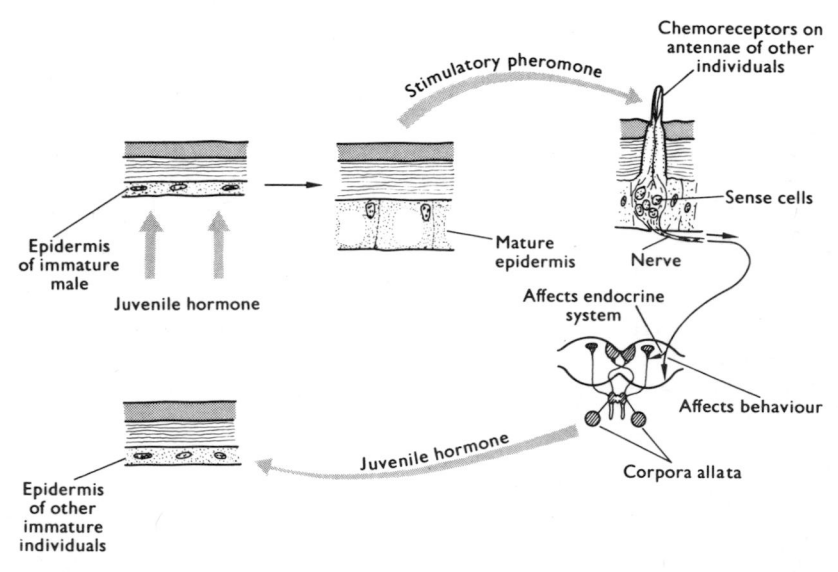

Fig. 13.5 The maturation pheromone of the desert locust, *Schistocerca gregaria*. Under the influence of juvenile hormone, the epidermis of one individual produces pheromone which stimulates the endocrine systems of other individuals probably by way of chemoreceptors and the nervous system. The consequent production of juvenile hormone in the recipients stimulates the production of more pheromone (see Loher[192]).

neurosecretory system. The pheromone can elicit an immediate behavioural response in the recipients (twitching of the palps of the mouthparts and a kicking of the hindlegs[192]), so the central nervous system is certainly involved in at least part of the effect of the pheromone.

The net result of the stimulatory pheromone in the desert locust is that a rapidly maturing individual hastens the development of the more slowly maturing members of a group. In the African migratory locust, *Locusta migratoria*, it seems probable that an inhibitory pheromone is produced by immature adults which suppresses the potentially more rapid development of some individuals. But in both *Schistocerca* and

Locusta the presence of the pheromones tends to synchronize reproductive development in all members of a group, although acting in quite opposite ways.

There are many obvious differences in morphology, physiology and behaviour between solitary and gregarious (swarming) locusts. Many of these differences can be traced to variation in the activities of parts of the endocrine system. The possibility that gregarization is caused by a specific pheromone[213b] will perhaps provide the best illustration of the relationships between endocrine glands and pheromones in insects.

Pheromones form only part of the many environmental influences which can affect development in insects, although in some instances their role is a major one. In social insects, where the colony may be considered a superorganism, the number and variety of pheromones may rival those of hormones in a highly developed animal. In many respects, the colony tends to isolate itself from the external environment, making contact at well defined points. The increased importance of pheromones in these situations is not surprising. Although it may not be wise to draw too close a parallel between the action of social pheromones upon the individuals of the colonial superorganism and that of hormones upon the cells and tissues of an individual, such a comparison does accentuate some common features of chemical communication systems in animals.

Pheromones and insect control

The use of pheromones in insect pest control has distinct possibilities. Like the hormonomimetic insecticides (p. 281), pheromones are active in minute quantities, have good species-specificity and are of low toxicity to other organisms. Many of the aggregation and sex pheromones are relatively simple, low molecular weight compounds which are comparatively easy to synthesize. Female sex pheromones can be used in initial surveys to determine the presence and increasing population of an insect pest, with eventual control being exercised by other means. Or the pheromones can be used themselves in eradication programmes, trapping and killing the males, or attracting them to chemosterilants for subsequent release as inviable mates. Sex attractants could also be used to saturate particular areas, obliterating response to the naturally produced substances and thus preventing mating. Other substances are known which mask or inhibit the attractiveness of the natural pheromones; their use would similarly prevent the normal response. But as with the hormonomimetic insecticides, greater knowledge of the biology and behaviour of each insect pest becomes essential when such 'natural' substances are employed. Some species react only to high concentrations of pheromones, others to low; some attractants are produced continuously, others in pulses of varying

duration; pheromones can also be aggregates of compounds, the precise proportion of each constituent being of the utmost importance in evoking a normal response. The behavioural reaction of the insect can depend upon the time of day, wind velocity, host plant, light intensity, previous exposure to a pheromone, previous mating experience, and a host of other factors. Once this information is available, the use of pheromones in insect control promises, at the least, that reliance upon conventional insecticides can be greatly reduced. The compounds are likely to find their brightest future as components in integrated control programmes.

14

Hormones and the Environment

The endocrine control of developmental and physiological processes in invertebrates has been amply demonstrated by the results of various kinds of experiment in the different phyla. It is clear that the diversity of invertebrate structure and habit imposes a corresponding variety upon their endocrine mechanisms. Among the vertebrates, the endocrine glands and hormone chemistry of a fish may profitably be compared with those of a mammal. A similar exercise even between related invertebrate phyla is difficult if not impossible. The diverse sources of hormones in the invertebrate groups are not amenable to any phylogenetic interpretation. But the comparative endocrinologists can illuminate the different ways in which similar developmental events in various animals are hormonally controlled. Thus gonadal differentiation in molluscs and crustaceans has features common to both groups, and although the endocrine mechanisms involved cannot be compared in terms of gland morphology or hormone chemistry, the overall *pattern* of control shows important similarities. Oocyte maturation and spawning in *Arenicola* has no possible phylogenetic relationship with that in starfish, yet these processes in the two groups show remarkable similarities in their endocrine control, although the hormones involved must be very different.

The unifying feature of endocrine controlled processes in invertebrates is the almost ubiquitous involvement of neurosecretory mechanisms. In the less highly organized invertebrates, neurosecretory hormones control directly the processes of growth, reproduction, water balance and so on. In the annelids, the neurosecretory control of growth and reproduction is also probably direct, although the possible intervention of other

secretory elements in the infracerebral gland has yet to be determined. The cephalopod molluscs are without a neurosecretory involvement in reproduction, perhaps because the relationship between visual stimuli and optic gland activity has become so close. In the arthropods, neurosecretions often have a direct effect upon particular targets—for example, the chromatophorotrophins of crustaceans, the diuretic hormones of insects, and neurosecretory hormones affecting synthetic mechanisms in the insect fatbody and the crustacean hepatopancreas. But particularly in their control of development, moulting and reproduction, arthropod neurosecretory hormones act through epithelial endocrine glands. In insects, ecdysone is secreted by the thoracic gland, the activity of which is stimulated by the neurosecretory thoracotrophic hormone; in crustaceans, crustecdysone secretion by the Y-organs is inhibited by the neurosecretory moult-inhibiting hormone. The insects' corpora allata are affected directly or indirectly by neurosecretory allatotrophic hormones: in the crustaceans, it is possible that metamorphosis is controlled directly by sinus gland hormones other than those which affect moulting in the post-metamorphic stages.[64a, 64b] But the endocrine activities of the crustacean ovaries and androgenic glands are again under neurosecretory control.

In the less highly organized invertebrates, therefore, neurosecretory hormones act directly to control growth and development, and other shorter-term processes too. In the arthropods, although some events are still controlled directly by neurosecretory hormones, others, particularly growth, moulting and differentiation, are controlled indirectly by these hormones, with epithelial endocrine glands intervening in the system. In the vertebrate animals, the neurosecretory hormones like oxytocin and vasopressin have *direct* effects upon physiological mechanisms, but hypothalamic neurosecretory hormones exert trophic effects upon the anterior pituitary whose own hormones control the activities of other endocrine glands. There is thus one step neurosecretory control in some invertebrates, two step control in more highly organized invertebrates, and three step control in the vertebrate animals[237] (Fig. 14.1). But direct neurosecretory control over *some* processes is still exercised even in the arthropods and vertebrates.

Where neurosecretions exert their effects indirectly through epithelial endocrine glands, it becomes possible to control the processes more efficiently by the introduction of feedback mechanisms between the glands and the neurosecretory systems. Such feedback controls are well documented in vertebrate endocrine systems, and the relationship between cerebral neurosecretory cells and the corpora allata in insects has been described (p. 167). But there is evidence that a feedback mechanism also exists between cerebral neurosecretion and the thoracic glands. In *Rhodnius prolixus*, ecdysone, whose production is stimulated by the thoracotrophic

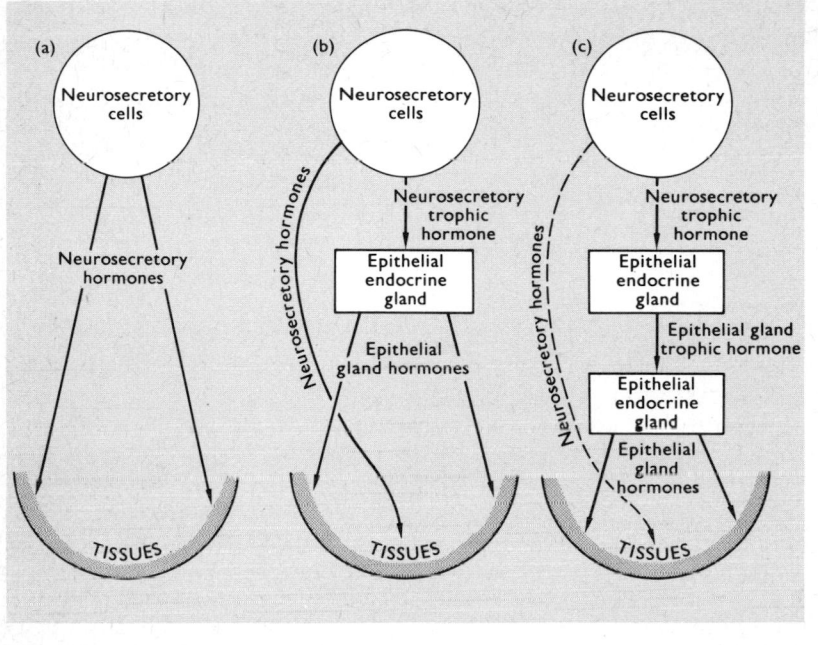

Fig. 14.1 (a) One step neurosecretory mechanism where neurosecretory hormones control developmental and physiological processes directly (e.g., coelenterates, annelids). (b) Two step neurosecretory mechanism where neurosecretory hormones exert *trophic* effects upon epithelial endocrine glands, hormones from which subsequently affect development and physiology. Note that elements of the one step system still occur (e.g., insects, crustaceans). (c) Three step neurosecretory mechanism, In which neurosecretory hormones exert trophic effects upon epithelial endocrine glands the hormones from which are themselves *trophic* to control the activity of yet further epithelial endocrine glands. Elements of the one and two step systems still occur (e.g., vertebrates). (See Scharrer and Scharrer[237] for a full discussion.)

hormone from the brain, inhibits the release of neurosecretion from the median neurosecretory cells and reduces the excitability of interneurones carrying nervous stimuli from the abdominal stretch receptors. Ecdysone also stimulates the synthesis of neurosecretory material, but the net result of ecdysone action is to 'inform' the brain that the thoracic glands are active and the intervention of the cerebral neurosecretory cells can be brought to an end.[250a]

There are, of course, many endocrine mechanisms which are without any neurosecretory control. Hormone production by the gut and its associated glands in vertebrate animals is a good example of such inde-

pendence. But in general, the endocrine control of growth and development in most animals seems ultimately to involve neurosecretory hormones. It has already been postulated that neurosecretory cells are the oldest hormone-producing cells in the animal kingdom, and that they function alongside purely nervous elements to coordinate responses in the individual with environmental change (Chapter 1). What is the evidence that neurosecretion functions in this way in modern animals?

Arguments in favour of the general hypothesis that neurosecretory mechanisms transform environmental stimuli to hormonal messages are often incomplete. Where a developmental process varies with the environmental situation and can be shown to be under neurosecretory control, it is often assumed that the neurosecretory mechanism is influenced by the environment. But in the vertebrates, reciprocal interactions between epithelial endocrine glands and hypothalamic neurosecretion are well established. In the insects, too, reciprocity between the corpora allata and cerebral neurosecretion is very likely (p. 167). Consequently, environmental effects upon developmental processes, such as the influence of temperature upon gonadal development in amphibians or insects, could *subsequently* cause changes in neurosecretory activity.

But there exist several very good examples of an immediate control over neurosecretory mechanisms by environmental stimuli. The best documented is perhaps the effect of photoperiod and temperature on the induction and termination of pupal diapause in insects (p. 178). There is now little doubt that the cerebral neurosecretory cells react to these environmental features and are the prime movers in the events that characterize diapause. The influence of direct and reflected light in controlling the release of chromatophorotrophins in crustaceans also indicates a fairly direct relationship between an environmental stimulus and neurosecretory activity (Fig. 11.14). The mammalian suckling reflex illustrates very well how an exogenous stimulus releases a neurosecretory hormone (oxytocin) which has a very rapid effect upon milk ejection from the mammary glands, but also a longer term effect in the continued secretion of prolactin for milk production (Fig. 14.2). Prolactin is an anterior pituitary hormone whose production and release is controlled by a neurosecretory hormone in the median eminence (Fig. 14.2).

In *Rhodnius*, nervous impulses from an abdomen stretched with food cause the rapid release of thoracotrophic hormone. In the desert locust, low frequency electrical stimulation of the central nervous system for a short time accelerates the movement of material within the cerebral neurosecretory system.[132] Continued low frequency stimulation, or high frequency stimulation for a shorter time, releases the material from the system.[133] Stimulation of sensory receptors by tumbling the locusts in a jar has an effect upon the neurosecretory system similar to that of electrical

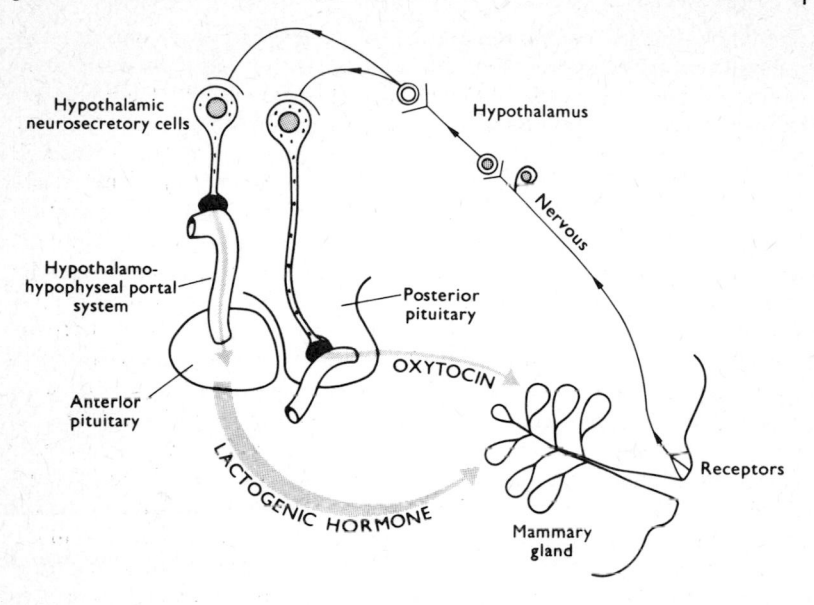

Fig. 14.2 The suckling reflex in a mammal. Both anterior and posterior pituitary hormones are secreted in response to stimulation of the nipple. Oxytocin release is very rapid and causes myoepithelial cells of the mammary alveoli to contract and force out the contained milk. Lactogenic hormone (among others) production and release is maintained by suckling stimuli so that milk synthesis is continued.

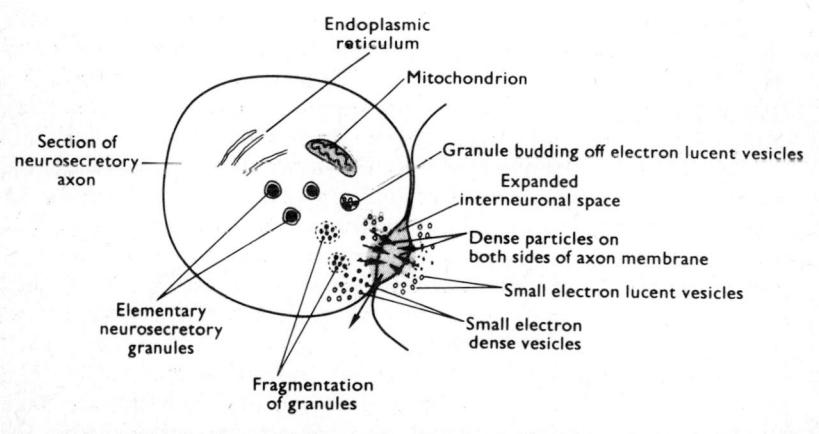

Fig. 14.3 Ultrastructure of a release site in a neurosecretory axon. The elementary neurosecretory granules appear to fragment and pass from the axon to the interneuronal space in diffuse form. (After Scharrer[236])

stimulation. These techniques mimic natural sensory stimulation, because copulation in female locusts previously kept without males,[141] oviposition,[133] feeding after previous starvation,[140, 148, 143b] and flying,[138] all bring about the rapid release of material from the neurosecretory system. In the vertebrates too, electrical stimulation of the hypothalamus releases both posterior and anterior pituitary hormones. That neurosecretory cells react to incoming nervous impulses by releasing hormones is beyond doubt. But *how* nervous activity accelerates the release of the elementary neurosecretory particles is unknown. Ultrastructural observations suggest that the particle of material is released from its surrounding membrane and passes through the axon terminal in a dispersed form (Fig. 14.3). But the relationship between this process and electrical activity in the axon is obscure.

A variety of environmental factors affects neurosecretory activity in insects (Figs. 14.4, 14.5); some, such as temperature and perhaps photoperiod, act directly upon the neurosecretory cell, others stimulate sensory receptors to control the cell through electrical activity within the central nervous system. The situation is similar in the crustaceans. In vertebrate animals, also, the environmental control of neurosecretory activity is well founded upon experiment and observation. In the less highly organized

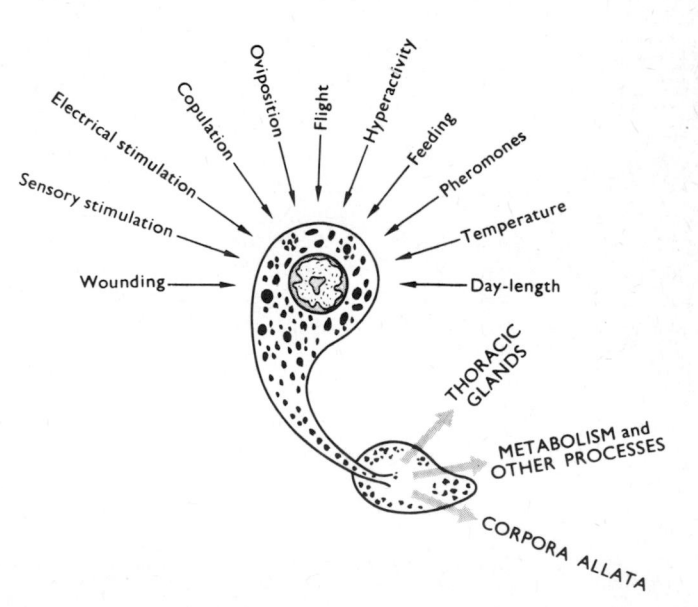

Fig. 14.4 Some of a variety of environmental and other factors which influence the release and/or synthesis of neurosecretions in insects. (After Highnam[136])

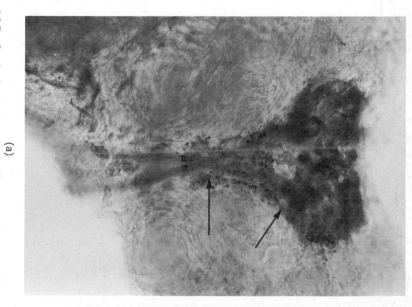

(a)

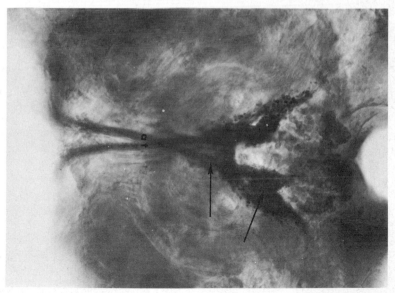

(b)

Fig. 14.5 Cerebral neurosecretory cells, neuropilar reservoir (arrowed) and axon tracts (at) of *Locusta migratoria* stained *in situ.* (a) The system in a normally maturing female : the axon tracts and reservoir contain little stainable neurosecretion ; (b) the system in a similar female starved for 4 days : the axon tracts and reservoir contain large amounts of stainable neurosecretion.

invertebrates, evidence for the exogenous control of neurosecretory mechanisms is more circumstantial. But at the least, there is now considerable evidence from a large number of quite disparate animals to suggest that neurosecretory mechanisms actually do have the general purpose of coordinating environmental change with developmental and physiological events within the animal. The environmental cues to which the neurosecretory system responds usually precede the situations which necessitate particular developmental and physiological changes. The animal is thus preadapted to take advantage of favourable or to avoid unfavourable conditions.

Of considerable interest in this connection is the part played by pheromones. These biologically very active molecules can be used to signal particular physiological states so that behavioural interactions will occur at appropriate times. The sex attractants of insects and other animals demonstrate the working of such a system. It is not surprising that both reproductive development and pheromone production are often coordinated by the endocrine system (p. 289).

But pheromones can also affect development itself. For example, the maturation pheromones of locusts control group reproduction, working either directly or indirectly through the endocrine systems of the recipients. The phenomenon is not confined to insects: in mice, for example, the male produces an olfactory pheromone which synchronizes oestrus cycles in groups of females. Moreover, the odour of a strange male can inhibit blastocyst implantation in recently mated females.[216] Here again, the pheromone is perceived by the recipient's olfactory receptors and ultimately acts through their endocrine systems. These pheromones may be considered as a kind of fine control over the effect produced by other internal and external forces. They assume overriding importance in the control of caste development in social insects.

Pheromones form one kind of chemical communication system in animals, hormones another. The effects of pheromones upon development illustrate how these two systems can interact. The metabolic machinery within a cell is a third kind of chemical communication system (p. 117), and there must be interaction between this and hormonal chemicals when the endocrine system controls metabolism and development. One way in which hormones can affect the intracellular system has already been described (p. 120). But hormones could act in many other ways: by altering membrane permeabilities, around and within the cell; by acting as co-enzymes; by activating pre-existing enzymes; and so on. It is likely that different hormones can affect the cell in various ways, and that they can have diverse effects upon different cells.

During embryogenesis, diffusable chemicals (organizer substances) are of the greatest importance in directing the development of particular

cells and tissues along specific paths. In some animals, this process of determination may begin even in the unfertilized egg: in others, its commencement is delayed, even up to gastrulation. The fertilized egg possesses all the genes of the eventual adult individual. The progress of differentiation into cell layers, tissues and organs must involve the suppression of the activities of certain genes, and allow the full expression of others (p. 121). In those larval Diptera in which giant chromosomes can be seen in a number of tissues, the same arrangement of bands is everywhere recognizable. But chromosomal puffs, representing active genes, occur on different bands in different tissues (Fig. 6.15).

Differentiation continues in the postembryonic stages in most animals, but is especially obvious in those with a metamorphosis from larval to adult form. It can be assumed that the sequence of gene suppressions and activations, initiated in the embryo, is continued into the postembryonic period. But now hormones take over the job of the embryonic organizer substances: consequently, at whatever level hormones act upon the cell it is to be inferred that ultimately the genome itself is affected.

Factors other than hormones can influence the developmental capacity of the cell. For example, an insect epidermal cell has the capacity to form a gland cell, a sensory cell, a bristle cell or to remain as an ordinary epithelial cell.[280] The developmental course that the cell takes is determined, among other things, by the proximity of other such structures. These perhaps drain some essential material from the surrounding cells, or produce a substance which inhibits the development of like within a certain area.

Hormones are thus only one of a number of chemical communication systems which act and interact during the development of an adult, reproductive individual from the fertilized egg. As animals become structurally more complicated, the endocrine system assumes greater importance as an integrating and coordinating mechanism. But it is not impossible that many of the hormones present in the more complicated animals have their effects duplicated in the simpler ones by chemicals produced by, and acting within, individual cells and tissues. This is to say that increasing structural complexity, reflected in the greater specialization of individual cells for specific functions, necessitates the removal of some of the erstwhile more general properties of the cell to tissues designed for the purpose. Differentiation and development in a coelenterate cell proceeds as regularly and as smoothly as in a mammalian cell.

Once the embryonic period has been passed, differentiation and development must be related to environmental events for greatest efficiency. Consequently, at least one system for the translation of environ-

mental stimuli into chemical messages must operate in all animals, and this system should exercise eventual control over the developmental machinery of the cell. It is suggested that the possession of neurosecretory mechanisms in animals as widely different in their structure and habits as coelenterates and man must reflect a common need to relate certain aspects of their development to particular environmental conditions.

References

The following references are to books and reviews which expand selected areas of invertebrate endocrinology. 1f, 5, 8, 8a, 13, 16a, 17l, 34, 41, 51, 55, 56, 58a, 67, 73, 86a, 86c, 88, 100, 115c, 115j, 118, 129, 134, 135, 136, 136a, 154a, 160a, 169, 171, 187, 191c, 196, 201, 227c, 237, 247b, 265j, 277, 280, 284, 284b.

1. ABELOOS, M. (1943). Éffets de la castration chez un mollusque, *Limax maximus*, L. *C.r. Acad. Sci. Paris*, **216**, 90–91.

1a. ABU-HAKIMA, R. and DAVEY, K. G. (1975). Two actions of juvenile hormone on the follicle cells of *Rhodnius prolixus* Stal. *Can. J. Zool.* **53**, 1187–1188.

1b. ADAMS, T. S. (1970). Ovarian regulation of the corpus allatum in the housefly, *Musca domestica. J. Insect Physiol.* **16**, 349–360.

1c. ADAMS, T. S., HINTZ, A. M. and POMONIS, J. G. (1968). Oöstatic hormone production in houseflies, *Musca domestica,* with developing ovaries. *J. Insect. Physiol.* **14**, 983–993.

1d. ADELUNG, D. (1967). Die workung von Ecdyson bei *Carcinus maenas L.* und der Crustecdysontiter während eines Häutengszyklus. *Zool. Anz.* **30**, Suppl. 264–272.

1e. ADELUNG, D. (1969). Die Ausschüttung und Funktion von Hautungshormon während eines zwischenhäutungs-Intervalls bei der Strandskrabbe *Carcinus maenas* L. *Z. Naturforsch. B.* **24**, 1447–1455.

1f. ADIYODI, K. G. and ADIYODI, R. G. (1970). Endocrine control of reproduction in decapod crustacea. *Biol. Rev.* **45**, 121–165.

1g. ADKISSON, P. L. (1964). Action of the photoperiod in controlling insect diapause. *Am. Nat.* **98**, 357–374.

1h. AJAMI, A. M. and RIDDIFORD, L. M. (1973). Comparative metabolism of the cecropia juvenile hormone. *J. Insect Physiol.* **19**, 635–645.

1i. ALEXANDROWICZ, J. S. (1964). The neurosecretory system of the vena cava in Cephalopoda. I. *Eledone cirrosa. J. mar. biol. Ass. U.K.* **44**, 111–132.

1j. ALEXANDROWICZ, J. S. (1965). The neurosecretory system of the vena cava in Cephalopoda. II. *Sepia officinalis* and *Octopus vulgaris. J. mar. biol. Ass. U.K.* **45**, 209–228.

2. ANDREWARTHA, H. G. (1952). Diapause in relation to the ecology of insects. *Biol. Rev.* **27**, 50–107.

2a. ANDREWARTHA, H. G., MIETHKE, P. M. and WELLS, A. (1974). Induction of diapause in the pupa of *Phalaenoides glycinae* by a hormone from the suboesophageal ganglion. *J. Insect Physiol.* **20**, 679–701.

2b. ANDREWS, P. M., COPELAND, D. E. and FINGERMAN, M. (1971). Ultrastructural study of the neurosecretory granules in the sinus gland of the blue crab, *Callinectes sapidus. Z. Zellforsch.* **113.** 461–471.

3. ANTHEUNISSE, L. J. (1963). Neurosecretory phenomena in the zebra mussel *Dreissena polymorpha* Palls. *Archs Néerl. Zool.* **16**, 237–314.

4. AOTO, T. and NISHIDA, H. (1956). Effects of removal of the eyestalk on the growth and maturation of the oocytes in a hermaphrodite prawn, *Pandalus kessleri. J. Fac. Sci. Hokkaido Univ.* (Ser. 6. Zool.) **12**, 412–424.

4a. ARVY, L., ECHALIER, G. and GABE, M. (1954). Modification de la gonade de *Carcinides maenas* L. après ablation bilaterale de l'organe Y. *C.R. Acad. Sci. Paris*, **239**, 1853–1855.

4b. ARVY, L., ECHALIER, G. and GABE, M. (1956). Organe Y et gonade chez *Carcinides maenas* L. *Ann. sci. nat. Zool. et biol. animale*, **18**, 263–267.

4c. ASTON, R. J. and WHITE, A. F. (1974). Isolation and purification of the diuretic hormone from *Rhodnius prolixus. J. Insect. Physiol.* **20**, 1673–1682.

5. BARRINGTON, E. W. J. (1964). Hormones and the control of colour. *The Hormones*, iv. Eds. PINCUS, G., THIMANN, K. V. and ASTWOOD, E. B. Academic Press, London and New York.

5a. BARTELL, C. K., RAO, R. K. and FINGERMAN, M. (1971). Comparison of the melanin-dispersing fractions in extracts prepared initially from ethanol, saline or distilled water from eyestalks of the fiddler crab, *Uca pugilator. Comp. Biochem. Physiol.* **38**, (A), 17–36.

6. BARTH, R. H. (1961). Hormonal control of sex attractant in the Cuban Cockroach. *Science, N.Y.* **133**, 1598–1599.

7. BARTH, R. H. (1965). Insect mating behaviour: endocrine control of a chemical communication system. *Science, N.Y.* **149**, 882–883.

8. BARTH, R. H. (1968). The comparative physiology of reproductive processes in cockroaches. *Advances in Reproductive Physiology*, vol. 3. Ed. MCLAREN, A. Logos Press, London.

8a. BARTH, R. H. and LESTER, L. J. (1973). Neurohumoral control of sexual behaviour in insects. *Ann. Rev. Ent.* **18**, 445–472.

8b. BASKIN, D. G. (1970). Studies on the infracerebral gland of the polychaet annelid, *Nereis limnicola*, in relation to reproduction, salinity and regeneration. *Gen. comp. Endocr.* **15**, 352–360.

8c. BASKIN, D. G. (1971). A possible neuroendocrine system in polynoid polychaetes. *J. Morph.* **133**, 93–104.

8d. BASKIN, D. G. and GOLDING, D. W. (1970). Experimental studies on the endocrinology and reproductive biology of the viviparous polychaete annelid *Nereis limnicola* Johnson. *Biol. Bull., Woods Hole*, **139**, 461–475.

8e. BECK, S. D., SHANE, J. L. and CALVIN, I. B. (1965). Proctodone production in the European corn borer, *Ostrinia nubilalis. J. Insect Physiol.* **11**, 297–303.

9. BECKER, H. J. (1962). Die Puffs der Speicheldrüsenchromosomen von *Drosophila melanogaster.* II. Die Auslösung der Puffbildung, ihre Spezifität und ihre Beziehung zur funktion der Ringdrüse. *Chromosoma* (Berl.), **13**, 341–384.

9a. BEENAKKERS, A. M. T. (1969). Carbohydrate and fat as a fuel for insect flight. A comparative study. *J. Insect Physiol.* **15**, 353–361.

10. BEERMANN, W. (1952). Chromomerenkonstanz und spezifische Modifikationen der Chromosomenstruktur in der Entwicklung und Organdifferenzierung von *Chironomus tentans*. *Chromosoma* (Berl.), **5**, 139–198.

11. BEERMANN, W. (1961). Ein Balbiani-ring als Locus einer Speicheldrüsenmutation. *Chromosoma* (Berl.), **12**, 1–25.

12. BELAMARICH, F. A. and TERWILLIGER, R. C. (1966). Isolation and identification of cardio-excitor hormone from the pericardial organs of *Cancer borealis*. *Am. Zool.* **6**, 101–106.

12a. BELL, W. J. (1969). Dual role of juvenile hormone in the control of yolk formation in *Periplaneta americana*. *J. Insect Physiol.* **15**, 1279–1290.

12b. BELL, W. J. (1971). Starvation induced oocyte resorption and yolk protein salvage in *Periplaneta americana*. *J. Insect Physiol.* **17**, 1099–1111.

12c. BELL, W. J. and BARTH, R. H. (1971). Evidence for juvenile-hormone initiated yolk deposition in a cockroach. *Nature N.B.* **230**, 220–221.

12d. BERLIND, A. and COOK, I. M. (1970). Release of a neurosecretory hormone as peptide by electrical stimulation of crab pericardial organs. *J. exp. Biol.* **53**, 679–686.

13. BERN, H. A. (1966). On the production of hormones by neurones and the role of neurosecretion in neuroendocrine mechanisms. *Symp. Soc. exp. Biol.* **20**, 325–344.

14. BERN, H. A., NISHIOKA, R. A. and HAGADORN, I. R. (1961). Association of clementary neurosecretory ganules with the Golgi complex. *J. Ultrastruct. Res.* **5**, 311–320.

15. BERN, H. A. and TAKASUGI, N. (1962). The caudal neurosecretory system of fishes. *Gen. comp. Endocr.* **2**, 96–110.

16. BERN, H. A., YAGI, K. and NISHIOKA, R. S. (1965). Structure and function of the caudal neurosecretory system of fishes. *Arch. Anat. microsc.* **54**, 217–238.

16a. BEROZA, M. (1970). *Chemicals controlling insect behaviour*. Academic Press. New York and London.

17. BERRIDGE, M. J. (1966). The physiology of excretion in the cotton stainer, *Dysdercus fasciatus* Signoret. IV. Hormonal control of excretion. *J. exp. Biol.* **44**, 553–566.

17a. BERRY, C. F. and COTTRELL, G. A. (1970). Neurosecretion in the vena cava of the cephalopod *Eledone cirrosa*. *Z. Zellforsch.* **104**, 107–115.

17b. DE BESSÉ, N. (1967). Neurosécrétion dans la chaîne nerveuse ventrale de deux Blattes, *Leucophaea maderae* (F) et *Periplaneta americana* (L). *Bull. Soc. zool. Fr.* **92**, 73–86.

17c. DE BESSÉ, N. and CAZAL, M. (1968). Action des extraits d'organes périsympathiques et de corpora cardiaca sur la diurèse de quelques insectes. *C.r. Acad. Sci. Paris*, **266**, 615–618.

17d. BEST, J. B., GOODMAN, A. B. and PIGON, A. (1969). Fissioning in planarians: control by the brain. *Science, N.Y.* **164**, 565.

17e. BIANCHI, S. (1969). The histochemistry of the neurosecretory system in *Cerebratulus marginatus* (Heteronemertini) *Gen. comp. Endocr.* **13**, 206–210.

17f. BIANCHI, S. (1969). On the neurosecretory system of *Cerebratulus marginatus* (Heteronemertini). *Gen. comp. Endocr.* **12**, 541–548.

17g. BIANCHI, S., ESPOSITO, V. and GRANATA, N. (1972). On the cerebral organs of *Cerebratulus marginatus* (Heteronemertini). *Gen. comp. Endocr.* **18**, 5–11.

17h. BIERL, B. A., BEROZA, M. and COLLIER, C. W. (1970). Potent sex attractant of the gypsy moth: its isolation, identification and synthesis. *Science*, *N.Y.* **170**, 87–89.

17i. BIERNE, J. (1964). Maturation sexuelle anticipée par décapitation de la femelle chez l'Hétéronémerte *Lineus ruber* Müller. *C.r. Acad. Sci. Paris*, **259**, 4841–4843.

17j. BIERNE, J. (1966). Localisation dans les ganglions cérébroides du centre régulateur de la maturation sexuelle chez la femelle de *Lineus ruber* Müller (Hétéronémertes). *C.r. Acad. Sci. Paris*, **262**, 1572–1575.

17k. BIERNE, J. (1973). Contrôle neuroendocrinien de la puberté chez le mâle de *Lineus ruber* Müller (Hétéronémerte). *C.r. Acad. Paris*, **276**, 363–366.

17l. BIRCH, M. C. (1974). *Pheromones*. North-Holland Publishing Company, Amsterdam and London.

17m. BLANCHI, D. (1969). Esperimenti sulla funzione de sisterm a neurosecretorio della vena cava nei Cefalopodi. Nota II. *Boll. Soc. Hal. Biol. Sper.* **45**, 1617–1618.

18. BLANCHI, D., NOVIELLO, L. and LIBONATI, M. (1973). A neurohormone of cephalopods with cardioexcitatory activity. *Gen. comp. Endocr.* **21**, 267–277.

19. BLISS, D. E. (1953). Endocrine control of metabolism in the land crab *Gecarcinus lateralis* (Fréminville). I. Differences in respiratory metabolism of sinus glandless and eyestalkless crabs. *Biol. Bull.*, *Woods Hole*, **104**, 275–296.

19a. BLISS, D. E. (1956). Neurosecretion and the control of growth in a decapod crustacean, p. 56–75. *Bertil Hanstrom, Zoological Papers in Honour of His 65th Birthday*. Ed. Wingstrand, K. G. Zool. Inst., Lund.

19b. BLISS, D. E. (1968). Transition from water to land in decapod crustaceans. *Am. Zool.* **8**, 355–392.

19c. BLISS, D. E. and BOYER, J. R. (1964). Environmental regulation of growth in the decapod crustacean *Gecarcinus lateralis*. *Gen. comp. Endocr.* **4**, 15–41.

20. BLISS, D. E., WANG, S. M. E. and MARTINEZ, E. A. (1966). Water balance in the land crab, *Gecarcinus lateralis*, during the intermoult cycle. *Am. Zool.* **6**, 197–212.

21. BOBIN, G. and DURCHON, M. (1952). Étude histologique du cerveau de *Perinereis cultrifera*. Mise en evidence d'une complèxe cérébrovasculaire. *Archs Anat. microsc. Morph. exp.* **41**, 25–40.

21a. BOER, H. H., SLOT, J. W. and ANDEL, J. VAN. (1968). Electron microscopical and histochemical observations on the relation between medio-dorsal bodies and neurosecretory cells in the basommatophoran snails *Lymnaea stagnalis*, *Ancylus fluviatilis*, *Australorbis glabratus* and *Planorbarius corneus*. *Z. Zellforsch.* **87**, 435–450.

21b. BOILLY-MARER, Y. (1969). Sur le determinisme des caractères sexuels secondaires chez *Nereis pelagica* L. (Annélide Polychète). *C.r. Acad. Sci. Paris*, **268** (D), 2462–2464.

21c. BOLLENBACHER, W. E., BORST, D. W. and O'CONNOR, J. D. (1972). Endocrine regulation of lipid synthesis in decapod crustaceans. *Am. Zool.* **12**, 381–384.

21d. BOWERS, W. S., FALES, H. M., THOMPSON, M. J. and UEBEL, E. C. (1966). Juvenile hormone: identification of an active compound from balsam fir. *Science, N.Y.* **154**, 1020–1021.

21e. BOYCOTT, B. B. and YOUNG, J. Z. (1956). The subpedunculate body and nerve and other organs associated with the optic tract of cephalopods. *Bertil Hanstrom, Zoological Papers in Honour of His 65th Birthday.* Ed. WINGSTRAND, K. G. Zool. Inst. Lund.

21f. BROOKES, V. J. (1969). The induction of yolk protein synthesis in the fat body of an insect, *Leucophaea maderae*, by an analog of the juvenile hormone. *Dev. Biol.* **20**, 459–471.

22. BROWN, F. A., FINGERMAN, M. and HINES, M. N. (1952). Alterations in the capacity for light and dark adaptations of the distal retinal pigment of *Palaemonetes*. *Physiol. Zoöl.* **25**, 230–239.

23. BROWN, F. A., FINGERMAN, M., SANDEEN, M. I. and WEBB, H. M. (1953). Persistent diurnal and tidal rhythms of colour change in the fiddler crab, *Uca pugnax*. *J. exp. Zool.* **123**, 29.

24. BROWN, F. A., HINES, M. N. and FINGERMAN, M. (1952). Hormonal regulation of the distal retinal pigment of *Palaemonetes*. *Biol. Bull., Woods Hole*, **102**, 212–225.

25. BROWN, F. A. and SANDEEN, M. I. (1948). Responses of the chromatophores of the fiddler crab, *Uca*, to light and temperature. *Physiol. Zoöl.* **21**, 361.

26. BROWN, F. A., WEBB, H. M. and SANDEEN, M. I. (1952). The action of two hormones regulating the red chromatophores of *Palaemonetes*. *J. exp. Zool.* **120**, 391.

27. BROWN, F. A., WEBB, H. M. and SANDEEN, M. (1953). Differential production of two retinal pigment hormones in *Palaemonetes* by light flashes. *J. cell. comp. Physiol.* **41**, 123–144.

28. BROWN, F. A. and WULFF, V. J. (1941). Chromatophore types in *Crangon* and their endocrine control. *J. cell. comp. Physiol.* **18**, 339–353.

28a. BRUSLÉ, J. (1969). Aspects ultrastructuraux de l'innervation des gonades chez l'etoile de mer *Asterina gibbosa*. *P. Z. Zellforsch.* **98**, 88–98.

29. BURNETT, A. L. and DIEHL, N. A. (1964). The nervous system of *Hydra*, I. Types, distribution and origin of nerve elements. *J. exp. Zool.* **157**, 217–226.

30. BURNETT, A. L., DIEHL, N. A. and DIEHL, F. (1964). The nervous system of *Hydra*. II. Control of growth and regeneration by neurosecretory cells. *J. exp. Zool.* **157**, 227–236.

31. BURNETT, A. L. and DIEHL, N. A. (1964). The nervous system of *Hydra*. III. The initiation of sexuality with special reference to the nervous system. *J. exp. Zool.* **157**, 237–250.

32. BUTENANDT, A. (1963). Bombycol, the sex-attractive substance of the silkworm, *Bombyx mori*. *J. Endocr.* **27**, ix–xvi.

33. BUTENANDT, A. and KARLSON, P. (1954). Über die Isolierung eines Metamorphose-Hormons der Insekten in kristellisierter Form. *Z. Naturf.* **9b**, 389–391.

34. BUTLER, C. G. (1967). Insect pheromones. *Biol. Rev.* **42**, 42–87.

35. BUTLER, C. G. and CALLOW, R. K. (1968). Pheromones of the honey bee (*Apis mellifera* L): the 'inhibitory scent' of the queen. *Proc. R. ent. Soc. Lond.* **43**, 62–65.

36. CARLISLE, D. B. (1953). Moulting hormones in *Leander* (Crustacea Decapoda). *J. mar. biol. Ass. U.K.* **32**, 289–296.

37. CARLISLE, D. B. (1953). Studies on *Lysmata seticaudata* Risso (Crustacea Decapoda). V. The ovarian inhibiting hormone and the hormonal inhibition of sex reversal. *Pubbl. Staz. zool. Napoli*, **24**, 355–372.

38. CARLISLE, D. B. (1956). On the hormonal control of water balance in *Carcinus*. *Pubbl. Staz. zool. Napoli*, **27**, 227–231.

39. CARLISLE, D. B. (1957). On the hormonal inhibition of moulting in decapod crustacea. II. The terminal anecdysis in crabs. *J. mar. biol. Ass. U.K.* **36**, 291–307.

40. CARLISLE, D. B. and DOHRN, P. F. R. (1952). Sulla presenza di un ormone d'accresciento in un crostaceo decapoda, la *Lysmata seticaudata* Risso. *Ric. sci. Torino*, **23**, 95–100.

41. CARLISLE, D. B. and KNOWLES, F. (1959). *Endocrine control in crustaceans.* Cambridge University Press.

42. CASANOVA, G. (1955). Influence du prostomium sur la régénération caudale chez *Platynereis massiliensis* (Moquin-Tandon). *C.r. Acad. Sci. Paris*, **240**, 1814–1816.

42a. CAZAL, M. and GIRARDIE, A. (1968). Contrôle humorale de l'équilibre hydrique chez *Locusta migratoria migratorioides*. *J. Insect Physiol.* **14**, 655–668.

43. CAZAL, P. (1948). Les glandes endocrines rétrocérébrales des insectes (étude morphologique). *Bull. biol. Fr. Belg.* (Suppl.), **32**, 1–227.

43a. CAZAUX, M. and ANDRÉ, F. (1972). Rôle des ganglions cérébroides dans la croissance pondérale du Lombricien *Eisenia foetida* Sav. *C.r. Acad. Sci. Paris*, **274**, 1550–1553.

43b. CAZAUX, M. and ANDRÉ, F. (1972). Rôle accélérateur et rôle inhibiteur des ganglions cérébroides sur la croissance du Lombricien *Eisenia foetida*. Sav., dans les conditions expérimentales. *C.r. Acad. Sci. Paris*, **274**, 2905–2908.

43c. CERNY, V., DOLEJS, L., LABLER, L., SORM, F. and SLAMA, K. (1967). Dehydrojuvabione—a new compound with juvenile hormone activity from balsam fir. *Coll. Czech. Chem. Comm.* **32**, 3926.

44. CHAET, A. B. (1966). The gamete-shedding substances of starfishes: a physiological biochemical study: *Am. Zool.* **6**, 263–271.

44a. CHAET, A. B. (1966). Neurochemical control of gamete release in starfish. *Biol. Bull.*, *Woods Hole*, **130**, 43–58.

44b. CHAISEMARTIN, C. (1968). Contrôle neuroendocrinien du renouvellement hydrosodique chez *Lymnaea limosa* L. *C.r. Soc. Biol.* **162**, 1994–1998.

44c. CHALAYE, D. (1967). Neurosécrétion au niveau de la chaîne nerveuse ventrale de *Locusta migratoria migratorioides* R et F. (Orthoptère Acridien). *Bull. Soc. Zool. Fr.* **92**, 87–108.

44d. CHAPRON, C. and CHAPRON, J. (1973). Induction de la régénération postérieure chez les Annélides du genre *Lumbricus*. *C.r. Acad. Sci. Paris*, **276** (D), 2641–2694.

45. CHARNIAUX-COTTON, H. (1952). Castration chirurgicale chez un Crustacé Amphipode (*Orchestia gammarella*) et déterminisme des caractères sexuel secondaires. Premier résultats. *C.r. Acad. Sci. Paris*, **234**, 2570–2572.

46 CHARNIAUX-COTTON, H. (1954). Découverte chez un crustacé amphipode (*Orchestia gammarella*) d'une gland endocrine responsible de la differentiation de caractères sexuel primaires et secondaires mâle. *C.r. Acad. Sci. Paris*, **239**, 780–782.

47. CHARNIAUX-COTTON, H. (1956). Existance d'un organ comparable à la 'glande androgène' chez un Pagure et un crabe. *C.r. Acad. Sci. Paris*, **243**, 1487–1489.

48. CHARNIAUX-COTTON, H. (1957). Croissance, régénération et déterminisme endocrinien des caractères sexuels d'*Orchestia gammarella* (Pallas) Crustacé Amphipode. *Ann. sci. nat. Zool. et biol. animale*, **19**, 411–559.

49. CHARNIAUX-COTTON, H. (1958). La glande androgène de quelques Crustacés Décapodes et particulièrement de *Lysmata seticaudata* espèce à hermaphrodisme protérandrique fonctionnel. *C.r. Acad. Sci. Paris*, **246**, 2817–2819.

50. CHARNIAUX-COTTON, H. (1958). Contrôle de la différenciation du sexe et de la reproduction chez les Crustacés supérieurs. *Bull. Soc. zool. Fr.* **83**, 314–336.

51. CHARNIAUX-COTTON, H. (1960). Sex determination. *Physiology of Crustacea* I. Ed. WATERMAN, T. H. Academic Press, London and New York.

51a. CHASSARD-BOUCHARD, C. and HUBERT, M. (1971). Premières données relatives à la présence de microtubules cytoplasmiques dans les chromatophores de Crustacés. *C.r. Acad. Sci. Paris*, **272** (D), 1405–1408.

51b. CHERBAS, L. and CHERBAS, P. (1970). Distribution and metabolism of α-ecdysone in pupae of the silkworm *Antheraea polyphemus*. *Biol. Bull., Woods Hole.* **138**, 115–128.

51c. CHOQUET, M. (1964). Culture organotypique de gonades de *Patella vulgata* L. (Mollusque Gastéropode Prosobranch). *C.r. Acad. Sci. Paris*, **258**, 1089–1091.

51d. CHOQUET, M. (1965). Recherche en culture organotypique sur la spermatogenèse chez *Patella vulgata* L. Rôle des ganglions cérébroides et des tentacules. *C.r. Acad. Sci. Paris*, **261**, 4521–4524.

51e. CHOQUET, M. (1971). Étude du cycle biologique et de l'inversion du sexe chez *Patella vulgata* L. (Mollusque Gastéropode Prosobranche). *Gen. comp. Endocr.* **16**, 59–73.

52. CLARK, R. B. (1956). On the origin of neurosecretory cells. *Ann. Sci. Nat. Zool. et biol. Animale.* **18**, 199–207.

53. CLARK, R. B. (1958). The micromorphology of the supra-oesophageal ganglion of *Nephtys*. *Zool. Jb.* (*Physiol.*), **68**, 261–299.

54. CLARK, R. B. (1959). The neurosecretory system of the supra-oesophageal ganglion of *Nephtys*. *Zool. Jb.* (*Physiol*), **68**, 395–424.

55. CLARK, R. B. (1961). The origin and formation of the heteronereis. *Biol. Rev.* **36**, 199–236.

56. CLARK, R. B. (1965). Endocrinology and reproductive biology of polychaetes. *Oceanography and Marine Biology, An Annual Review*, **3**, Ed. BARNES, H. Allen and Unwin, London.

57. CLARK, R. B. and BONNEY, D. G. (1960). Influence of the supra-oesophageal ganglion on posterior regeneration in *Nereis diversicolor*. *J. Embryol. exp. Morph.* **8**, 112–118.

58. CLARK, R. B. and EVANS, S. M. (1961). The effects of delayed brain extirpation and replacement on caudal regeneration in *Nereis diversicolor*. *J. Embryol. exp. Morph.* **9**, 97–105.

58a. CLARK, R. B. and OLIVE, P. J. W. (1973). Recent advances in polychaete endocrinology and reproductive biology. *Oceanography and Marine Biology, An Annual Review*, **11**, Ed. BARNES, H. Allen and Unwin, London.

59. CLEVER, U. (1961). Genaktivitäten in den Reisenchromosomen von *Chironomus tentans* und ihre Beziehungen zur Entwicklung. I. Genaktivierungen durch Ecdyson. *Chromosoma* (Berl.), **12**, 607–675.

60. CLEVER, U. (1962). Genaktivitäten in den Riesenchromosomen von *Chironomus tentans* und ihre Beziehungen zur Entwicklung. II. Das verhalten der Puffs. während des letzten Larvenstadiums und der Puppenhäutung. *Chromosoma* (Berl.), **13**, 385–486.

61. CLEVER, U. (1964). Actinomycin and puromycin: effects on sequential gene activation by ecdysone. *Science N. Y.*, **146**, 794–795.

61a. CLIFT, A. D. (1971). Control of germarial activity and yolk deposition in nonterminal oocytes of *Lucilia cuprina*. *J. Insect Physiol.* **17**, 601–606.

61b. COCHRAN, R. C. and ENGELMANN, F. (1972). Echinoid spawning induced by a radial nerve factor. *Science N. Y.* **178**, 423–424.

61c. COGGESHALL, R. E. (1970). A cytologic analysis of the bag cell control of egg laying in *Aplysia*. *J. Morph.* **132**, 461–486.

62. COLES, G. C. (1965). Haemolymph proteins and yolk formation in *Rhodnius prolixus*. Stal. *J. exp. Biol.* **43**, 425–431.

63. COLES, G. C. (1965). Studies on the hormonal control of metabolism in *Rhodnius prolixus* Stal. I. The adult female. *J. Insect Physiol.* **11**, 1325–1330.

64. COOKE, I. M. (1966). The site of action of pericardial organ extract and 5-hydroxytryptamine in the Decapod Crustacean heart. *Am. Zool.* **6**, 107–121.

64a. COSTLOW, J. D. (1963). The effect of eyestalk extirpation on metamorphosis of megalops of the blue crab, *Callinectes sapidus* Rathsum. *Gen. comp. Endocr.* **3**, 120–130.

64b. COSTLOW, J. D. (1966). The effect of eyestalk extirpation on larval development of the mud crab, *Rhithropanopeus harrisii* (Gould). *Gen. comp. Endocr.* **7**, 255–274.

64c. COTTRELL, C. B. (1962). The imaginal ecdysis of blowflies. The control of cuticular hardening and darkening. *J. exp. Biol.* **39**, 395–411.

64d. COTTRELL, C. B. (1962). The imaginal ecdysis of blowflies. Detection of the blood-borne darkening factor and determination of some of its properties. *J. exp. Biol.* **39**, 413–430.

64e. CRAWFORD, M. A. and WEBB, K. L. (1972). An ultrastructural study of strobilation in *Chrysaora quinquecirrha* with special reference to neurosecretion. *J. exp. Zool.* **182**, 251–270.

65. DAHM, K. H., ROLLER, H. and TROST, B. M. (1968). The juvenile hormone. IV. Stereochemistry of juvenile hormone and biological activity of some of its isomers and related compounds. *Life Sci.* **7**, 129–138.

65a. DALY, J. M. (1970). The Breeding Biology and Behaviour of *Harmothoë imbricata* (L). Ph.D. Thesis, University of Newcastle upon Tyne.

66 DANIEL, P. M. (1966). Blood supply of hypothalamus and pituitary gland. *Br. med. Bull.* **22**, 202–208.

67. DANILEVSKII, A. S. (1965). *Photoperiodism and seasonal development of insects.* Oliver and Boyd, Edinburgh and London.

68. DAVEY, K. G. (1958). The migration of spermatozoa in the female of *Rhodnius prolixus* Stal. *J. exp. Biol.* **35**, 694–701.

69. DAVEY, K. G. (1959). Spermatophore production in *Rhodnius prolixus*. *Q. Jl. microsc. Sci.* **100**, 221–230.

70. DAVEY, K. G. (1961). The mode of action of the heart accelerating factor from the corpus cardiacum of insects. *Gen. comp. Endocr.* **1**, 24–29.

71. DAVEY, K. G. (1961). The release by feeding of a pharmacologically active factor from the corpus cardiacum of *Periplaneta americana*. *J. Insect Physiol.* **8**, 205–208.

72. DAVEY, K. G. (1962). The nervous pathway involved in the release by feeding of a pharmacologically active factor from the corpus cardiacum of *Periplaneta*. *J. Insect Physiol.* **8**, 579–583.

73. DAVEY, K. G. (1965). *Reproduction in the insects.* Oliver and Boyd, Edinburgh and London.

74. DAVEY, K. G. (1966). Neurosecretion and moulting in some parasitic nematodes. *Am. Zool.* **6**, 243–250.

75. DAVEY, K. G. (1967). Some consequences of copulation in *Rhodnius prolixus*. *J. Insect Physiol.* **13**, 1629–1636.

75a. DAVEY, K. G. and HUEBNER, E. (1974). The response of the follicle cells of *Rhodnius prolixus* to juvenile hormone and antigonadotropin *in vitro. Can. J. Zool.* **52**, 1407–1412.

75b. DAVIDSON, J. (1973). Population growth in planaria *Dugesia tigrina* (Gerard). Regulation by the absolute number in the population. *J. gen. Physiol.* **61**, 767–785.

75c. DAVIS, L. E. (1972). Ultrastructural evidence for the presence of nerve cells in the gastrodermis of Hydra. *Z. Zellforsch.* **123**, 1–17.

75d. DAVIS, L. E., BURNETT, A. L. and HAYNES, J. F. (1968). Histological and ultrastructural study of the muscular and nervous system in *Hydra*. II. Nervous system. *J. exp. Zool.* **167**, 295–332.

76. DELPHIN, G. (1965). The histology and possible functions of neurosecretory cells in the ventral ganglia of *Schistocerca gregaria* Forskål (Orthoptera: Acrididae). *Trans. R. ent. Soc. Lond.* **117**, 167–214.

77. DEMEUSY, N. (1958). Recherches sur la mue de puberté du decapode brachyoure *Carcinus maenas* Linné. *Archs Zool. exp. gén.* **95**, 253–491.

77a. DHAINAUT, A. (1967). Étude de la vitellogenèse chez *Nereis diversicolor* O. F. Muller (Annélide Polychète) par autoradiographie à haute résolution. *C.r. Acad. Sci. Paris*, **265** D, 434–436.

77b. DHAINAUT, A. (1970). Étude cytochimique et ultrastructurale de l'évolution ovocytaire de *Nereis pelagica* L. Evolution expérimentale en l'absence d'hormone cérébrale. *Z. Zellforsch.* **104**, 390–404.

77c. DHAINAUT, A. and PORCHET, M. (1967). Evolution ovocytaire en l'absence d'hormone cérébrale chez *Perinereis cultrifera*. G. C.r. *Acad. Sci. Paris*, **264**, 2807–2810.

77d. DILLY, P. N. and MESSENGER, J. B. (1972). The branchial gland: a site of hemocyanin synthesis in *Octopus*. *Z. Zellforsch.* **132**, 193–201.

77e. DOWNER, R. G. H. and STEELE, J. E. (1969). Hormonal control of lipid concentration in fatbody and hemolymph of the American cockroach *Periplaneta americana. Proc. ent. Soc. Ontario*, **100**, 113–116.

77f. DOWNER, R. G. H. and STEELE, J. E. (1972). Hormonal stimulation of lipid transport in the American cockroach, *Periplaneta americana. Gen. comp. Endocr.* **19**, 259–265.

77g. DOWNER, R. G. H. and STEELE, J. E. (1973). Haemolymph lipase activity in the American cockroach, *Periplaneta americana. J. Insect Physiol.* **19**, 523–532.

77h. DUBOIS, F. S. and LENDER, T. (1956). Corrélations humorales dans la régenération des planaires paludicoles. *Ann. Sci. Nat. Zool. et biol. Animale.* **18**, 223–230.

77i. DUFOUR, D., TASKAR, S. P. and PERRON, J. M. (1970). Ontogenesis of a female specific protein from the locust, *Schistocerca gregaria. J. Insect Physiol.* **16**, 1369–1377.

78. DUPONT-RAABE, M. (1966). Étude des phénomènes de neurosécrétion au niveau de la chaîne nerveuse ventrale des phasmides. *Bull. Sci. zool. Fr.* **90**, 631–654.

79. DURCHON, M. (1956). Influence du cerveau sur les processus de régénération caudale chez Néréidiens (Annélides Polychètes). *Archs Zool. exp. gén.* **94** (N and R), 1–9.

80. DURCHON, H. (1960). L'endocrinologie chez les annélides polychètes. *Bull. Soc. zool. Fr.* **85**, 275–301.

81. DURCHON, M., MONTREUIL, J. and BOILLY-MARER, Y. (1963). Résultats préliminaires sur la nature chimique de l'hormone inhibitrice du cerveau des Néréidiens (Annélides Polychètes). *C.r. Acad. Sci. Paris*, **257**, 1807–1808.

81a. DURCHON, M. and PORCHET, M. (1971). Premières données quantitative sur l'activité endocrine du cerveau des Néréidiens au cours de leur cycle sexuel. *Gen. comp. Endocr.* **16**, 555–565.

81b. DURCHON, M. and RICHARD, A. (1967). Étude, en culture organotypique, du rôle endocrine de la glande optique dans la maturation ovarienne chez *Sepia officinalis* L. (Mollusque Cephalopode). *C.r. Acad. Sci. Paris*, **264**, 1497–1500.

81c. DUTKOWSKI, A. B. and OBERLANDER, H. (1973). The influence of larval fatbody on wing disk development *in vitro. J. Insect Physiol.* **19**, 2155–2162.

81d. DUTKOWSKI, A. B. and OBERLANDER, H. (1974). Interactions between beta-ecdysone and fatbody during wing disk development *in vitro. J. Insect Physiol.* **20**, 743–749.

82. ECHALIER, G. (1954). Recherches expérimèntales sur le rôle de 'l'organe Y' dans la mue de *Carcinus maenas* (L). Crustacé decapode). *C.r. Acad. Sci. Paris*, **238**, 523–525.

83. ECHALIER, G. (1955). Rôle de l'organe Y dans la déterminisme de la mue de *Carcinides (Carcinus) maenas* L. (Crustacé decapode). Expérience d'implantation. *C.r. Acad. Sci. Paris*, **240**, 1581–1583.

84. EDSTROM, J. E. and BEERMANN, W. (1962). The base composition of nucleic acids in chromosomes, puffs, nucleoli and cytoplasm of *Chironomus* salivary gland cells. *J. cell. Biol.* **14**, 371–379.

84a. ELOFSSON, R. and KAURI, T. (1971). The ultrastructure of the chromatophores of *Crangon* and *Pandalus* (Crustacea). *J. ultrastructure Res.* **36**, 263–270.

84b. ELSE, J. G. and JUDSON, C. L. (1972). Enforced egg retention and its effects on vitellogenesis in the mosquito, *Aedes aegypti. J. med. Ent.* 9, 527–530.

84c. EMMERICH, H. and HARTMANN, R. (1973). A carrier lipoprotein for juvenile hormone in the haemolymph of *Locusta migratoria. J. Insect Physiol.* 19, 1663–1675.

85. ENGELMANN, F. (1960). Mechanisms controlling reproduction in two viviparous roaches (Blattaria). *Ann. N.Y. Acad. Sci.* 89, 516–536.

86. ENGELMANN, F. (1965). The mode of regulation of the corpus allatum in adult insects. *Archs Anat. microsc. Morph. exp.* 54, 387–404.

86a. ENGELMANN, F. (1968). Endocrine control of reproduction in insects. *Ann. Rev. Ent.* 13, 1–26.

86b. ENGELMANN, F. (1969). Food-stimulated synthesis of intestinal proteolytic enzymes in the cockroach, *Leucophaea maderae. J. Insect Physiol.* 15, 217–235.

86c. ENGELMANN, F. (1970). *The physiology of insect reproduction.* Pergamon Press, Oxford.

86d. ENGELMANN, F. (1972). Juvenile hormone induced RNA and specific protein synthesis in an adult insect. *Gen. comp. Endocr.* Suppl. 3, 168–173.

86e. ENGELMANN, F., HILL, L. and WILKENS, J. L. (1971). Juvenile hormone control of female specific protein synthesis in *Leucophaea maderae, Schistocerca vaga,* and *Sarcophaga bullata. J. Insect Physiol.* 17, 2179–2191.

86f. ENGELMANN, F. and WILKENS, J. L. (1969). Synthesis of digestive enzymes in the fleshfly *Sarcophaga bullata* stimulated by food. *Nature, Lond.,* 222, 798.

86g. FALLON, A. M., HAGEDORN, H. H., WYATT, G. R. and LAUFER, H. (1974). Activation of vitellogenin synthesis in the mosquito *Aedes aegypti* by ecdysone. *J. Insect Physiol.* 20, 1815–1923.

86h. FAUX, A., HORN, D. H. S., MIDDLETON, E. J., FALES, H. M. and LOWE, M. E. (1969). Moulting hormone of a crab during ecdysis. *Chem. Comm.* (4), 175–176.

86i. FERGUSON, J. (1964). Nutrient transport in starfish. II. Uptake of nutrients by isolated organs. *Biol. Bull., Woods Hole,* 126, 391–406.

86j. FERNLUND, P. (1970). Chromactivating hormones of *Pandalus borealis*: isolation and purification of a light-adapting hormone. *Biochim. Biophys. Acta,* 237, 519–529.

86k. FERNLUND, P. and JOSEFSSON, L. (1972). Crustacean color-change hormone: amino acid sequence and chemical synthesis. *Science, N.Y.* 177, 173–175.

87. FINGERMAN, M. (1959). Comparison of the chromatophorotropins of two crayfish with special reference to electrophoretic behaviour. *Tulane Stud. Zool.* 7, 21.

88. FINGERMAN, M. (1963). *The control of chromatophores.* Pergamon Press, Oxford.

89. FINGERMAN, M. (1966). Neurosecretory control of pigmentary effectors in crustaceans. *Am. Zool.* 6, 169–179.

89a. FINGERMAN, M. (1973). Comparison of the effects of partially purified eyestalk extracts of the shrimp *Crangon septemspinosa* on its black, red, and white chromatophores. *Physiol. Zoöl.* 46, 173–179.

89b. FINGERMAN, M. (1973). Behaviour of chromatophores of the fiddler crab *Uca pugilator* and the dwarf crayfish *Cambarellus shufeldtii* in response to synthetic *Pandulus* red concentrating hormone. *Gen. comp. Endocr.* **20**, 589–592.

90. FINGERMAN, M. and BARTELL, C. K. (1973). Comparison of the pigmentary effector tropins in the eyestalks and abdominal nerve cord of the prawn *Palaemonetes vulgaris*. *Biol. Bull., Woods Hole,* **144**, 276–288.

91. FINGERMAN, M., DOMINIEZAK, T., MIYAWAKI, M. and OGURO, C. (1967). Neuroendocrine control of the hepatopancreas in the crayfish *Procambarus clarki*. *Physiol. Zoöl.* **40**, 23–30.

91a. FINGERMAN, M. and FINGERMAN, S. W. (1972). Evidence for a substance in the eyestalks of brachyurans that darkens the shrimp *Crangon septemspinosa*. *Comp. Biochem. Physiol.* **43** (A), 37–46.

91b. FINGERMAN, M., FINGERMAN, S. and HAMMOND, R. D. (1974). Comparison of red pigment-concentrating hormones from the eyestalks of the Fiddler Crab, *Uca pugilator*, and the Prawn, *Palaemonetes vulgaris*, with synthetic red pigment-concentrating hormone of *Pandalus borealis*. *Gen. comp. Endocr.* **23**, 124–126.

91c. FINGERMAN, M., KRASNOW, R. A. and FINGERMAN, S. W. (1971). Separation, assay and properties of the distal retinal pigment light-adapting and dark-adapting hormones in the eyestalks of the prawn *Palaemonetes vulgaris*. *Physiol. Zoöl.* **44**, 119–128.

92. FINGERMAN, M., SANDEEN, M. I. and LOWE, M. E. (1958). Experimental analysis of the red chromatophore system of the prawn *Palaemonetes vulgaris*. *Physiol. Zoöl.* **32**, 128–149.

93. FINGERMAN, M. and TINKLE, D. W. (1956). Responses of the white chromatophores of two species of prawns (*Palaemonetes*) to light and temperature. *Biol. Bull., Woods Hole,* **110**, 144.

93a. FOGAL, W. and FRAENKEL, G. (1969). The role of bursicon in melanization and endocuticle formation in the adult fleshfly, *Sarcophaga bullata*. *J. Insect Physiol.* **15**, 1235–1247.

93b. FONTAINE, A. R. (1962). Neurosecretion in the ophiuroid *Ophiopholis aculeata*. *Science, N.Y.* **138**, 908–909.

94. FRAENKEL, G. (1935). A hormone causing pupation in the blowfly, *Calliphora erythrocephala*. *Proc. R. Soc.* (B), **118**, 1–12.

94a. FRAENKEL, G. and HSIAO, C. (1962). Hormonal and nervous control of tanning in the fly. *Science, N.Y.* **138**, 27–29.

94b. FRAENKEL, G. and HSIAO, C. (1963). Tanning in the adult fly: a new function of neurosecretion in the brain. *Science, N.Y.* **141**, 1057–1058.

95. FRAENKEL, G. and HSIAO, C. (1965). Bursicon, a hormone which mediates tanning of the cuticle in the adult fly and other insects. *J. Insect Physiol.* **11**, 513–526.

95a. FRAENKEL, G. and HSIAO, C. (1967). Calcification, tanning, and the role of ecdysone in the formation of the puparium of the facefly *Musca autumnalis*. *J. Insect Physiol.* **13**, 1387–1394.

95b. FRAENKEL, G., HSIAO, C. and SELIGMAN, M. (1966). Properties of bursicon: an insect protein hormone that controls cuticular tanning. *Science, N.Y.* **151**, 91–93.

95c. FRISTROM, J. W. (1972). The biochemistry of imaginal disk development. In *The Biology of Imaginal Disks* (URSPRUNG, H. and NOTHIGER, R. Eds.), pp. 109–154. Springer-Verlag, New York.

95d. FROESCH, D. (1974). The subpedunculate lobe of the Octopus brain: evidence for dual function. *Brain Res.* **75**, 277–285.

96. FUKAYA, M. (1962). The inhibitory action of farnesol on the development of the rice stem borer in post-diapause. *Jap. J. appl. Ent. Zool.* **6**, 298.

97. FUKUDA, S. (1963). Déterminisme hormonal de la diapause chez le ver á soie. *Bull. Soc. zool. Fr.* **88**, 151–179.

98. GABE, M. (1955). Données histologiques sur la neurosécrétion chez les Arachnides. *Archs Anat. microsc.* **44**, 351–383.

99. GABE, M. (1956). Histologie comparé de la glande de mue (organe Y) des Crustacés malacostracés. *Ann. Sci. nat. (Zool.)*, **18**, 145–152.

100. GABE, M. (1966). *Neurosecretion*. Pergamon Press, Oxford.

100a. GALBRAITH, M. N., HORN, D. H. S., HOCKS, P., SCHULZ, G. and HOFF-MEISTER, H. (1967). The identity of the 20-hydroxy-ecdysones from various sources. *Naturwissenschaften*, **54**, 471–472.

101. GALBRAITH, M. N., HORN, D. H. S., MIDDLETON, E. J. and HACKNEY, R. J. (1968). Structure of deoxycrustecdysone, a second crustacean moulting hormone. *Chem. Comm.* **2**, 83–85.

101a. GALBRAITH, M. N., HORN, D. H. S., MIDDLETON, E. J., THOMSON, J. A., SIDDALL, J. B. and HAFFERL, W. (1969). Catabolism of crustecdysone in the blowfly, *Calliphora stygia*. *Chem. Comm.* 1969, 1134–1135.

102. GALLISSIAN, A. (1963). Action des ganglions cérèbroîdes sur la diapause et la régénération d'*Eophila dollfusi* Tetry. (Lumbricide). *C.r. Acad. Sci. Paris*, **256**, 1158.

102a. GALLISSIAN, A. (1973). Rôle des cellules neurosécrétrices dans le determinisme de la diapause et de la régénération postérieure chez le Lombricidae *Eophila dollfusi*. Tétry. *C.r. Acad. Sci. Paris*, **276**(D), 2721–2723.

103. GANGULY, D. N. and RAY, A. K. (1961). Caudal neurosecretory system in the vertebrates. *Sci. Cult.* **27**, 585–586.

103a. GAST, R. (1967). Untersuchungen über die Einfluss der Koniginnen substanz auf die Entwicklung der endokrinen Drüsen bei der Arbeiterin der Honigbiene (*Apis mellifica*). *Insectes Soc.* **14**, 1–12.

104. GELDIAY, S. (1967). Hormonal control of adult reproductive diapause in the Egyptian grasshopper, *Anacridium aegyptium* L. *J. Endocr.* **37**, 63–71.

104a. GERAERTS, W. P. M. (1974). Effects on growth of endocrine centres in the cerebral ganglia of *Lymnaea stagnalis*. *Gen. comp. Endocr.* **22**, 377.

104b. GERSCH, M. (1969). Neurosecretory phenomena in invertebrates. *Gen. comp. Endocr.* Suppl. **2**, 553–564.

104c. GERSCH, M. (1972). Experimentelle Untersuchungen zum Freisetzungsmechanismus von Neurohormonen nach elektrischer Reizung der Corpora cardiaca von *Periplaneta americana in vitro*. *J. Insect Physiol.* **18**, 2425–2439.

104d. GERSCH, M. and BRAUER, R. (1974). *In vitro* Stimulation der Prothoracaldrüsen von Insekten als Testsystem (Prothoracaldrusen test). *J. Insect Physiol.* **20**, 735–741.

105. GERSCH, M. and STÜRZEBECHER, J. (1968). Weitere Untersuchungen zur Kernzeichnung des Aktivations hormons der Insektenhautung. *J. Insect Physiol.* **14**, 87–96.

106. GERSCH, M. and STÜRZEBECKER, J. (1971). Über eine Synthese von Ecdyson-^{3}H und Ecdysteron-^{3}H aus Cholesterin-^{3}H in geschnürten Abdomina von *Mamestra brassicae*—Raupen. *Experientia*, **27**, 1475–1476.

107. GILBERT, L. I. and GOODFELLOW, R. D. (1965). Endocrinological significance of sterols and isoprenoids in the metamorphosis of the American Silkmoth *Hyalophora cecropia. Zool. Jb. (Physiol.)*, **8**, 718–726.

108. GILBERT, L. I. and SCHNEIDERMAN, H. A. (1959). Prothoracic gland stimulation by juvenile hormone extracts of insects. *Nature, Lond.* **184**, 171–173.

109. GILBERT, L. I. and SCHNEIDERMAN, H. A. (1960). The development of a bioassay for the juvenile hormone of insects. *Trans. Am. microsc. Soc.* **74**, 38–67.

110. GILBERT, L. I. and SCHNEIDERMAN, H. A. (1961). The content of juvenile hormone and lipid in Lepidoptera: sexual differences and developmental changes. *Gen. comp. Endocr.* **1**, 453–472.

111. GILGAN, M. W. and IDLER, D. R. (1967). The conversion of androstenedione to testosterone by some lobster (*Homarus americanus* Milne Edwards) tissues. *Gen. comp. Endocr.* **9**, 319–324.

112. GIRARDIE, A. (1964). Action de la pars intercérèbralis sur le dévelopment de *Locusta migratoria* L. *J. Insect Physiol.* **10**, 599–609.

112a. GIRARDIE, A. (1966). Contrôle de l'activité génitale chez *Locusta migratoria*. Mise en évidence d'un facteur gonadotrope et d'un facteur allatrope dans la pars intercerebralis. *Bull. Soc. zool. Fr.* **91**, 423–439.

113. GIRARDIE, A. (1967). Controle neuro-hormonal de la métamorphose et de la pigmentation chez *Locusta migratoria cinerascens* (Orthoptère). *Bull. biol. Fr. Belg.* **101**, 79–114.

113a. GOLDBARD, G. A., SAUER, J. R. and MILLS, R. R. (1970). Hormonal control of excretion in the cockroach—II. Preliminary purification of a diuretic and antidiuretic hormone. *Comp. gen. Pharmac.* **1**, 82–86.

114. GOLDING, D. W. (1967). Regeneration and growth control in *Nereis*. I. Growth and regeneration. *J. Embryol. exp. Morph.* **18**, 67–77.

115. GOLDING, D. W. (1967). Regeneration and growth control in *Nereis*. II. An axial gradient in growth potentiality. *J. Embryol. exp. Morph.* **18**, 79–90.

115a. GOLDING, D. W. (1970). The infracerebral gland in *Nephtys*—a possible neuroendocrine complex. *Gen. comp. Endocr.* **14**, 114–126.

115b. GOLDING, D. W. (1972). Studies in the comparative neuroendocrinology of polychaete reproduction. *Gen. comp. Endocr.* Suppl. **3**, 580–590.

115c. GOLDING, D. W. (1974). A survey of neuroendocrine phenomena in nonarthropod invertebrates. *Biol. Rev.* **49**, 161–224.

155d. GOLDING, D. W., BASKIN, D. W. and BERN, H. A. (1968). The infracerebral gland of *Nereis. J. Morphol.* **124**, 187–216.

115e. GOLDSWORTHY, G. J. (1969). Hyperglycaemic factors from the corpus cardiacum of *Locusta migratoria. J. Insect Physiol.* **15**, 2131–2140.

115f. GOLDSWORTHY, G. J. (1970). The action of hyperglycaemic factors from the corpus cardiacum of *Locusta migratoria* on glycogen phosphorylase. *Gen. comp. Endocr.* **14**, 78–85.

115g. GOLDSWORTHY, G. J., COUPLAND, A. J. and MORDUE, W. (1973). The effects of corpora cardiaca on tethered flight in the locust. *J. comp. Physiol.* **82**, 339–346.

115h. GOLDSWORTHY, G. J., JOHNSON, R. A. and MORDUE, W. (1972). *In vivo* studies on the release of hormones from the corpora cardiaca of locusts. *J. comp. Physiol.* **79**, 85–96.

115i. GOLDSWORTHY, G. J., MORDUE, W. and GUTHKELCH, J. (1972). Studies on insect adipokinetic hormones. *Gen. comp. Endocr.* **18**, 545–551.

115j. GOLDSWORTHY, G. J. and MORDUE, W. (1974). Neurosecretory hormones in insects. *J. Endocr.* **60**, 529–558.

116. GOMEZ, R. (1965). Acceleration of development of the gonads by implantation of brain in the crab *Paratelphusa hydrodromus*. *Naturwissenschaften*, **52**, 216.

116a. GOMOT, L. and GUYARD, A. (1964). Evolution en culture *in vitro* de la glande hermaphrodite de jeunes escargots de l'espèces *Helix aspersa* Mull. *C.r. Acad. Sci. Paris*, **258**, 2902–2905.

117. GOODFELLOW, R. D. and GILBERT, L. I. (1963). Sterols and terpenes in the Cecropia silkmoth. *Am. Zool.* **3**, 137.

118. GORBMAN, A. and BERN, H. A. (1962). *A textbook of comparative endocrinology*. John Wiley, New York.

118a. GORELL, T. A. and GILBERT, L. I. (1969). Stimulation of protein and RNA synthesis in the crayfish hepatopancreas by crustecdysone. *Gen. comp. Endocr.* **13**, 308–310.

118b. GORELL, T. A. and GILBERT, L. I. (1971). Protein and nucleic acid synthesis during the crayfish molt cycle. *Z. vergl. Physiol.* **73**, 345–356.

118c. GORELL, T. A., GILBERT, L. I. and SIDDALL, J. S. (1972). Binding proteins for an ecdysone metabolite in the crustacean hepatopancreas. *Proc. natn. Acad. Sci.* **69**, 812–815.

118d. GORELL, T. A., GILBERT, L. I. and SIDDALL, J. B. (1972). Studies on hormone target tissues. *Am. Zool.* **12**, 347–356.

118e. GORELL, T. A., GILBERT, L. I. and TASH, J. (1972). The uptake and conversion of α-ecdysone by the pupal tissues of *Hyalophora cecropia*. *Insect Biochem.* **2**, 94–106.

118f. GOTTFRIED, H. and DORFMAN, R. I. (1970). Steroids of invertebrates. IV. On the optic tentacle-gonadal axis in the control of the male phase ovotestis in the slug (*Ariolimax californicus*). *Gen. comp. Endocr.* **15**, 101–119.

118g. GOTTFRIED, A., DORFMAN, R. I. and WALL, P. E. (1967). Steroids of invertebrates: production of estrogens by an accessory reproductive tissue of the slug *Arion ater rufus* (Linn). *Nature, Lond.* **215**, 409–410.

118h. GRASSO, M. and QUAGLIA, A. (1970). Studies on neurosecretion in planarians. I. Neurosecretory fibres near the testes of *Dugesia lugubris*. *J. submicr. Cytol.* **2**, 119–125.

118i. GRASSO, M. and QUAGLIA, A. (1970). Studies on neurosecretion in planarians. II. Observations on the ovaries of *Dugesia lugubris*. *J. submicr. Cytol.* **2**, 127–132.

118j. GRASSO, M. and QUAGLIA, A. (1971). Studies on neurosecretion in planarians. III. Neurosecretory fibres near the testes and ovaries of *Polycelis nigra*. *J. submicr. Cytol.* **3**, 171–180.

118k. GUYARD, A. (1967). Féminisation de la glande hermaphrodite juvenile d'*Helix aspersa* Müll. associée *in vitro* au ganglion cérébroïde d'Escargot adulte ou de Paludione femelle. *C.r. Acad. Sci. Paris*, **265**, 147–149.

118l. GUYARD, A. (1969). Autodifferenciation femelle de l'ebauche gonodique de l'escargot *Helix aspersa* Müll. cultivée sur milieu anhormonal. *C.r. Acad. Sci. Pars*, **268**, 966–969.

118m. GWADZ, R. W. and SPIELMAN, A. (1973). Corpus allatum control of ovarian development in *Aedes aegypti. J. Insect Physiol.* **19**, 1441–1448.

119. HAGADORN, I. R. (1958). Neurosecretion and the brain of the rhynchobdellid leech *Theromyzon rude.* (Baird, 1869). *J. Morph.* **102**, 55–90.

120. HAGADORN, I. R. (1966). Neurosecretion in the Hirudinea and its possible role in reproduction. *Am. Zool.* **6**, 251–262.

120a. HAGEDORN, H. H. (1974). The control of vitellogenesis in the Mosquito *Aedes aegypti. Am. Zool.* **14**, 1207–1217.

120b. HAGEDORN, H. H. and FALLON, A. M. (1973). Ovarian control of vitellogenin synthesis by the fat body in *Aedes aegypti. Nature, Lond.* **244**, 103–105.

120c. HAGEDORN, H. H., FALLON, A. M. and LAUFER, H. (1973). Vitellogenin synthesis by the fatbody of the mosquito *Aedes aegypti*: evidence for transcriptional control. *Dev. Biol.* **31**, 285–294.

120d. HAGEDORN, H. H. and JUDSON, C. L. (1972). Purification and synthesis of *Aedes aegypti* yolk proteins, *J. exp. Zool.* **182**, 367–377.

121. HAMPSHIRE, F. and HORN, D. H. S. (1966). Structure of crustecdysone, a crustacean moulting hormone. *Chem. Comm.* **2**, 37–38.

122. HANSTRÖM, B. (1954). On the transformation of ordinary nerve cells into neurosecretory cells. *K. fysiogr. Sällsk. Lund. Förh.* **24**, 1–8.

123. HARRIS, G. W., REED, M. and FAWCETT, C. P. (1966). Hypothalamic releasing factors and the control of anterior pituitary function. *Br. med. Bull.* **22**, 266–272.

123a. HARTMAN, H. B. and SUDA, M. (1973). Pheromone production and mating behaviour by allatectomized males of the cockroach *Nauphoeta cinerea. J. Insect Physiol.* **19**, 1417–1422.

123b. HASEGAWA, K. (1964). Studies on the mode of action of the diapause hormone in the silkworm, *Bombyx mori* L.—II. Content of diapause hormone in the suboesophageal ganglion. *J. exp. Biol.* **41**, 855–863.

123c. HASEGAWA, K., ISOBE, M. and GOTO, T. (1972). Highly purified diapause hormone from the silkworm. *Naturwissenschaften* **59**, 364–365.

124. HAUENSCHILD, C. (1960). Lunar periodicity. *Cold Spring Harb. Symp. quant. Biol.* **25**, 491–497.

125. HAUENSCHILD, C. (1964). Postembryonale Entwicklungssteuerung durch ein Gehirn-Hormon bei *Platynereis dumerilii. Zool. Anz.* Suppl. **27**, 111–120.

126. HAUENSCHILD, C. and FISCHER, A. (1962). Neurosecretory control of development in *Platynereis dumerilii. Mem. Soc. Endocr.* **12**, 297–312.

126a. HAUENSCHILD, C., FISCHER, A. and HOFMANN, D. K. (1968). Untersuchungen am pacifischen Palolowurm *Eunice viridis* (Polychaeta) in Samoa. *Helgol. wiss. Meeresunters*, **18**, 254–295.

126b. HECHTER, O. and HALKERSTON, I. D. K. (1965). Effects of steroid hormones on gene regulation and cell metabolism. *Ann. Rev. Physiol.* **27**, 133–162.

126c. HEINRICH, G. and HOFFMEISTER, H. (1967). Ecdyson als Begleitsubstanz des Ecdysterons in *Polypodium vulgare* L. *Experientia*, **23**, 995.

126d. HEINRICH, G. and HOFFMEISTER, H. (1968). 5β-hydroxyecdysterone, ein Pflanzansteroid mit Häutungshormon—aktivitat aus *Polypodium vulgare* L. *Tetrahedron Letters*, **58**, 6063–6064.

126e. HEKSTRA, G. P. and LEVER, J. (1960). Some effects of ganglion extirpation in *Lymnaea stagnalis*. *Proc. kon. ned. Akad. Wet. C.* **63**, 271–282.

126f. HERLANT-MEEWIS, H. (1956). Croissance et neurosécrétion chez *Eisenia foetida*. *Annls. Soc. r. Zool. Belg.* **87**, 151–183.

127. HERLANT-MEEWIS, H. (1957). Réproduction et neurosécrétion chez *Eisenia foetida* (Sav). *Annls Soc. r. zool. Belg.* **87**, 151–183.

128. HERLANT-MEEWIS, H. (1959). Phénomènes neurosécrétoires et sexualité chez *Eisenia foetida*. *C.r. Acad. Sci. Paris*, **248**, 1405–1407.

129. HERLANT-MEEWIS, H. (1964). Regeneration in annelids. *Advances in Morphogenesis*, vol. 4. Eds. ABERCROMBIE, M. and BRACHET, J. Academic Press, London and New York.

130. HEYMONS, R. (1895). Die Embryonalentwicklung von Dermapteren und Orthopteren, unter besonderer Berücksichtigung der Keimblätterbildung. Jena.

131. HIGHNAM, K. C. (1961). The histology of the neurosecretory system of the adult female desert locust, *Schistocerca gregaria*. *Q. Jl. microsc. Sci.* **102**, 27–38.

132. HIGHNAM, K. C. (1961). Induced changes in the amounts of material in the neurosecretory system of the desert locust. *Nature, Lond.* **191**, 199–200.

133. HIGHNAM, K. C. (1962). Neurosecretory control of ovarian development in *Schistocerca gregaria*. *Q. Jl. microsc. Sci.* **103**, 57–72.

134. HIGHNAM, K. C. (1964). Endocrine relationships in insect reproduction. In *Insect Reproduction*, pp. 26–42. Symp. No. 2. Royal Ent. Soc. Lond.

135. HIGHNAM, K. C. (1964). Hormones and behaviour in insects. *Viewpoints in Biology*, **3**, 219–260.

136. HIGHNAM, K. C. (1965). Some aspects of neurosecretion in arthropods. *Zool. Jb. (Physiol.)*, **71**, 558–582.

136a. HIGHNAM, K. C. (1970). Mode of action of Arthropod steroid and other hormones. *Advances in steroid biochemistry and pharmacology*. I. Ed. BRIGGS, M. H. Academic Press, London and New York.

137. HIGHNAM, K. C. (1971). Estimates of neurosecretory activity during maturation in female locusts. *Insect Endocrines*. Eds. NOVAK, V. J. A. and SLAMA, K. Czechoslovak Academy of Sciences.

138. HIGHNAM, K. C. and HASKELL, P. T. (1964). The endocrine systems of isolated and crowded *Locusta* and *Schistocerca* in relation to oocyte growth, and the effects of flying upon maturation. *J. Insect Physiol.* **10**, 849–864.

139. HIGHNAM, K. C., HILL, L. and GINGELL, D. (1965). Neurosecretion and water balance in the male desert locust (*Schistocerca gregaria*). *J. Zool.* **147**, 201–215.

140. HIGHNAM, K. C., HILL, L. and MORDUE, W. (1966). The endocrine system and oocyte growth in *Schistocerca* in relation to starvation and frontal ganglionectomy. *J. Insect Physiol.* **12**, 977–994.

141. HIGHNAM, K. C. and LUSIS, O. (1962). The effect of mature males on the neurosecretory control of ovarian development in the desert locust. *Q. Jl. microsc. Sci.* **103**, 73–83.

142. HIGHNAM, K. C., LUSIS, O. and HILL, L. (1963). The role of the corpora allata during oocyte growth in the desert locust, *Schistocerca gregaria* Forsk. *J. Insect Physiol.* **9**, 587–596.

143. HIGHNAM, K. C., LUSIS, O. and HILL, L. (1963). Factors affecting oocyte resorption in the desert locust, *Schistocerca gregaria*. *J. Insect Physiol.* **9**, 827–837.

143a. HIGHNAM, K. C. and MORDUE (LUNTZ), A. J. (1970). Estimates of neurosecretory activity by an autoradiographic method in adult female *Schistocerca gregaria* (Forsk). *Gen. comp. Endocr.* **15**, 31–38.

143b. HIGHNAM, K. C. and MORDUE (LUNTZ), A. J. (1974). Induced changes in neurosecretory activity of adult female *Schistocerca gregaria* in relation to feeding. *Gen. comp. Endocr.* **22**, 519–525.

144. HILL, L. (1962). Neurosecretory control of haemolymph protein concentration during ovarian development in the desert locust. *J. Insect Physiol.* **8**, 609–619.

145. HILL, L. (1965). The incorporation of C^{14}-glycine into the proteins of the fatbody of the desert locust during ovarian development. *J. Insect Physiol.* **11**, 1605–1615.

146. HILL, L. and GOLDSWORTHY, G. J. (1968). Growth, feeding activity and the utilization of reserves in larvae of *Locusta*. *J. Insect. Physiol.* **14**, 1085–1098.

146a. HILL, L. and IZATT, M. E. G. (1973). Hormonal control of lipid and protein metabolism in the adult desert locust. *J. Endocr.* **57**, l–li.

146b. HILL, L. and IZATT, M. E. G. (1974). The relationships between corpora allata and fat body and haemolymph lipids in the adult female desert locust. *J. Insect Physiol.* **20**, 2143–2156.

147. HILL, L., LUNTZ, A. J. and STEELE, P. A. (1968). The relationships between somatic growth, ovarian growth, and feeding activity in the adult desert locust. *J. Insect Physiol.* **14**, 1–20.

148. HILL, L., MORDUE, W. and HIGHNAM, K. C. (1966). The endocrine system, frontal ganglion, and feeding during maturation in the female desert locust. *J. Insect Physiol.* **12**, 1197–1208.

148a. HIRAI, S. and KANATANI, H. (1971). Site of production of meiosis-inducing substance in ovary of starfish. *Exptl. Cell Res.* **67**, 224–227.

148b. HIRAI, S., KUBOTA, J. and KANATANI, H. (1971). Induction of cytoplasmic maturation by 1-methyladenine in starfish oocytes after removal of the germinal vesicle. *Exptl. Cell Res.* **68**, 137–143.

148c. HITCHO, P. J. and THORSON, R. E. (1971). Possible molting and maturation controls in *Trichinella spiralis*. *J. Parasit.* **57**, 787–793.

148d. HOCKS, P., SCHULTZ, G. and KARLSON, P. (1967). Die struktur des β-Ecdysones. *Naturwissenschaften*, **54**, 44–45.

148e. HOFFMANN, J. A., KOOLMAN, J., KARLSON, P. and JOLY, P. (1974). Molting hormone titer and metabolic fate of injected ecdysone during the fifth larval instar and in adults of *Locusta migratoria* (Orthoptera). *Gen. comp. Endocr.* **22**, 90–97.

149. HOFFMEISTER, H. (1966). Ecdysteron, eine neues Häutungshormon der Insekten. *Angew. Chem.* **78**, 269–270.

150. HOFFMEISTER, H., NAKANISHI, K., KOREEDA, M. and HSU, H. Y. (1968). The moulting hormone activity of ponasterones in the *Calliphora* test. *J. Insect Physiol.* **14**, 53–54.

150a. HOWIE, D. I. D. (1959). The spawning of *Arenicola marina* (L). I. The breeding season. *J. mar. Biol. Ass. U.K.* **38**, 395–406.

151. HOWIE, D. I. D. (1963). Experimental evidence for the humoral stimulation of ripening of the gametes and spawning in the polychaete *Arenicola marina* (L). *Gen. comp. Endocr.* **3**, 660–668.

151a. HOWIE, D. I. D. (1966). Further data relating to the maturation hormone and its site of secretion in *Arenicola marina*. *Gen. comp. Endocr.* **6**, 347–360.

151b. HOWIE, D. I. D. and MCCLENAGHAN, C. M. (1965). Evidence for a feed-back mechanism influencing spermatogonial divisions in the lug-worm (*Arenicola marina* L). *Gen. comp. Endocr.* **5**, 40–44.

151c. HUBL., H. (1956). Über die Beziehungen der Neurosekretion zum Regenerationsgeschehen bei Lumbriciden nebst Beschreibung eines neuertigen neurosekretorischen Zelltyps im Unterschlundganglion. *Arch. Entw. Mech.* **149**, 73–87.

151d. HUEBNER, E. and DAVEY, K. G. (1973). An antigonadotropin from the ovaries of the insect *Rhodnius prolixus* Stål. *Can. J. Zool.* **51**, 113–120.

151e. HUTCHINGS, P. A. (1973). Gametogenesis in a Northumberland popula-tion of the Polychaete *Melinna cristata*. *Mar. Biol.* **18**, 199–211.

152. ICHIKAWA, M. (1962). Brain and metamorphosis of Lepidoptera. *Gen. comp. Endocr.* Suppl. **1**, 331–336.

153. ICHIKAWA, M. and ISHIZAKI, H. (1961). Brain hormone of the silkworm, *Bombyx mori*. *Nature, Lond.* **191**, 933–934.

153a. IKEGAMI, S., TAMURA, S. and KANATANI, H. (1967). Starfish gonad: action and chemical identification of spawning inhibitor. *Science, N.Y.* **158**, 1052–1053.

153b. IMAI, S., FUJIOKA, S., NAKANISHI, K., KOREEDA, M. and KUROKAWA, T. (1967). Extraction of ponasterone A and ecdysterone from Podo-carpaceae and related plants. *Steroids*, **10**, 557–565.

153c. IMAI, S., IIORI, M., FUJIOKA, S., MURATA, E., GOTO, M. and NAKANISHI, K. (1968). Isolation of four new phytoecdysones, makisterone A, B, C, D and the structure of makisterone A, a C_{28} steroid. *Tetrahedron Letters*, (36), 3883–3886.

153d. ISHIZAKI, H. and ICHIKAWA, M. (1967). Purification of the brain hormone of the silkworm *Bombyx mori*. *Biol. Bull., Woods Hole*, **133**, 355–368.

153e. ISOBE, M., HASEGAWA, K. and GOTO, T. (1973). Isolation of the diapause hormone from the silkworm, *Bombyx mori*. *J. Insect Physiol.* **19**, 1221–1239.

154. ITO, H. (1918). On the glandular nature of the Corpora allata of the Lepidoptera. *Bull. imp. seric. Coll., Tokyo*, **1**, 64–103.

154a. JACOBSON, M. (1972). *Insect Sex Pheromones*. Academic Press, New York and London.

155. JACOBSON, M., BEROZA, M. and JONES, W. A. (1960). Isolation, identifica-tion and synthesis of the sex attractant of the gypsy moth. *Science N.Y.* **139**, 48–49.

155a. JOHANSSON, A. S. (1958). Relation of nutrition to endocrine-reproduc-tive functions in the milkweed bug *Oncopeltus fasciatus* (Dallas) (Heteroptera: Lygaeidae). *Nytt. Mag. Zool.* **7**, 1–132.

156. JOHNSON, B. (1963). A histological study of neurosecretion in aphids. *J. Insect Physiol.* **9**, 727–739.

156a. JOHNSON, R. A. and HILL, L. (1973). The activity of the corpora allata in the fourth and fifth larval instars of the migratory locust. *J. Insect Physiol.* **19**, 1921–1932.

156b. JOHNSON, R. A. and HILL, L. (1973). Quantitative studies on the activity of the corpora allata in adult male *Locusta* and *Schistocerca*. *J. Insect Physiol.* **19**, 2459–2467.

157. JOLY, P. (1945). La fonction ovarienne et son contrôle humorale chez les dytiscides. *Archs. Zool. exp. gén.* **84**, 47–164.

158. JOOSSE, J. (1964). Dorsal bodies and dorsal neurosecretory cells of the cerebral ganglia of *Lymnaea stagnalis*. *Archs. Néerl. Zool.* **16**, 1–103.

158a. JOOSSE, J. (1972). Endocrinology of reproduction in molluscs. *Gen. comp. Endocr.* Suppl. **3**, 591–601.

158b. JOOSSE, J. and GERAERTS, W. J. (1969). On the influence of the dorsal bodies and the adjacent neurosecretory cells on the reproduction and metabolism of *Lymnaea stagnalis*. *Gen. comp. Endocr.* **13**, 540.

158c. JOOSSE, J., GERAERTS, W. P. M., BOHLKEN, S. and FLOOR, A. (1974). Some effects of endocrine centres in the cerebral ganglia of *Lymnaea stagnalis* on protein synthesis and the biochemical composition of various organs. *Gen. comp. Endocr.* **22**, 386.

158d. JOOSSE, J. and REITZ, D. (1969). Functional anatomical aspects of the ovotestis of *Lymnaea stagnalis*. *Malacologia*, **9**, 101–109.

158e. JUDY, K. J., SCHOOLEY, D. A., HALL, M. S., BERGOT, B. J. and SIDDALL, J. B. (1973). Chemical structure and absolute configuration of a juvenile hormone from grasshopper corpora allata *in vitro*. *Life Sci.* **13**, 1511–1516.

158f. KAI, H. and HASEGAWA, K. (1971). Studies on the mode of action of the diapause hormones with special reference to the protein metabolism in the silkworm, *Bombyx mori* L. I. The diapause hormone and the protein soluble in ethanol containing trichloroacetic acid in mature eggs of adult ovaries. *J. seric. Sci. Tokyo*, **40**, 199–208.

158g. KAI, H. and HASEGAWA, K. (1972). Electrophoretic protein patterns and esterase zymograms in ovaries and mature eggs of *Bombyx mori* in relation to diapause. *J. Insect Physiol.* **18**, 133–142.

158h. KAI, H. and HASEGAWA, K. (1972). Studies on the mode of action of the diapause hormone with special reference to the protein metabolism in the silkworm, *Bombyx mori* L. III. Effects of acid-treatment and detergents on 'esterase A' in diapause eggs. *J. seric. Sci. Tokyo*, **41**, 253–262.

158i. KAI, H. and HASEGAWA, K. (1973). An esterase in relation to yolk cell lysis at diapause termination in the silkworm, *Bombyx mori*. *J. Insect Physiol.* **19**, 799–810.

158j. KAMBYSELLIS, M. P. and WILLIAMS, C. M. (1971). *In vitro* studies of insects. 1. A macromolecular factor prerequisite for silkworm spermatogenesis. *Biol. Bull., Woods Hole*, **141**, 527–540.

158k. KAMBYSELLIS, M. P. and WILLIAMS, C. M. (1971). *In vitro* development of insect tissues. II. The role of ecdysone in the spermatogenesis of silkworms. *Biol. Bull., Woods Hole*, **141**, 541–552.

158l. KAMBYSELLIS, M. P. and WILLIAMS, C. M. (1972). Spermatogenesis in cultured testes of the cynthia silkworm: effects of ecdysone and of prothoracic glands. *Science N.Y.* **175**, 769–770.

158m. KAMEMOTO, F. I. (1964). The influence of the brain on osmotic and ionic regulation in earthworms. *Gen. comp. Endocr.* **4**, 420–426.

158n. KAMEMOTO, F. I. and TULLIS, R. E. (1972). Hydromineral regulation in Decapod Crustacea. *Gen. comp. Endocr.* Suppl. **3**, 299–307.

158o. KANATANI, H. (1969). Mechanism of starfish spawning: action of neural substance on the isolated ovary. *Gen. comp. Endocr.* Suppl. **2**, 582–589.

158p. KANATANI, H. and HIRAMOTO, Y. (1970). Site of action of 1-methyl-adenine in inducing oocyte maturation in starfish. *Exptl. Cell Res.* **61**, 280–284.

158q. KANATANI, H. and OHGURI, M. (1966). Mechanism of starfish spawning. I. Distribution of active substance responsible for maturation of oocytes and shedding of gametes. *Biol. Bull., Woods Hole,* **131**, 104–114.

158r. KANATANI, H. and SHIRAI, H. (1967). *In vitro* production of meiosis inducing substances by nerve extracts in ovary of starfish. *Nature, Lond.* **216**, 284–286.

158s. KANATANI, H. and SHIRAI, H. (1969). Mechanism of starfish spawning. II. Some aspects of action of a neural substance obtained from radial nerve. *Biol. Bull., Woods Hole,* **137**, 297–311.

158t. KANATANI, H. and SHIRAI, H. (1970). Mechanism of starfish spawning. III. Properties and action of meiosis-inducing substance produced in gonad under influence of gonad stimulating substances. *Dev. Growth & Differentiation,* **12**, 119–140.

158u. KANATANI, H. and SHIRAI, H. (1971). Chemical structural requirements for induction of oocyte maturation and spawning in starfishes. *Dev. Growth & Differentiation,* **13**, 53–64.

158v. KANATANI, H. and SHIRAI, H. (1972). On the maturation-inducing substance produced in the starfish gonad by neural substance. *Gen. comp. Endocr.* Suppl. **3**, 571–579.

158w. KANATANI, H., SHIRAI, H., NAKANISHI, K. and KUROKAWA, T. (1969). Isolation and identification of meiosis inducing substance in starfish, *Asterias amurensis, Nature, Lond.* **221**, 273–274.

158x. KAPLANIS, J. N., THOMPSON, M. J., ROBBINS, W. E. and BRYCE, B. M. (1967). Insect hormones: alpha ecdysone and 20-hydroxyecdysone in bracken fern. *Science, N.Y.* **157**, 1436–1438.

159. KAPLANIS, J. N., THOMPSON, M. J., YAMAMOTO, R. T., ROBBINS, W. E. and LOULOUDES, S. J. (1966). Ecdysones from the pupa of the tobacco hornworm, *Manduca sexta* (Johannson). *Steroids,* **8**, 605–623.

160. KARLSON, P. (1963). Chemie und Biochemie der Insektenhormone. *Angew. Chem.* **75**, 257–265.

160a. KARLSON, P. (1967). The effects of ecdysone on giant chromosomes, RNA metabolism and enzyme induction. *Mem. Soc. Endocr.* **15**, 67–76.

161. KARLSON, P. and HOFFMEISTER, H. (1963). Zur Biogenese des Ecdysons. I. Umwandlung von Cholesterin in Ecdyson. *Hoppe-Seylers. Z. physiol. Chem.* **331**, 298–300.

162. KARLSON, P., HOFFMEISTER, H., HOPPE, W. and HUBER, R. (1963). Zur Chemie des Ecdysons. *Justus Liebigs Annl. or. Chem.* **662**, 1–20.

163. KARLSON, P. and LUSCHER, M. (1959). 'Pheromones': a new term for a class of biologically active substances. *Nature, Lond.* **183**, 55–56.

164. KARLSON, P. and SCHMIALEK, H. (1959). Nachreis der Exkretion von Juvenilhormon. *Z. Naturf.* **146**, 821.

165. KARLSON, P. and SCHWEIGER, A. (1961). Zum Tyrosinstoffwechsel der Insekten. IV. Mitteilung Das Phenoloxydase-System von *Calliphora* und seine Beeinflussung durch das Hormon Ecdyson. *Hoppe-Seylers Z. physiol. Chem.* **323**, 199–210.

166. KARLSON, P. and SEKERIS, C. (1962). Zum Tyrosinstoffwechsel der Insekten IX. Kontrolle des Tyrosinstoffwechsels durch Ecdyson. *Biochim. biophys. Acta.* **63**, 489–495.

167. KARLSON, P. and SEKERIS, C. (1966). Ecdysone, an insect steroid hormone, and its mode of action. *Rec. Progr. Horm. Res.* **22**, 473–502.

167a. KELLER, R. and ADELUNG, D. (1970). Vergleichende morphologische und physiologische Untersuchungen der Integumentgewebes und Häutungshormongehaltes beim Fluszkrebs *Orconectes limosus* während eines Häutungszyklus. *Wilhelm Roux Arch. Entwicklungsmech. Organismen.* **164**, 209–221.

167b. KELLER, R. and ANDREW, E. M. (1973). The site of action of the crustacean hyperglycaemic hormone. *Gen. comp. Endocr.* **20**, 572–578.

168. KENNEDY, J. S. (1965). Coordination of successive activities in an aphid. Reciprocal effects of settling on flight. *J. exp. Biol.* **43**, 489–509.

169. KENNEDY, J. S. (1967). Behaviour as physiology. *Insects and physiology.* Eds. BEAMENT, J. W. L. and TREHERNE, J. E. Oliver and Boyd, Edinburgh and London. 1967.

169a. KENNEDY, J. S. (1972). The emergence of behaviour. *J. Aust. ent. Soc.* **11**, 168–176.

169b. KING, D. S. (1972). Metabolism of α-ecdysone and possible immediate precursors by insects *in vivo* and *in vitro*. *Gen. comp. Endocr.* Suppl. **3**, 221–227.

169c. KING, D. S. and SIDDALL, J. B. (1969). Conversion of α-ecdysone to β-ecdysone by crustaceans and insects. *Nature, Lond.* **221**, 955–956.

170. KLEINHOLZ, L. H. (1936). Crustacean eyestalk hormones and retinal pigment migration. *Biol. Bull., Woods Hole,* **70**, 159–184.

171. KLEINHOLZ, L. H. (1961). Pigmentary effectors. *Physiology of Crustacea.* II. Ed. WATERMAN, T. H. Academic Press, London and New York.

171a. KLEINHOLZ, L. H. (1972). Comparative studies of crustacean melanophore-stimulating hormones. *Gen. comp. Endocr.* **19**, 473–483.

172. KLEINHOLZ, L. H. and KELLER, R. (1973). Comparative studies in crustacean neurosecretory hyperglycaemic hormones. *Gen. comp. Endocr.* **21**, 554–564.

173. KOBAYASHI, M. and KIRIMURA, J. (1958). The 'brain' hormone in the silkworm, *Bombyx mori. Nature, Lond.* **181**, 1217.

174. KOBAYASHI, M., KIRIMURA, J. and SAITO, M. (1962). The 'brain' hormone in an insect *Bombyx mori* L. (Lepidoptera). *Mushi*, **36**, 85–92.

175. KOBAYASHI, M., TAKEMOTO, T., OGAWA, S. and NISHIMOTO, N. (1967). The moulting hormone activity of ecdysterone and inokosterone isolated from *Achyranthis radix. J. Insect Physiol.* **13**, 1395–1399.

175a. KOBAYASHI, M. and YAMAZAKI, M. (1966). The proteinic brain hormone in an insect, *Bombyx mori* L. (Lepidoptera: Bombycidae). *Appl. Entomol. & Zool.* **1**, 53–60.

176. KOLLER, G. (1928). Versuch über die inkretorischen Vorgange bien Garneelfarbweichsel. *Z. vergl. Physiol.* **8**, 601.

177. KOPEC, S. (1922). Studies on the necessity of the brain for the inception of insect metamorphosis. *Biol. Bull., Woods Hole,* **42**, 323–342.

178. KRISHNAKUMARAN, A., OBERLANDER, H. and SCHNEIDERMAN, H. A. (1965). Rates of DNA and RNA synthesis in various tissues during a larval moult cycle of *Samia cynthia ricini* (Lepidoptera). *Nature, Lond.* **205**, 1131–1133.

178a. KRISHNAKUMARAN, A. and SCHNEIDERMAN, H. A. (1968). Chemical control of moulting in arthropods. *Nature, Lond.* **220**, 601–603.

178b. KRISHNAKUMARAN, A. and SCHNEIDERMAN, H. A. (1969). Induction of molting in Crustacea by an insect moulting hormone. *Gen. comp. Endocr.* **12**, 515–518.

178c. KUPFERMANN, I. (1967). Stimulation of egg laying: possible neuro-endocrine function of bag cells of abdominal ganglion of *Aplysia californica. Nature, Lond.* **216**, 814–815.

178d. KUPFERMANN, I. (1970). Stimulation of egg laying by extracts of neuro-endocrine cells (bag cells) of abdominal ganglion of *Aplysia. J. neurophysiol.* **33**, 877–881.

179. KUPFERMAN, I. and KANDEL, E. R. (1970). Electrophysiological properties and functional interconnections of two symmetrical neurosecretory clusters (bag cells) in abdominal ganglion of *Aplysia. J. neurophysiol.* **33**, 865–876.

180. LANE, N. J. (1962). Neurosecretory cells in the optic tentacles of certain pulmonates. *Q. Jl. microsc. Sci.* **103**, 211–223.

181. LANE, N. J. (1964). Elementary neurosecretory granules in neurones of the snail, *Helix aspersa. Q. Jl. microsc. Sci.* **105**, 31–34.

182. LAUFER, H. (1965). Developmental studies of the Dipteran salivary gland. III. Relationships between chromosomal puffing and cellular function during development. *Developmental and metabolic control mechanisms and Neoplasia.* Williams & Wilkins Co., Baltimore.

183. LAUFER, H. and GOLDSMITH, M. (1965). Ultrastructural evidence for a protein transport system in *Chironomus* salivary glands and its implications for chromosomal puffing. *J. cell. Biol.* **27**, 57A.

184. LAUFER, H. and NAKASE, Y. (1965). Developmental studies of the Dipteran salivary gland. II. DNAase activity in *Chironomus thummi. J. cell. Biol.* **25**, 97–102.

185. LAUFER, H. and NAKASE, Y. (1965). Salivary gland secretion and its relation to chromosomal puffing in the Dipteran *Chironomus thummi. Proc. natn. Acad. Sci.* **53**, 511–516.

185a. LAVIOLETTE, P. (1954). Rôle de la gonade dans le determinisme glandulaire du tractus génital chez quelques gastéropodes arionidae et limacidae. *Bull. biol. Fr. et Belg.* **88**, 310–332.

185b. LAWRENCE, J. M., CRAIG, J. V. and CLOUGH, D. (1972). The presence of a hyperglycemic factor in the suprapharyngeal ganglia of *Lumbricus terrestris. Gen. comp. Endocr.* **18**, 260–267.

185c. LEA, A. O. (1963). Some relationships between environment, corpora allata and egg maturation in aedine mosquitoes. *J. Insect Physiol.* **9**, 793–809.

185d. LEA, A. O. (1967). The medial neurosecretory cells and egg maturation in mosquitoes. *J. Insect Physiol.* **13**, 419–429.

185e. LEA, A. O. (1970). Endocrinology of egg maturation in autogenous and anautogenous *Aedes taeniorhynchus. J. Insect Physiol.* **16**, 1689–1696.

185f. LEA, A. O. (1972). Regulation of egg maturation in the mosquito by the neurosecretory system: the role of the corpus cardiacum. *Gen. comp. Endocr.* Suppl. **3**, 602–608.

185g. LEA, A. O. and THOMSEN, E. (1969). Size independent secretion by the corpus allatum of *Calliphora erythrocephala. J. Insect Physiol.* **15**, 477–482.

186. LECHENAULT, H. (1962). Sur l'existence de cellules neurosécrétrices dans les ganglions cérébroides des Lineidae (Hétéronémertes). *C.r. Acad. Sci. Paris,* **255,** 194–196.

186a. LECHENAULT, H. (1965). Neurosécrétion et osmorégulation chez les Lineidae (Hétéronémertes). *C.r. Acad. Sci. Paris,* **261,** 4868–4871.

187. LEES, A. D. (1955). *The physiology of diapause in arthropods.* Cambridge University Press.

188. LEGENDRE, R. (1959). Contribution à l'étude du systéme nerveux des Aranéides. *Ann. Sci. nat. Zool.* **1,** 339–474.

188a. LENDER, T. (1964). Mise en evidence et role de la neurosécrétion chez les Planaires d'eau douce (Turbellaries, Triclades). *Ann. Endocrinol.* **25,** 61–65.

189. LENDER, TH. and KLEIN, N. (1961). Misé en evidence de cellules sécrétrices dans le cerveau de la Planaire *Polycelis nigra.* Variations de leur nombre au cours de la regénération postérieure. *C.r. Acad. Sci. Paris,* **253,** 331–333.

189a. LENDER, T. and ZGHAL, F. (1968). Influence du cerveau et de la neurosécrétion sur la scissiparite de la Planaire *Dugesia gonocephala. C.r. Acad. Sci. Paris,* **267,** 2008–2009.

189b. LENTZ, T. L. (1966). Histochemical localization of neurohumors in a sponge. *J. exp. Zool.* **162,** 171–175.

189c. LEONE, D. E. (1970). The maturation of *Hydroides dianthus. Biol. Bull., Woods Hole,* **138,** 306–315.

189d. LEVER, J. (1958). On the relation between the medio-dorsal bodies and the cerebral ganglia in some pulmonates. *Arch. Néerl. Zool.* **13,** 194–201.

190. LEVER, J., DEVRIES, C. M. and JAGER, J. C. (1965). On the anatomy of the central nervous system and the location of neurosecretory cells in *Australorbis glabratus. Malacologia,* **2,** 219–230.

191. LEVER, J., JANSEN, J. and DE VLIEGER, T. A. (1961). Pleural ganglia and water balance in the freshwater pulmonate *Lymnaea stagnalis. Proc. K. ned. Akad. Wet.* Section C. **64,** 532–542.

191a. LING, E. A. (1970). Further investigations on the structure and function of cephalic organs of a nemertine *Lineus ruber. Tissue and Cell,* **2,** 569–588.

191b. LOCKE, M. (1970). The molt/intermolt cycle in the epidermis and other tissues of an insect, *Calpodes ethlius* (Lepidoptera, Hesperiidae). *Tissue and Cell,* **2,** 197–223.

191c. LOCKE, M. (1970). The control of activities in insect epidermal cells. *Mem. Soc. Endocr.* **18,** 285–299.

191d. LOCKSHIN, R. A. (1969). Programmed cell death. Activation of lysis by a mechanism involving the synthesis of protein. *J. Insect Physiol.* **15,** 1505–1516.

191e. LOCKSHIN, R. A. and WILLIAMS, C. M. (1965). Programmed cell death. 1. Cytology of degeneration in the intersegmental muscles of the Pernyi silkmoth. *J. Insect Physiol.* **11,** 123–133.

191f. LOCKSHIN, R. A. and WILLIAMS, C. M. (1965). Programmed cell death. III. Neural control of the breakdown of the intersegmental muscles of silkmoths. *J. Insect Physiol.* **11,** 601–610.

191g. LOCKSHIN, R. A. and WILLIAMS, C. M. (1965). Programmed cell death. IV. The influence of drugs on the breakdown of the intersegmental muscles of silkmoths. *J. Insect Physiol.* **11,** 803–809.

192. LOHER, W. (1960). The chemical acceleration of the maturation process and its hormonal control in the male of the desert locust. *Proc. R. Soc.* (B), **153**, 380–397.

192a. LOHER, W. (1962). Die Kontrolle des Weibchengesanges von *Gomphocerus rufus* L. (Acridinae) durch die Corpora allata. *Naturwissenschaften*, **49**, 406.

193. LOHER, W. and HUBER, F. (1966). Nervous and endocrine control of sexual behaviour in a grasshopper (*Gomphocerus rufus* L., Acridinae). *Symp. Soc. exp. Biol.* **20**, 381–400.

193a. DE LOOF, A. and DE WILDE, J. (1970). The relation between haemolymph proteins and vitellogenesis in the Colorado beetle *Leptinotarsa decemlineata. J. Insect Physiol.* **16**, 157–169.

193b. DE LOOF, A. and DE WILDE, J. (1970). Hormonal control of synthesis of vitellogenic female protein in the Colorado beetle, *Leptinotarsa decemlineata. J. Insect Physiol.* **16**, 1455–1466.

193c. LOWE, M. F., HORN, D. H. S. and GALBRAITH, M. N. (1968). The role of crustecdysone in the moulting crayfish. *Experientia*, **24**, 518–519.

194. LUBET, P. (1955). Cycle neurosécretoire de *Chlamys varia* L. et *Mytilus edulis* L. *C.r. Acad. Sci. Paris*, **241**, 119 121.

195. LUBET, P. (1956). Effects de l'ablation des centres nerveux sur l'émission des gametes chez *Mytilus edulis* L. et *Chlamys varia* L. *Ann. Sci. natur. Zool.* **18**, 175–183.

196. LÜSCHER, M. (1962). In *Insect Polymorphism*. Social control of polymorphism in termites. Ed. KENNEDY, J. S. Symp. No. 1. Royal Ent. Soc. Lond. 57 67.

197. LÜSCHER, M. and KARLSON, P. (1958). Experimentelle Auslösung von Häutungen bei der Termite *Kalotermes flavicollis. J. Insect Physiol.* **1**, 341–345.

198. LUSIS, O. (1963). The histology and histochemistry of development and resorption in the terminal oocytes of the desert locust *Schistocerca gregaria. Q. Jl. microsc. Sci.* **104**, 57–68.

199. MADDRELL, S. H. P. (1963). Excretion in the blood-sucking bug, *Rhodnius prolixus* Stal. 1. The control of diuresis. *J. exp. Biol.* **40**, 247–256.

199a. MADDRELL, S. H. P. (1964). Excretion in the blood-sucking bug, *Rhodnius prolixus* Stål. III. The control of the release of the diuretic hormone. *J. exp. Biol.* **41**, 459–472.

200. MADHAVAN, K. and SCHNEIDERMAN, H. A. (1968). Effects of ecdysone on epidermal cells in which DNA synthesis has been blocked. *J. Insect Physiol.* **14**, 777–781.

201. MANNING, A. (1967). *An introduction to animal behaviour*. Edward Arnold, London.

201a. MANTEL, L. H. (1968). The foregut of *Gecarcinus lateralis* as an organ of salt and water balance. *Am. Zool.* **8**, 433–442.

201b. MARTIN, R. (1968). Fine structure of the vena cava in *Octopus. Brain Res.* **8**, 201–205.

201c. MARTOJA, M. (1965). Existence d'un organe juxtaganglionnaire chez *Aplysia punctata* Cuv. (Gastropode Opisthobranche). *C.r. Acad. Sci. Paris*, **260**, 4615–4617.

201d. MARTOJA, M. (1965). Données relatives a l'organe juxtaganglionnaire des Prosobranches Diotocardes. *C.r. Acad. Sci. Paris*, **261**, 3195–3196.

202. MASNER, P., SLAMA, K. and LANDA, V. (1968). Sexually spread insect sterility induced by the analogues of juvenile hormone. *Nature, Lond.* **219**, 395–396.

202a. MAYER, R. J. and CANDY, D. J. (1969). Control of haemolymph lipid concentration during locust flight: an adipokinetic hormone from the corpora cardiaca. *J. Insect Physiol.* **15**, 611–620.

203. MAYNARD, D. M. (1960). Circulation and heart function. *Physiology of Crustacea*, I. Ed. WATERMAN, T. H. Academic Press, London and New York.

203a. MCCAFFERY, A. R. (1973). Hormonal interactions during development in locusts. Ph.D. Thesis, University of Sheffield.

203b. MCCAFFERY, A. R. and HIGHNAM, K. C. (1975). Effects of corpus allatum hormone and its mimics on the activity of the cerebral neurosecretory system of *Locusta migratoria migratorioides* R & F. *Gen. comp. Endocr.* **25**, 358–372.

203c. MCCAFFERY, A. R. and HIGHNAM, K. C. (1975). Effects of corpora allata on the activity of the cerebral neurosecretory system of *Locusta migratoria migratorioides* R & F. *Gen. comp. Endocr.* **25**, 373–386.

203d. MCCURRY, P. M. (1972). Terpenoid routes to the synthesis of homojuvenates. *In Insect Juvenile Hormones: chemistry and action.* pp. 237–247. (J. J. Mann and M. Beroza, Eds.) Academic Press, New York and London.

204. MCWHINNIE, M. A., CAHOON, M. O. and JOHANNECK, R. (1969). Hormonal effects on calcium metabolism in Crustacea. *Am. Zool.* **9**, 841–855.

205. MCWHINNIE, M. A., KIRCHENBERG, R. J., URBANSKI, R. J. and SCHWARZ, J. E. (1972). Crustecdysone mediated changes in crayfish. *Am. Zool.* **12**, 357–372.

206. MECHELKE, F. (1953). Reversible Strukturmodificationen der Speicheldrüsenchromosomen von *Acricotopus lucidus*. *Chromosoma* (Berl.), **5**, 511.

206a. MEENAKSHI, V. R. and SCHEER, B. T. (1969). Regulation of galactogen synthesis in the slug *Ariolimax columbianus*. *Comp. Biochem. Physiol.* **29**, 841–845.

206b. MEOLA, R. and LEA, A. O. (1972). Humoral inhibition of egg development in mosquitoes. *J. med. Ent.* **9**, 99–103.

206c. MESSENGER, J. B., MUZII, E. O., NARDI, G. and STEINBERG, H. (1974). Haemocyanin synthesis and the branchial gland of *Octopus*. *Nature, Lond.* **250**, 154–155.

206d. MEYER, A. S., SCHNEIDERMAN, H. A., HANZMANN, E. and KO, J. H. (1968). The two juvenile hormones from the cecropia silkmoth. *Proc. natn. Acad. Sci.* **60**, 853–860.

206e. MILLS, R. R., MATHUR, R. B. and GUERRA, A. A. (1965). Studies on the hormonal control of tanning in the American cockroach. 1. Release of an activation factor from the terminal abdominal ganglion. *J. Insect Physiol.* **11**, 1047–1053.

207. MILLS, R. R. and NEILSEN, D. J. (1967). Hormonal control of tanning in the American cockroach. V. Some properties of the purified hormone. *J. Insect Physiol.* **13**, 273–280.

207a. MILLS, R. R. and WHITEHEAD, D. L. (1970). Hormonal control of tanning in the American cockroach: changes in blood cell permeability during ecdysis. *J. Insect Physiol.* **16**, 331–340.

208. MINKS, A. (1967). Biochemical aspects of juvenile hormone action in the adult *Locusta migratoria. Arch. Néerl. Zool.* **17**, 175–257.

208a. MORDUE (LUNTZ), A. J. and HIGHNAM, K. C. (1973). Incorporation of cysteine into the cerebral neurosecretory system of adult desert locusts. *Gen. comp. Endocr.* **20**, 351–357.

208b. MORDUE, W. (1967). The influence of feeding upon the activity of the neuro-endocrine system during oocyte growth in *Tenebrio molitor. Gen. comp. Endocr.* **9**, 406–415.

209. MORDUE, W. (1969). Hormonal control of Malpighian tube and rectal function in the desert locust, *Schistocerca gregaria. J. Insect Physiol.* **15**, 273–285.

209a. MORDUE, W. (1970). Evidence for the existence of diuretic and anti-diuretic hormones in locusts. *J. Endocr.* **46**, 119–120.

209b. MORDUE, W. (1971). The hormonal control of excretion and water balance in locusts. *Endocr. Exp.* **5**, 79–84.

209c. MORDUE, W. (1972). Hormones and excretion in locusts. *Gen. comp. Endocr. Suppl.* **3**, 289–298.

210. MORDUE, W. and GOLDSWORTHY, G. J. (1969). The physiological effects of corpus cardiacum extracts in locusts. *Gen. comp. Endocr.* **12**, 360–369.

210a. MORIYAMA, H., NAKANISHI, K., KING, D. S., OKAUCHI, T., SIDDALL, J. B. and HAFFERL, W. (1970). Origin and metabolic fate of α-ecdysone in insects. *Gen. comp. Endocr.* **15**, 80–87.

211. NAISSE, J. (1965). Contrôle endocrinien de la différenciation sexuelle chez les insectes. *Arch. d'anat. micr. et exper.* **54**, 417–428.

211a. NAKANISHI, K., MORIYAMA, H., OKAUCHI, T., FUJIOKA, S. and KOREEDA, M. (1972). Biosynthesis of α- and β-ecdysones from cholesterol outside the prothoracic gland in *Bombyx mori. Science, N.Y.* **176**, 51–52.

212. NAYAR, K. K. (1958). Studies on the neurosecretory system of *Iphita limbata* Stål. V. Probable endocrine basis of oviposition in the female insect. *Proc. Indian Acad. Sci.* **47**, 233–251.

213. NELSON, J. A. (1915). *The embryology of the honey bee.* Princeton University Press.

213a. NISHIITSUTSUJI-UWO, J. (1972). Purification and some properties of insect brain hormone extracted from silkworm heads. *Botyu-Kagaku,* **37**, 93–102.

213b. NOLTE, D. J., EGGERS, S. H. and MAY, I. R. (1973). A locust pheromone: locustol. *J. Insect Physiol.* **19**, 1547–1554.

213c. NORMANN, T. C. (1972). Heart activity and its control in the adult blowfly, *Calliphora erythrocephala. J. Insect Physiol.* **18**, 1793–1810.

213d. NOUMURA, T. and KANATANI, H. (1962). Induction of spawning by radial nerve extracts in some starfish. *J. Fac. Sci. Univ. Tokyo,* **9**, 397–402.

213e. OBERLANDER, H. (1969). Effects of ecdysone, ecdysterone and inokosterone on the *in vitro* initiation of metamorphosis of wing disks of *Galleria mellonella. J. Insect Physiol.* **15**, 297–304.

213f. OBERLANDER, H. (1969). Ecdysone and DNA synthesis in cultured wing disks of the wax moth, *Galleria mellonella. J. Insect Physiol.* **15**, 1803–1806.

213g. OBERLANDER, H., LEACH, C. E. and TOMBLIN, C. (1973). Cuticle deposition in imaginal disks of three species of Lepidoptera: effects of ecdysones *in vitro*. *J. Insect Physiol.* **19**, 993–998.

213h. OBERLANDER, H. and TOMBLIN, C. (1972). Cuticle deposition in imaginal disks: effects of juvenile hormone and fatbody. *Science, N.Y.* **177**, 441–442.

214. ODHIAMBO, T. R. (1966). The fine structure of the corpus allatum of the sexually mature male of the desert locust. *J. Insect Physiol.* **12**, 819–828.

214a. O'CONNOR, J. D. and GILBERT, L. I. (1968). Aspects of lipid metabolism in crustaceans. *Am. Zool.* **8**, 529–539.

214b. O'CONNOR, J. D. and GILBERT, L. I. (1969). Alterations in lipid metabolism associated with premoult activity in a land crab and fresh water crayfish. *Comp. Biochem. Physiol.* **29**, 889–904.

214c. O'DOR, R. K. and WELLS, M. J. (1973). Yolk protein synthesis in the ovary of *Octopus vulgaris* and its control by the optic gland gonadotropin. *J. exp. Biol.* **59**, 665–674.

214d. OHMORI, K. (1974). Effect of ecdysterone contact period on development of wing disks of the fleshfly, *Sarcophaga peregrina*, *in vitro*. *J. Insect Physiol.* **20**, 1697–1706.

214e. OHMORI, K. and OHTAKI, T. (1973). Effects of ecdysone analogues on development and metabolic activity of wing disks of the fleshfly, *Sarcophaga peregrina*, *in vitro*. *J. Insect Physiol.* **19**, 1199–1210.

214f. OHTAKI, T., MILKMAN, R. D. and WILLIAMS, C. M. (1967). Ecdysone and ecdysone analogues: their assay on the fleshfly, *Sarcophaga peregrina*. *Proc. natn. Acad. Sci.* **58**, 981–984.

214g. OHTAKI, T. MILKMAN, R. D. and WILLIAMS, C. M. (1968). Dynamics of ecdysone secretion and action in the fleshfly, *Sarcophaga peregrina*. *Biol. Bull., Woods Hole*, **135**, 322–334.

214h. OHTAKI, T. and WILLIAMS, C. M. (1970). Inactivation of α-ecdysone and cyasterone by larvae of the fleshfly, *Sarcophaga peregrina*, and pupae of the silkworm, *Samia cynthia*. *Biol. Bull., Woods Hole*, **138**, 326–333.

214i. OLIVE, P. W. B. (1970). Reproduction of a Northumberland population of the polychaete *Cirratulus cirratus*. *Mar. Biol.* **5**, 259–273.

214j. OLIVE, P. J. W. (1972). Regulation and kinetics of spermatogonial proliferation in *Arenicola marina* (Annelida, Polychaeta). I. The annual cycle of mitotic index in the testis. *Cell Tissue Kinet.* **5**, 245–253.

214k. OLIVE, P. J. W. (1972). Regulation and kinetics of spermatogonial proliferation in *Arenicola marina* (Annelida, Polychaeta). 2. Kinetics. *Cell Tissue Kinet.* **5**, 255–267.

214l. OLIVE, P. J. W. (1973). The regulation of ovary function in *Cirratulus cirratus* (Polychaeta). *Gen. comp. Endocr.* **20**, 1–15.

214m. PAN, M. L., BELL, W. J. and TELFER, W. H. (1969). Vitellogenic blood protein synthesis by insect fat body. *Science, N.Y.* **165**, 393–394.

215. PANOUSE, J. B. (1943). Influence de l'ablation du pedoncule oculaire sur la croissance de l'ovaire chez la crevette *Leander serratus*. *C.r. Acad. Sci. Paris*, **217**, 553–555.

216. PARKES, A. S. and BRUCE, H. M. (1962). Olfactory stimuli in mammalian reproduction. *Science, N.Y.* **134**, 1049–1054.

217. PASSANO, L. M. (1953). Neurosecretory control of moulting in crabs by the X-organ sinus gland complex. *Physiologia comp. Oecol.* **3**, 155–189.

218. PASSANO, L. M. (1960). Moulting and its control. *Physiology of Crustacea*, I. Ed. WATERMAN, T. H. Academic Press, London and New York.

218a. PASSANO, L. M. and JYSSUM, S. (1963). The role of the Y-organ in crab proecdysis and limb regeneration. *Comp. Biochem. Physiol.* **9**, 195–213.

218b. PAYNE, T. L., SHOREY, H. H. and GASTON, L. K. (1970). Sex pheromones of noctuid moths: factors influencing antennal responsiveness in males of *Trichoplusia ni. J. Insect Physiol.* **16**, 1043–1055.

219. PELLING, C. (1965). The mechanism of activation in Dipteran salivary chromosomes. II. Characteristics of puffing in giant chromosomes. *Archs Anat. microsc. Morph. exp.* **54**, 645–647.

220. PELLUET, D. and LANE, N. J. (1961). The relation between neurosecretion and cell differentiation in the ovotestis of slugs. *Can. J. Zool.*, **39**, 789–805.

221. PÉREZ-GONZÁLEZ, M. D. (1957). Evidence for hormone-containing granules in sinus glands of the fiddler crab (*Uca pugilator*). *Biol. Bull.*, Woods Hole, **113**, 426.

222. PERKINS, E. B. (1928). Colour change in crustaceans, especially in *Palaemonetes. J. exp. Zool.* **50**, 71–195.

223. PHILLIPS, J. E. (1964). Rectal reabsorption in the desert locust, *Schistocerca gregaria* Forskal. I. Water. *J. exp. Biol.* **41**, 15–38.

224. PIEPHO, H. (1939). Raupenhäutungen bereits verpuppter Hautstücke bei der Wachsmotte *Galleria mellonella* L. *Naturwissenschaften*, **27**, 301–302.

224a. PORCHET, M. (1967). Rôle des ovocytes submatures dans l'arrêt de l'inhibition cérébrale chez *Perinereis cultrifera. C.r. Acad. Sci. Paris*, **265**, 1394–1396.

224b. PORCHET, M. (1970). Relations entre le cycle hormonal cérébral et l'évolution ovocytaire chez *Perinereis cultrifera* Grube (annélide polychète). *Gen. comp. Endocr.* **15**, 220–231.

224c. PORCHET, M. (1972). Variation de l'activité endocrine des cerveaux en fonction de l'espèce, du sexe et du cycle vital chez quelques Néréidiens (Annélides Polychètes). *Gen. comp. Endocr.* **18**, 276–284.

224d. PORCHET, M. and DHAINAUT, A. (1969). Formation et migration des mucopolysaccharides au cours de l'ovogenèse de *Perinereis cultrifera* Grube et *Nereis pelagica* L. (Annélides Polychètes). Etude histochimique et ultrastructurale. *C.r. Seanc. Soc. Biol.* **163**, 418–420.

224e. POST, L. C. (1972). Bursicon: its effects on tyrosine permeation into insect haemocytes. *Biochim. Biophys. Acta*, **290**, 424–428.

225. PRABHU, V. K. K. (1961). The structure of the cerebral glands and connective bodies of *Jonespeltis splendidus* Verhoeff. (Myriapoda: Diplopoda.) *Z. fur Zellforsch.* **54**, 717–733.

225a. PRATT, G. E. and DAVEY, K. G. (1972). The corpus allatum and oogenesis in *Rhodnius prolixus* (Stål). I. The effects of allatectomy. *J. exp. Biol.* **56**, 201–214.

225b. PRATT, G. E. and DAVEY, K. G. (1972). The corpus allatum and oogenesis in *Rhodnius prolixus* (Stål). III. The effect of mating. *J. exp. Biol.* **56**, 223–237.

225c. PRATT, G. E. and TOBE, S. S. (1974). Juvenile hormones radiobiosynthesized by corpora allata of adult female locusts *in vitro*. *Life Sci.* **14**, 575–586.

226. RAE, C. A. (1955). Possible new elements in the endocrine complex of cockroaches. *Aust. J. Sci.* **18**, 33–34.

227. RAE, C. A. and O'FARRELL, A. F. (1959). The retrocerebral complex and ventral glands of the primitive orthopteroid *Grylloblatta campodeiformis* with a note on the homology of the muscle core of the 'prothoracic gland' in Dictyoptera. *Proc. R. ent. Soc. Lond. A.* **34**, 76–82.

227a. RAO, K. R., FINGERMAN, S. W. and FINGERMAN, M. (1973). Effects of exogenous ecdysones on the molt cycles of fourth and fifth stage American lobsters, *Homarus americanus. Comp. Biochem. Physiol.* **44(A)**, 1105–1120.

227b. RICHARD, A. (1971). Action qualitative de la lumière dans le déterminisme du cycle sexuel chez le céphalopode *Sepia officinalis* L. *C.r. Acad. Sci. Paris*, **272**, (D), 106–109.

227c. RIDDIFORD, L. M. (1974). The role of hormones in the reproductive behaviour of female wild silkmoths. *Experimental Analysis of Insect Behaviour*. Ed. BARTON BROWNE, L. Springer-Verlag, Berlin and New York.

227d. RIDDIFORD, L. M., AJAMI, A. M., CAREY, E. J., YAMAMOTO, H. and ANDERSON, J. E. (1971). Synthetic imino analogs of *Cecropia* juvenile hormones as potentiates of juvenile hormone activity. *J. Am. Chem. Soc.* **93**, 1815–1816.

227e. RIDDIFORD, L. M. and ASHENHURST, J. B. (1973). The switchover from virgin to mated behaviour in female cecropia moths: the role of the bursa copulatrix. *Biol. Bull., Woods Hole*, **144**, 162–171.

227f. RIDDIFORD, L. M. and WILLIAMS, C. M. (1967). Volatile principle from oak leaves: role in sex life of the Polyphemus moth. *Science, N.Y.* **155**, 589–590.

227g. RIDDIFORD, L. M. and WILLIAMS, C. M. (1971). Role of the corpora cardiaca in the behaviour of saturniid moths. 1. Release of sex pheromone. *Biol. Bull., Woods Hole*, **140**, 1–7.

227h. ROBINSON, N. L. and GOLDSWORTHY, G. J. (1974). The effects of locust adipokinetic hormone on flight muscle metabolism *in vivo* and *in vitro. J. comp. Physiol.* **89**, 369–377.

227i. ROGERS, D. C. (1969). Fine structure of the epineural connective tissue sheath of the suboesophageal ganglion in *Helix aspersa. Z. Zellforsch.* **102**, 99–112.

227j. ROGERS, W. P. (1973). Juvenile and moulting hormones from nematodes. *Parasitology*, **67**, 105–113.

228. RÖLLER, H., DAHM, K. H., SWEELY, C. C. and TROST, B. M. (1967). The structure of the Juvenile Hormone. *Angew. Chem.* **6**, 179–180.

229. ROONWAL, M. L. (1937). Studies on the embryology of the African migratory locust, *Locusta migratoria* R. and F. (Orthoptera, Acrididae). II. Organogeny. *Phil. Trans. R. Soc.* (B), **227**, 175–244.

229a. ROUBOS, E. W. (1973). Regulation of neurosecretory activity in the freshwater pulmonate *Lymnaea stagnalis* (L). A quantitative electron microscopical study. *Z. Zellforsch.* **146**, 177–205.

229b. ROUSSEL, J. P. and CAZAL, M. (1969). Action, *in vivo*, d'extraits de corpora cardiaca sur le rhythme cardiaque de *Locusta migratoria* L. *C.r. Acad. Sci. Paris*, **268**, 581–583.

229c. RUH, M. F., RUH, T. S., DeWERT, W. and DUENAS, V. (1974). Ecdysterone-induced protein synthesis *in vitro*. *J. Insect Physiol.* **20**, 1729–1736.

229d. RUNHAM, N. W. and HUNTER, P. J. (1970). *Terrestrial Slugs*. Hutchinson, London.

229e. SAMOILOFF, M. R. (1973). Nematode morphogenesis: localization of controlling regions by laser microbeam surgery. *Science, N.Y.* **180**, 976–977.

229f. SANBURG, K. J., KRAMER, K. J., KEZDY, F. J., LAW, J. H. and OBERLANDER, H. (1975). Role of juvenile hormone esterases and carrier proteins in insect development. *Nature, Lond.* **253**, 266–267.

229g. SANCHEZ, S. and SABLIER, H. (1962). Histophysiologie neuro-hormonale chez quelques mollusques gastéropodes. II. Corrélations hormonales. *Bull. Soc. zool. Fr.* **87**, 319–330.

230. SAROJINI, S. (1963). Comparison of the effects of androgenic hormone and testosterone propionate on the female Ocypod crab. *Curr. Sci.* **32**, 411–412.

231. SCHARRER, B. (1936). Über 'Drusennvenzellen' im Gehirn von *Nereis virens* Sars. *Zool. Anz.* **25**, 131–138.

232. SCHARRER, B. (1941). Neurosecretion. III. The cerebral organ of the Nemerteans. *J. comp. Neurol.* **74**, 109–130.

233. SCHARRER, B. (1941). Neurosecretion. 4. Localization of neurosecretory cells in the central nervous system of *Limulus*. *Biol. Bull., Woods Hole,* **81**, 96–104.

234. SCHARRER, B. (1952). Neurosecretion. XI. The effects of nerve section on the intercerebralis-cardiacum-allatum system of the insect *Leucophaea maderae*. *Biol. Bull. Woods Hole,* **102**, 261–272.

235. SCHARRER, B. (1964). The ultrastructure of the corpus allatum of *Blaberus craniifer* (Blattariae). *Am. Zool.* **4**, 327–328.

236. SCHARRER, B. (1967). The neurosecretory neurone in neuroendocrine regulatory mechanisms. *Am. Zool.* **7**, 161–169.

237. SCHARRER, E. and SCHARRER, B. (1963). *Neuroendocrinology*. Columbia University Press, New York and London.

238. SCHEER, B. T. and SCHEER, M. A. R. (1951). The hormonal regulation of metabolism in crustacea. I. Blood sugar in spiny lobsters. *Physiol. comp.* **2**, 198–209.

239. SCHEFFEL, H. (1965). Der Einfluss von Dekapitation und Schnürung auf die Häutung und die Anamorphose der Larven von *Lithobius forficatus* L. (Chilopoda). *Zool. Jb.* (*Physiol.*), **71**, 359–370.

241. SCHMIALEK, P. (1961). Die Identifizierung zweier im *Tenebrio* kot und in der Hefe vorkommender Substanzen mit Juvenilhormonwirkung. *Z. Naturf.* **16b**, 461–464.

242. SCHMIALEK, P. (1963). Über die Bildung von Juvenilhormonen in Wildseidenspinnern. *Z. Naturf.* **186**, 462–465.

243. SCHNEIDERMAN, H. A. and GILBERT, L. I. (1964). Control of growth and development in insects. *Science N.Y.*, **143**, 325–333.

244. SCHNEIDERMAN, H. A., KRISHNAKUMARAN, A., KULKARNI, V. G. and FRIEDMAN, L. (1965). Juvenile hormone activity of structurally un-related compounds. *J. Insect Physiol.* **11**, 1641–1649.

244a. SCHROEDER, P. C. (1971). Studies on oogenesis in the polychaete annelid *Nereis grubei* (Kinberg). II Oocyte growth rates in intact and hormone-deficient animals. *Gen. comp. Endocr.* **16**, 312–322.

244b. SCHUETZ, A. W. (1969). Chemical properties and physiological actions of a starfish radial nerve factor and ovarian factor. *Gen. comp. Endocr.* **12**, 209–221.

244c. SCHUETZ, A. W. and BIGGERS, J. D. (1967). Regulation of germinal vesicle breakdown in starfish oocytes. *Exptl. Cell Res.* **46**, 624–628.

245. SCULLY, U. (1964). Factors influencing the secretion of regeneration-promoting hormone in *Nereis diversicolor*. *Gen. comp. Endocr.* **4**, 91–98.

245a. SELIGMAN, I. M., FRIEDMAN, S. and FRAENKEL, G. (1969). Bursicon mediation of tyrosine hydroxylation during tanning of the adult cuticle of the fly, *Sarcophaga bullata*. *J. Insect Physiol.* **15**, 553–562.

245b. SILVERTHORNE, S. U. (1973). Respiration in eyestalkless *Uca* (Crustacea: Decapoda) acclimated to two temperatures. *Comp. Biochem. Physiol.* **45**(A), 417–420.

245c. SILVERTHORNE, S. U. (1975). Hormonal involvement in thermal acclimation in the fiddler crab *Uca pugilator* (Bosc) I. Effect of eyestalk extracts on whole animal respiration. *Comp. Biochem. Physiol.* **50**(A), 281–283.

246. SIMPSON, L., BERN, H. A. and NISHIOKA, R. S. (1966). Survey of the evidence for neurosecretion in gastropod molluscs. *Am. Zool.* **6**, 123–138.

246a. SIMPSON, M. (1962). Reproduction of the Polychaete *Glycera dibranchiata* at Solomons, Maryland. *Biol. Bull., Woods Hole*, **123**, 396–410.

246b. SKORKOWSKI, E. F. (1971). Isolation of three chromatophorotropic hormones from the eystalk of the shrimp *Crangon crangon*. *Mar. Biol.* **8**, 220–223.

246c. SKORKOWSKI, E. F. (1972). Separation of three chromatophorotropic hormones from the eyestalk of the crab *Rhithropanopeus harrisi* (Gould). *Gen. comp. Endocr.* **18**, 329–334.

246d. SKORKOWSKI, E. F. and KLEINHOLZ, L. H. (1973). Comparison of white pigment concentrating hormone from (1973) *Crangon* and *Pandalus*. *Gen. comp. Endocr.* **20**, 595–597.

246e. SLADE, M. and ZIBITT, C. H. (1972). Metabolism of cecropia juvenile hormone in insects and in mammals. In *Insect Juvenile Hormones: chemistry and action* pp. 155–177. (J. J. Mann and M. Beroza Eds.). Academic Press, New York.

247. SLAMA, K. (1964). Hormonal control of respiratory metabolism during growth, reproduction and diapause in female adults of *Pyrrhocoris apterus* L. *J. Insect Physiol.* **10**, 283–304.

247a. SLAMA, K., ROMANUK, M. and SORM, F. (1969). Natural and synthetic materials with insect hormone activity. 2. Juvenile hormone activity of some derivatives of farnesenic acid. *Biol. Bull., Woods Hole*, **136**, 91–95.

247b. SLAMA, K., ROMANUK, M. and SORM, F. (1974). *Insect hormones and bioanalogues*. Springer-Verlag, Wien and New York.

248. SLAMA, K. and WILLIAMS, C. M. (1965). Juvenile hormone activity for the bug, *Pyrrhocoris apterus*. *Proc. natn. Acad. Sci.* **54**, 411–414.

249. SLAMA, K. and WILLIAMS, C. M. (1966). The juvenile hormone. V. The sensitivity of the bug, *Pyrrhocoris apterus*, to a hormonally, active factor in American paper-pulp. *Biol. Bull., Woods Hole*, **130**, 235–246.

249a. SMITH, G. N. and GREENBERG, M. J. (1973). Chemical control of the evisceration process in *Thyone briareus*. *Biol. Bull., Woods Hole*, **144**, 421–436.

249b. SPAZIANA, E. and KATER, S. B. (1973). Uptake and turnover of cholesterol-14C in Y-organs of the crab *Hemigrapsus* as a function of the molt cycle. *Gen. comp. Endocr.* **20**, 534–549.

250. STAMM, M. D. (1959). Estudios sobre hormonas de invertebrados. II. Aislamiento de hormonas de la metamorfosis en el ortoptero *Dociotaurus marocannus*. *Espan. Fis. Quim.* (Madrid), **55B**, 171–178.

250a. STEEL, C. G. H. (1975). A neuroendocrine feedback mechanism in the insect moult cycle. *Nature, Lond.* **253**, 267–269.

251. STEELE, J. E. (1963). The site of action of insect hyperglycaemic hormone. *Gen. comp. Endocr.* **3**, 46–52.

251a. STEVENS, M. (1970). Procedures for induction of spawning and meiotic maturation of starfish oocytes by treatment with 1-methyladenine. *Expl. Cell Res.* **59**, 482–484.

251b. STEVENSON, J. R. (1972). Changing activities of the crustacean epidermis during the molting cycle. *Am. Zool.* **12**, 373–380.

251c. STEVENSON, J. R. and TSCHANTZ, J. A. (1973). Acceleration by ecdysterone of premoult substages in the crayfish. *Nature, Lond.* **242**, 133–134.

252. STRANGWAYS-DIXON, J. (1961). The relationship between nutrition, hormones and reproduction in the blowfly *Calliphora erythrocephala* Meig. I. Selective feeding in relation to the reproductive cycle, the corpus allatum volume and fertilization. *J. exp. Biol.* **38**, 225–235.

253. STRANGWAYS-DIXON, J. (1961). The relationship between nutrition, hormones and reproduction in the blowfly *Calliphora erythrocephala* Meig. II. The effect of removing the ovaries, the corpus allatum and the median neurosecretory cells upon selective feeding and the demonstration of the corpus allatum cycle. *J. exp. Biol.* **38**, 637–646.

254. STRANGWAYS-DIXON, J. (1962). The relationship between nutrition, hormones and reproduction in the blowfly *Calliphora erythrocephala* Meig. III. The corpus allatum in relation to nutrition, the ovaries, innervation and the corpus cardiacum. *J. exp. Biol.* **39**, 293–306.

254a. STREIFF, W. (1967). Étude endocrinologique du cycle sexuel chez un Mollusque hermaphrodite protandre *Calyptraea sinensis* L. III. Mise en évidence par culture *in vitro* de facteurs hormonaux conditionnant l'évolution de la gonade. *Ann. Endocrinol.* **28**, 641–656.

254b. STREIFF, W., LEBRETON, J. and SILBERZAHN, N. (1970). Non spécificité des facteurs hormonaux responsables de la morphogenèse et du cycle du tractus génital mâle chez les Mollusques Prosobranches. *Ann. Endocrinol.* **31**, 548–556.

255. STRICH-HALBWACHS, M. C. (1954). Rôle de la glande ventrale chez *Locusta migratoria*. *C.r. Séanc. Soc. Biol.* **148**, 2087–2090.

256. STRICH-HALBWACHS, M. C. (1959). Contrôle de la mue chez *Locusta migratoria*. *Ann. Sci. nat. Zool.* Ser. 12, **1**, 483–570.

256a. STRUMWASSER, F., JACKLET, J. W. and ALVAREZ, R. B. (1969). A seasonal rhythm in the neural extract induction of behavioural egg laying in *Aplysia*. *Comp. Biochem. Physiol.* **29**, 197–206.

257. STUMM-ZOLLINGER, E. (1957). Histological study of regenerative processes after transection of the nervi corporis cardiaci in transplanted brains of the cecropia silkworm (*Platysamia cecropia* L). *J. exp. Zool.* **134**, 315–326.

257a. SUCHY, M., SLAMA, K. and SORM, F. (1968). Insect hormone activity of p-(1,5-dimethyl-hexyl) benzoic acid derivatives in *Dysdercus* species. *Science, N.Y.* **162**, 582–583.

258. TAKI, I. (1964). On the morphology and physiology of the branchial gland in Cephalopoda. *J. Fac. Fish. Anim. Husb. Hiroshima Univ.* **5**, 345–417.

259. THOMAS, P. J. and BHATNAGAR-THOMAS, P. L. (1968). Use of a juvenile hormone analogue as insecticide for pests of stored grain. *Nature, Lond.* **219**, 949.

259a. THOMPSON, M. J., KAPLANIS, J. N., ROBBINS, W. E. and YAMAMOTO, R. T. (1967). 20, 26-dihydroxyecdysone, a new steroid with moulting hormone activity from the tobacco hornworm, *Manduca sexta* (Johannson). *Chem. Comm.* 1967 (3), 650–653.

260. THOMSEN, E. (1949). Influence of the corpus allatum on the oxygen consumption of adult *Calliphora erythrocephala* Meig. *J. exp. Biol.* **26**, 137–149.

261. THOMSEN, E. (1952). Functional significance of the neurosecretory brain cells and corpus cardiacum in the female blowfly, *Calliphora erythrocephala* Meig. *J. exp. Biol.* **29**, 137–172.

262. THOMSEN, E. and HAMBURGER, K. (1955). Oxygen consumption of castrated females of the blowfly, *Calliphora erythrocephala* Meig. *J. exp. Biol.* **32**, 692–699.

263. THOMSEN, E. and MÖLLER, I. (1963). Influence of neurosecretory cells and of corpus allatum on intestinal protease activity in the adult *Calliphora erythrocephala* Meig. *J. exp. Biol.* **40**, 301–321.

264. THOMSEN, M. (1965). The neurosecretory system of the adult *Calliphora erythrocephala*. II. Histology of the neurosecretory cells of the brain and some related structures. *Z. Zellforsch.* **67**, 693–717.

264a. TOEVS, L. A. and BRACKENBURY, R. W. (1969). Bag cell-specific proteins and the humoral control of egg-laying in *Aplysia californica*. *Comp. Biochem. Physiol.* **29**, 207–216.

264b. TOOLE, B., SCHUETZ, A. W. and BOYLAN, E. (1974). Uptake and cellular localization of ^{3}H-1-methyladenine and ^{3}H-adenine in the starfish gonad and oocytes. *Gen. comp. Endocr.* **22**, 199–208.

265. TOYAMA, K. (1902). Contributions to the study of silkworms. I. On the embryology of the Silkworm. *Bull. Coll. Agric. Tokyo, imp. Univ.* **5**.

265a. TRAUTMANN, K. H. (1972). *In vitro* Studium der Trägerproteine von ^{3}H-markierten juvenilhormonwirksamen Verbindungen in der Hämolymphe von *Tenebrio moliter* L. Larnen. *Z. Naturf.* **276**, 263–273.

265b. TROST, B. M. (1972). The origin of juvenile hormone chemistry. *Insect Juvenile Hormones: Chemistry and Action*. Eds. MANN, J. J. and BEROZA, M. Academic Press, New York and London.

265c. TRUMAN, J. W. (1970). The eclosion hormone: its release by the brain, and its action on the central nervous system of silkmoths. *Amer. Zool.* **10**, 511–512.

265d. TRUMAN, J. W. (1971). Physiology of insect ecdysis. 1. The eclosion behaviour of Saturniid moths and its hormonal release. *J. exp. Biol.* **54**, 805–814.

265e. TRUMAN, J. W. (1972). Physiology of Insect Rhythms. II. The silkmoth brain as the location of the biological clock controlling eclosion. *J. comp. Physiol.* **81**, 99–114.

265f. TRUMAN, J. W. (1972). Involvement of a photoreversible process in the circadian clock controlling silkmoth eclosion. *Z. vergl. Physiologie*, **76**, 32–40.

265g. TRUMAN, J. W. (1973). Physiology of insect ecdysis. III. Relationships between the hormonal control of eclosion and of tanning in the tobacco hornworm, *Manduca sexta. J. exp. Biol.* **58**, 821–829.

265h. TRUMAN, J. W. and RIDDIFORD, L. M. (1970). Neuroendocrine control of ecdysis in silkmoths. *Science, N.Y.* **167**, 1624–1626.

265i. TRUMAN, J. W. and RIDDIFORD, L. M. (1971). Role of the corpora cardiaca in the behavior of saturniid moths. II. Oviposition. *Biol. Bull., Woods Hole*, **140**, 8–14.

265j. TRUMAN, J. W. and RIDDIFORD, L. M. (1973). Hormonal mechanisms underlying insect behaviour. *Adv. Insect Physiol.* **10**, 297–352.

265k. TULLIS, R. E. and KAMEMOTO, F. I. (1974). Separation and biological effects of CNS factors affecting water balance in the decapod crustacean *Thalamita crenata. Gen. comp. Endocr.* **23**, 19–28.

266. VAN DER KLOOT, W. G. (1955). The control of neurosecretion and diapause by physiological changes in the brain of the cecropia silkworm. *Biol. Bull., Woods Hole*, **109**, 276–294.

266a. VENDRIX, J. J. (1963). Existence de cellules neurosecrétrices chez *Polycelis nigra* Ehrenberg et *Dugesia gonocephala* Dugés (Triclades Paludicoles). Caractéristiques cytologiques et histochimiques. *Bull. Soc. Sci. Liège*, **32**, Annee 34, 293–305.

266b. VICENTE, N. (1969). Etude histologique et histochimique du système nerveux central, des rhinopores et de la gonade chez les gastéropodes opisthobranches. *Téthys.* **1**, 833–874.

266c. VICENTE, N. (1969). Corrélations neuroendocrines chez *Aplysia rosea* ayant subi l'ablation de divers ganglions nerveux. *Téthys.* **1**, 875–900.

266d. VICENTE, N. (1969). Contribution a l'étude des gastéropodes opisthobranches du Golfe de Marseille. II. Histophysiologie du système nerveux. Etude des phéromènes neurosécrétoires. *Rec. Trav. St. Mar. Endocrine.* **62**, 13–121.

266e. VICENTE, N. (1970). Observations sur l'ultrastructure d'un organe juxtacommissural dans le système nerveux du Chiton (Mollusque Polyplacophore). *C. R. soc. Biol.* **164**, 601–607.

266f. VICENTE, N. and GASQUET, M. (1970). Etude du système nerveux et de la neurosécrétion chez quelques mollusques polyplacophores. *Téthys.* **2**, 515–546.

266g. VINCENT, J. F. V. (1971). The effects of bursicon on cuticular properties in *Locusta migratoria migratorioides. J. Insect Physiol.* **17**, 625–636.

266h. WARNER, A. C. and STEVENSON, J. R. (1972). The influence of ecdysones and eyestalk removal on the molt cycle of the crayfish *Orconectes obscurus. Gen. comp. Endocr.* **18**, 454–462.

266i. WEIR, S. B. (1970). Control of moulting in an insect. *Nature, Lond.* **228**, 580–581.

267. WEISMANN, A. (1864). Die nachembryonale Entwicklung der Musciden nach Beobachtungen an *Musca vomitoria* und *Sarcophaga carnaria*. *Z. wiss. Zool.* **14**, 187–336.

268. WELLS, M. J. and WELLS, J. (1959). Hormonal control of sexual maturity in *Octopus*. *J. exp. Biol.* **36**, 1–33.

268a. WENDELAAR BONGA, S. E. (1970). Ultrastructure and histochemistry of neurosecretory cells and neurohaemal areas in the pond snail *Lymnaea stagnalis* (L). *Z. Zellforsch.* **108**, 190–224.

268b. WENDELAAR BONGA, S. E. (1971). Osmotically induced changes in the activity of neurosecretory cells located in the pleural ganglia of the fresh water snail *Lymnaea stagnalis* (L), studied by quantitative electron microscopy. *Neth. J. Zool.* **21**, 127–158.

268c. WENDELAAR BONGA, S. E. (1971). Formation, storage and release of neurosecretory material studied by quantitative electron microscopy in the fresh water snail *Lymnaea stagnalis* (L). *Z. Zellforsch.* **113**, 490–517.

268d. WENDELAAR BONGA, S. E. (1972). Neuroendocrine involvement in osmoregulation in a freshwater mollusc, *Lymnaea stagnalis*. *Gen. comp. Endocr.* Suppl. **3**, 308–316.

268e. WESTFALL, J. A. (1973). Ultrastructural evidence for a granule-containing sensory-motor-interneuron in *Hydra littoralis*. *J. Ultrastruct. Res.* **42**, 268–282.

268f. WHITE, A. F. (1972). Metabolism of the juvenile hormone analogue methyl farnesoate 10,11-epoxide in two insect species. *Life Sci.* **11**, 201–210.

268g. WHITMORE, E. and GILBERT, L. I. (1972). Haemolymph lipoprotein transport of juvenile hormone. *J. Insect Physiol.* **18**, 1153–1167.

268h. WHITMORE, D., GILBERT, L. I. and ITTYCHERIAH, P. I. (1974). The origin of hemolymph carboxylesterases 'induced' by the insect juvenile hormone. *Mol. Cell. Endocr.* **1**, 37–54.

268i. WHITMORE, D., WHITMORE, E. and GILBERT, L. I. (1972). Juvenile hormone induction of esterases: a mechanism for the regulation of juvenile hormone titer. *Proc. natn. Acad. Sci.* **69**, 1592–1595.

269. WIGGLESWORTH, V. B. (1936). The function of the corpus allatum in the growth and reproduction of *Rhodnius prolixus* (Hemiptera). *Q. Jl. microsc. Sci.* **79**, 91–121.

270. WIGGLESWORTH, V. B. (1939). Häutung bei Imagines von Wanzen. *Naturwissenschaften*, **27**, 301.

271. WIGGLESWORTH, V. B. (1940). The determination of characters at metamorphosis in *Rhodnius prolixus* (Hemiptera). *J. exp. Biol.* **17**, 201–222.

272. WIGGLESWORTH, V. B. (1940). Local and general factors in the development of 'pattern' in *Rhodnius prolixus* (Hemiptera). *J. exp. Biol.* **17**, 180–200.

273. WIGGLESWORTH, V. B. (1943). The fate of haemoglobin in *Rhodnius prolixus* (Hemiptera) and other blood-sucking arthropods. *Proc. R. Soc. Series B*, **131**, 313–339.

274. WIGGLESWORTH, V. B. (1948). The functions of the corpus allatum in *Rhodnius prolixus* (Hemiptera). *J. exp. Biol.* **25**, 1–14.

275. WIGGLESWORTH, V. B. (1952). The thoracic gland in *Rhodnius prolixus* (Hemiptera) and its role in moulting. *J. exp. Biol.* **29**, 561–570.

276. WIGGLESWORTH, V. B. (1953). Determination of cell function in an insect. *J. Embryol. exp. Morph.* **1**, 269–277.

277. WIGGLESWORTH, V. B. (1954). *The physiology of insect metamorphosis.* Cambridge University Press.

278. WIGGLESWORTH, V. B. (1955). The breakdown of the thoracic gland in the adult insect, *Rhodnius prolixus. J. exp. Biol.* **32**, 485–491.

279. WIGGLESWORTH, V. B. (1957). The action of growth hormones in insects. *Symp. Soc. exp. Biol.* **11**, 204–227.

280. WIGGLESWORTH, V. B. (1959). *The control of growth and form.* Cornell University Press.

281. WIGGLESWORTH, V. B. (1961). Some observations on the juvenile hormone effect of farnesol in *Rhodnius prolixus* Stål. *J. Insect Physiol.* **7**, 73–78.

282. WIGGLESWORTH, V. B. (1963). The juvenile hormone effect of farnesol and some related compounds: quantitative experiments. *J. Insect Physiol.* **9**, 105–119.

283. WIGGLESWORTH, V. B. (1963). The action of moulting hormone and juvenile hormone at the cellular level in *Rhodnius prolixus. J. exp. Biol.* **40**, 231–245.

284. WIGGLESWORTH, V. B. (1964). The hormonal regulation of growth and reproduction in insects. *Adv. in Insect Physiology*, **2**, 247–336.

284a. WIGGLESWORTH, V. B. (1966). The hormonal regulation of growth and reproduction in insects. *Adv. Insect Physiol.* **2**, 248–336.

284b. WIGGLESWORTH, V. B. (1970). *Insect Hormones.* Oliver and Boyd, Edinburgh.

285. DE WILDE, J., DUINTJER, C. S. and MOOK, L. (1959). Physiology of diapause in the adult Colorado beetle. I. The photoperiod as a controlling factor. *J. Insect Physiol.* **3**, 75–85.

285a. WILKENS, J. L. (1968). The endocrine and nutritional control of egg maturation in the fleshfly, *Sarcophaga bullata. J. Insect Physiol.* **14**, 927–943.

285b. WILKENS, J. L. (1969). The endocrine control of protein metabolism as related to reproduction in the fleshfly, *Sarcophaga bullata. J. Insect Physiol.* **15**, 1015–1024.

286. WILLIAMS, C. M. (1946). Physiology of insect diapause: the role of the brain in the production and termination of pupal dormancy in the giant silkworm *Platysamia cecropia. Biol. Bull., Woods Hole*, **90**, 234–243.

287. WILLIAMS, C. M. (1952). Physiology of insect diapause. IV. The brain and prothoracic glands as an endocrine system in the cecropia silkworm. *Biol. Bull., Woods Hole*, **103**, 120–238.

288. WILLIAMS, C. M. (1961). The juvenile hormone. II. Its role in the endocrine control of molting, pupation and adult development in the cecropia silkworm. *Biol. Bull., Woods Hole*, **121**, 572–585.

289. WILLIAMS, C. M. (1963). The juvenile hormone. III. Its accumulation and storage in the abdomens of certain male moths. *Biol. Bull., Woods Hole*, **124**, 355–367.

289a. WILLIAMS, C. M. (1967). The present status of the brain hormone. In *Insects and Physiology*, pp. 133–139. (Eds. J. W. L. Beament & J. E. Treherne). Oliver and Boyd, Edinburgh.

289b. WILLIAMS, C. M. (1968). Ecdysone and ecdysone-analogues: their assay and action on diapausing pupae of the cynthia silkworm. *Biol. Bull.*, *Woods Hole*, **134**, 344–355.

290. WILLIAMS, C. M. and ADKISSON, P. L. (1964). Physiology of insect diapause. XIV. An endocrine mechanism for the photoperiodic control of pupal diapause in the oak silkworm, *Antheraea pernyi*. *Biol. Bull.*, *Woods Hole*, **127**, 511–525.

291. WILLIAMS, C. M., ADKISSON, P. C. and WALCOTT, C. (1965). Physiology of insect diapause. XV. The transmission of photoperiod signals to the brain of the oak silkworm, *Antheraea pernyi*. *Biol. Bull.*, *Woods Hole*, **128**, 497–507.

291a. WILLIAMS, E. E. and NAZER, I. O. (1975). Personal communication.

291b. WILSON, E. O. (1963). Pheromones. *Sci. Amer.* **208**, 100–114.

291c. WISSOCQ, J. C. (1966). Rôle du proventricule dans le determinisme de la stolonisation de *Syllis amica*. *Quatrefages* (Annélide polychète). *C.r. acad. Sci. Paris*, **262**, 2605–2608.

291d. WYATT, S. S. and WYATT, G. R. (1971). Stimulation of RNA and protein synthesis in silkmoth pupal wing tissue by ecdysone *in vitro*. *Gen. comp. Endocr.* **16**, 369–374.

292. YAGI, K. and BERN, H. A. (1965). Electrophysiologic analysis of the response of the caudal neurosecretory system of *Tilapia massambica* to osmotic manipulations. *Gen. comp. Endocr.* **5**, 509–526.

293. YAGI, K., BERN, H. A. and HAGADORN, I. R. (1963). Action potentials of neurosecretory neurones in the leech, *Theromyzon rude*. *Gen. comp. Endocr.* **3**, 490–495.

293a. YAMASHITA, O., HASEGAWA, K. and SEKI, M. (1972). Effect of the diapause hormone on trehalase activity in pupal ovaries of the silkworm, *Bombyx mori* L. *Gen. comp. Endocr.* **18**, 515–523.

293b. YAMAZAKI, M. and KOBAYASHI, M. (1969). Purification of the proteinic brain hormone of the silkworm, *Bombyx mori*. *J. Insect Physiol.* **15**, 1981–1990.

Index

347